Concepts of Medical Physics

Concepts of Medical Physics

First Edition

Dr Shahid Majid

FRAS, MAAPT, MAPS

Lecturer in Physics

Globular Publications
London

Globular Publications
Kingston-Upon-Thames
Surrey   KT1

www.globular.org.uk

enquiry@globular.org.uk

No part of this publication may be reproduced or transmitted in any form or by any means, electronic or mechanical, including photocopy, recording, or any information storage and retrieval system, without permission in writing from the author.

No responsibility is assumed by the publisher or author for any injury and/or damage to persons or property as a matter of products liability, negligence or otherwise, or from any use or operation of any methods, products, instructions or ideas contained in the material herein. Because of rapid advances in the medical sciences, in particular, independent verification of diagnoses and drug dosages should be made.

ISBN 978-1-291-30701-6

© Shahid Majid 2013

First Edition 2013

British Library Cataloguing in Publication Data
A catalogue record for this book is available from the British Library

Printed and bound in UK by LULU

Contents

Introduction vi

Chapter 1 - Energy

Types of energy 1
Gravitational potential energy 1
Elastic potential energy 2
Kinetic energy 3
Thermal energy 4
Chemical energy 5
Sound energy 6
Electrical energy 8
Nuclear energy 9
Radiant energy 10
The law of conservation of energy 11
Energy conversion flow diagram 12
Sankey energy diagram 13
Exercise – 1 14

Chapter 2 - Waves

Transverse waves 16
Longitudinal waves 16
Wave anatomy 17
The speed of waves 18
How energy is transported on waves 20
Refraction of waves at plane boundaries 21
The law of refraction 21
The speed of light and index of refraction 23
The law of reflection 25
Total internal reflection 26
Use of total internal reflection 27
Exercise – 2 29

Chapter 3 - Electromagnetic Spectrum

Nature of electromagnetic waves 31
Thermal emission 31
Non-thermal emission 32
Line emission of electromagnetic spectra 32
Creating an electromagnetic wave 37
Energy in electromagnetic wave 38
Regions of the electromagnetic spectra 38
Radio waves 39
Micro waves 40

Infra-red waves	41
Visible region	42
Ultra-violet waves	42
X-rays	43
Gamma rays	45
Exercise – 3	46

Chapter 4 - Optics

Type of lenses	49
Construction of ray diagram-convex lens	50
Magnification	54
Lens equation	55
Sign convention	56
Power of lens	58
Exercise – 4	59

Chapter 5 – Radioactivity   61

Alpha decay	63
Beta decay	65
Gamma decay	70
Detecting radioactivity	73
How to distinguish radioactive source	73
Law of radioactive decay	74
Half-life	76
Radiocarbon dating	77
Exercise – 5	79

Chapter 6 - Radiopharmaceutical (Nuclear Medicine)   83

Production of radiopharmaceutical	84
Cyclotron	84
Nuclear reactor	86
Neutron activation	87
Generator	87
Separation technique for radionuclide	89
Molybdenum/Technetium generator	91
Equilibrium	93
Secular equilibrium	94
Transient equilibrium	95
Physical half-life	96
Biological half-life	97
Effective half-life	97
Measurement of radiation	98
Radiation absorbed dose units	99
Equivalent dose	99
Exposure	100
Diagnostic applications of radiopharmaceutical	101
Therapeutic applications of radiopharmaceutical	102

Gamma camera	104
Collimator	104
Scintillation crystal	106
Photomultiplier tube	107
Detection circuitry	108
Exercise – 6	110

Chapter 7 - Body lever system

First class lever system	114
Second class lever system	115
Third class lever system	117
Lever arm length	119
Mechanical advantage	119
The muscular system	119
Types of muscle contractions	121
Tendons and ligaments	122
The elbow	123
The hip	126
The vertebral column	128
Standing	129
Bending and lifting	131
Low-back biomechanics of bending and lifting	133
Exercise – 7	135

Chapter 8 - Energy expenditure

Sugars	138
Fats	140
Protein	141
Basal metabolic rate	141
Harris-Benedict equation	143
BMR and weight loss	145
Factors that affect basal metabolism	145
How does human body regulate temperature?	149
Daily variations in human body temperature	150
Cooling of human body	151
Conduction	152
Convection	152
Radiation	154
Evaporation of perspiration	155
Exercise – 8	156

Chapter 9 - The human eye and vision system

Human vision	160
Human vision system	163

Eye anatomy	164
The Cornea	164
Diseases and disorders affecting cornea	165
Keralitis	165
Keratocornus	165
Ocular herpes	166
Herpes zoster	166
Mop-dot-fingerprint dystrophy	166
Fuchs' dystrophy	166
Pink eye	167
The Conjunctiva	167
Diseases and disorders affecting conjunctiva	168
Pinguecula	168
Pterygium	168
The Iris	169
Diseases and disorders affecting iris	169
Iritis	169
The Pupil	170
The lens	170
Diseases and disorders affecting lens	170
Cataract	170
Aqueous humour	172
Vitreous humour	172
Diseases and disorders affecting vitreous humour	172
Posterior vitreous detachment	172
Retinal detachment	172
Choroid	174
Choroidal detachment and haemorrhage	174
Choroidal nevus	174
Sclera	175
The retina	175
Retinal scan	178
Diseases and disorders affecting retina	178
Retinitis pigmentosa	178
Macular degeneration	179
Retinoblastoma	180
Fovea	181
Blind spot	181
Accommodation of the eye	183
Power of accommodation	184
Visual acuity	184
The resolution of the human eye	186
Depth of field	188
Depth of focus	189
Contrast	189
Chromatic aberration	190
Optical detection	191
Myopia	191
Hypermetropia	196
Astigmatism	197
Exercise – 9	199

Chapter 10 – The human Ear and hearing system

Transmission of sound in a medium	202
Sound intensity and pressure	205
What is decibel, Phons and Sones	208
Sound Power	210
The anatomy of the ear	212
Outer ear, Pinna	212
The auditory canal	213
Eardrum	214
Middle ear	216
Eustachian tube	217
Inner ear	218
Cochlea	218
Hearing loss	222
Conductive hearing loss	222
Sensorineural hearing loss	223
Congenital	223
Acquired	224
Hearing aid	224
Cochlear implant	225
Tinnitus	227
Exercise 10	228

Chapter 11 – Biopotential and measurements

Action Potential	231
Quantitative model of the action potential	233
Hodgkin – Huxley model	234
The cardiovascular system	243
Electrocardiogram	247
Interpretation of ECG trace	255
QRS complex	256
Normal ECG and its interval	256
P wave	257
P-R interval	257
QRS interval	257
ST segment	258
T wave	258
QT interval	258
Hemodynamics	259
Flow velocity	263
Flow resistance	264
Turbulent flow	267
Pressure and blood vessel wall	269
Measuring blood velocity using ultrasound	269
Exercise 11	271

Appendix A 276

Solutions to Exercise-1 276
Solutions to Exercise-2 277
Solutions to Exercise-3 278
Solutions to Exercise-4 281
Solutions to Exercise-5 284
Solutions to Exercise-6 286
Solutions to Exercise-7 289
Solutions to Exercise-8 291
Solutions to Exercise-9 291
Solutions to Exercise-10 294
Solutions to Exercise-11 298

Index 303

Acknowledgement 310

Preface

This book provides an introduction to medical physics. It is aimed primarily at students who are studying various medical physics related courses, such as access to HE diploma in science or allied subjects, BTEC, level three or first year graduate courses. Medical physics cover a very wide range of subjects, not all covered in this book. An attempt is made to cover the main topics of physics which are relevant to this discipline. Students should, however, find this book a useful introduction to the main principles, which can help to build their foundation for further study in this field.

The book has been written in recognising that most students have not studied medical physics before, and therefore I have tried to write in a simple accessible style, with a clear explanation for any technical terminology. Interspersed throughout each chapter are numerous examples. Students are strongly encouraged to attempt these questions. They are designed to make sure that students have absorbed all the material and now they are ready to apply their understanding. This will also encourage students to think carefully about the topic they have just covered and link it with the activity at hand and may design one their own. The answer to each example is provided immediately after the question and they should be covered up while you are attempting to find the solution. Once you are satisfied with your answer you can then compare with the one given.

At the end of each chapter, there is an exercise. The questions in these exercises are more demanding and comprehensive. These questions are designed to check your understanding of the entire chapter so you could see your level of understanding of the topic. You should consult to text for refreshing your memory when you are ready to attempt these questions. The full detailed answers to each question is given at the end of this book in Appendix A. You must attempt each question on your own thoroughly before referring to these solutions. If you have any difficulty with these questions you must go over the relevant topic in the chapter.

I accept full responsibility for any error. I would welcome any suggestions as to how this book might yet be improved.

Shahid Majid
2013

1. ENERGY

In physics, energy is the ability of a physical system to do work on another physical system. Work is defined as force acting on an object through a distance. Energy is a scalar quantity. In the SI units energy is measured in Joules (J). Therefore, if something has energy it can do work. When any action in nature occurs is accompanied by a reduction in the overall quality of energy while the first law of thermodynamic entail that energy obeys the principle of conservation of energy which states that energy can not be created neither destroyed, but transformed into other form(s) of energy.

We can write a mathematical equation of work done:

$$W = F.d \qquad (1.1)$$

where, F is force in Newton (N) and d is the distance in meters (m).

Example: A 250 N force is applied to move a 12 kg block a horizontal distance of 5 meters at constant velocity. Calculate the work done on the block?

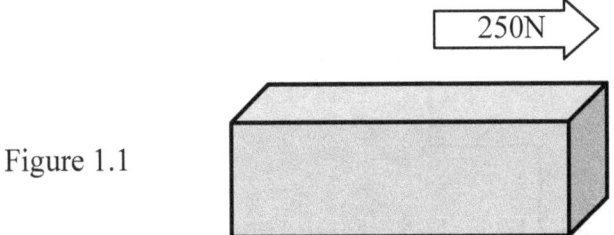

Figure 1.1

Box is moved in the direction of applied force.

$$W = F.d = 250 \times 5$$
$$= 1250 \text{ J}$$

TYPES OF ENERGY

There are just two types of energies, Potential energy and Kinetic energy and many forms of energies like chemical, heat and sound etc.

Potential Energy – This is the energy stored in object due to its position. We can look at two forms of potential energy.

Gravitational Potential Energy (GPE) - When object is raised above the ground level it stores this energy. Consider that you raise two objects one twice as heavy as the other to the same height. Which object has more GPE? Think that you have applied force to raise each object and this mean you have done some work on each object. So twice heavy requires twice as much force so it must store twice as mush energy. So GPE is directly proportional to mass. Now consider this that you raise two objects of identical mass but one raised twice as high compared to the other object. This means that the object which raised twice as high, you must have done twice as mush work, so it too store twice as much energy. GPE is also directly proportional to height. These relationships can be expressed by the following equation:

1

$$GPE = m.g.h \qquad (1.2)$$

where; m = mass in kg, g = gravitational acceleration in 9.81 kg m/s, h = height in meters.

Example: An object of mass 2 kg is dropped from a tall building. It losses a 750 J of GPE when it hit the ground. Find the height of the building?

Rearranging equation 1.2 $\quad h = \dfrac{GPE}{m.g}$

$$= \dfrac{750}{2 \times 9.81} = 38.23 m$$

GPE is directly proportional to its height above the datum level.

Example: Determine the GPE at each point in the diagram.

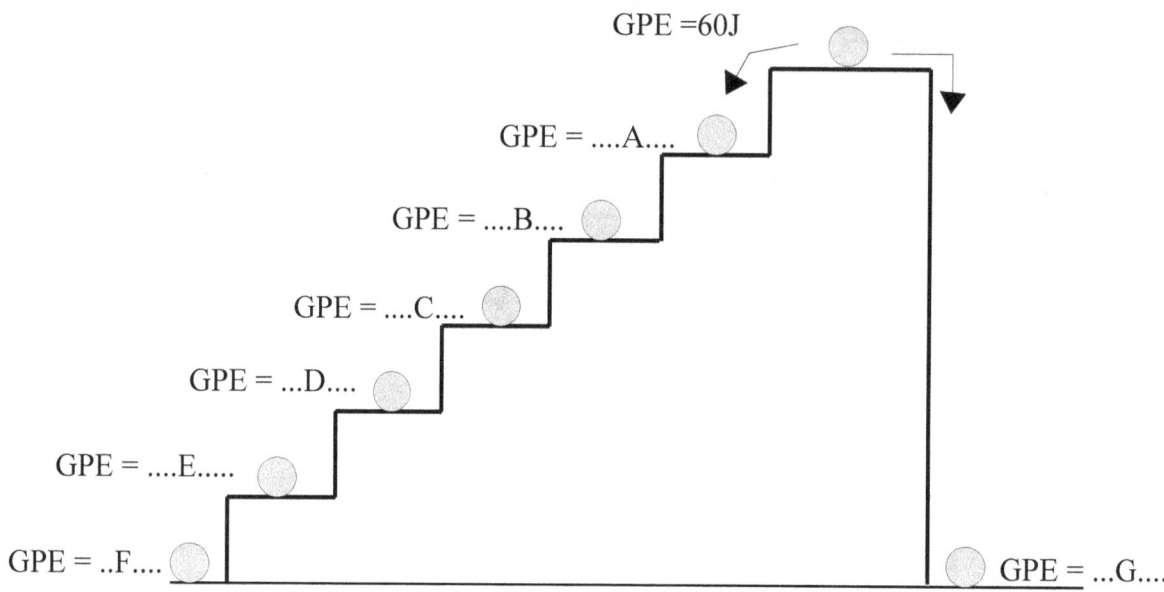

Figure 1.2

$A = 50J$ (5/6th of the height. 5/6 x 60 = 50)
$B = 40J$ (4/6th of the height. 4/6 x 60 = 40)
$C = 30J$ (3/6th of the height. 3/6 x 60 = 30)
$D = 20J$ (2/6th of the height. 2/6 x 60 = 20)
$E = 10J$ (1/6th of the height. 1/6 x 60 = 10)
$F = G = 0J$ (datum level)

Elastic Potential Energy - This is the energy stored in elastic material as the result of their stretching or compressing. It is equal to work done to stretch the spring. According to Hook's law, the force required to stretch the spring is directly proportional to the amount of stretch.

$$F = -kx \qquad (1.3)$$

Since the change in the potential energy of an object (i.e. spring in the above case) between two positions (natural length of spring and stretched spring) is equal to the work that must be done to move the object from one point to the other. The calculation of work involves an integral.

$$\text{Work} = \text{Energy} = \int_0^x kx\, dx = \tfrac{1}{2} \cdot kx^2 \qquad (1.4)$$

EPE is directly proportional to the amount of compression or stretch. Springs can store elastic potential energy. The amount of EPE stored in a spring is related with this equation.

This energy stored can also be visualized as the area under the force curve:

Area under the graph = ½ base x height = ½ F.x

substitute equation 1.3 for F in the above equation.

Area under the graph = ½ (k.x).x = ½ k.x²

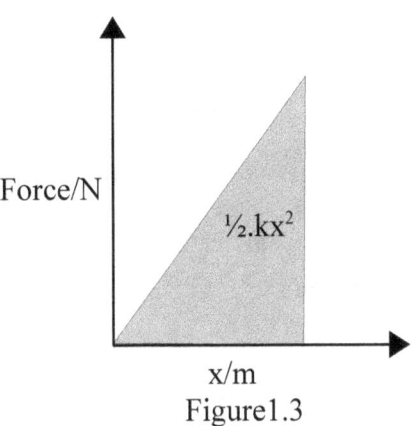

Figure1.3

Example: A spring-loaded toy dart gun is used to shoot a dart straight up in the air, and the dart reaches a maximum height of 36 m. The same dart is shot straight up a second time from the same gun, but this time the spring is compressed only half as far as before firing. How far up does the dart go this time, neglecting friction and assuming an ideal spring?

All the elastic potential energy of the spring is converted into gravitational potential energy of dart.

½ kx² = mgh; mass, spring constant and g is constant. Hence x² = h

So if compression is halved then the height will be one quarter. [x = ½ = x² = ¼]

h = ¼ x 36 = 9m

Kinetic Energy – The kinetic energy of an object is the energy it contains because of its motion. This energy arises from the net work done on an object moving from one point to another. Let say a force is applied to move a stationary object.

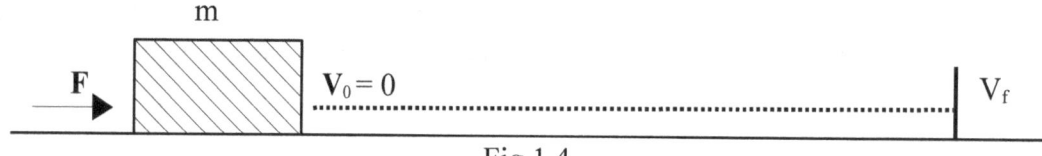

Fig.1.4

displacement ' d ' at time t

$F = ma$ Newton's second law.   Average velocity $\bar{U} = \dfrac{V_f + V_0}{2} = \dfrac{V_f}{2}$

$$\text{displacement } d = \bar{U} \times t = \dfrac{V_f}{2} t$$

$$\text{acceleration} = \dfrac{V_f - 0}{t} = \dfrac{V_f}{t}$$

$$\text{work = kinetic energy} = Fd = m \times a \times d = m \times a \times \dfrac{V_f}{2} t = m \left(\dfrac{V_f}{t}\right) \cdot \left(\dfrac{V_f}{2} t\right)$$

$$\mathbf{E_k = \tfrac{1}{2} m V_f^2 = \tfrac{1}{2} m V^2} \qquad (1.5)$$

Kinetic energy is a scalar quantity; it does not have a direction. The standard metric unit of measurement of kinetic energy is the Joule.

Example: A car is travelling at 72 km/h and has mass of 1350 kg. Calculate the kinetic energy of the car.

First we must convert 72 km/h velocity into m/s. For this we divide by a factor of 3.6

$$72 \text{km/h} = \dfrac{72}{3.6} = 20 \text{ m/s}$$

$$E_k = \tfrac{1}{2}(1350) \times 20^2 = 270 \text{kJ}$$

Thermal Energy – Thermal energy is produced by the movement of atoms and molecules in matter. It is a type of kinetic energy due to the random motion of atoms and molecules in matter.

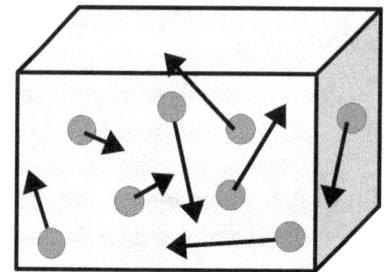

Figure 1.5 – Random motion in matter

Increase in the velocity of atoms and molecule will result in increase of temperature. Thermal energy can be considered as the part of the total, internal energy of a thermodynamic system which results in that system's temperature. When two such systems are brought into contact, they exchange energy in the form of heat, which can be defined as a transfer of thermal energy from system with higher temperature to the system with lower temperature. You must not say that heat is an energy. Heat is a characteristic of a process but it's not property of matter. Heat can only be absorbed or produced in thermal energy exchange.

Thermal energy can be defined in terms of the ideal gas (which is approximated by a monoatomic gas at a low pressure). The kinetic energy of single atom in closed container is:

$$E_k = \tfrac{1}{2} m \langle V^2 \rangle \qquad (1.5a)$$

$\langle V^2 \rangle$ is the speed associated with the average kinetic energy. This useful average is **root-mean-square speed**, which has a symbol $\sqrt{\langle V^2 \rangle}$. This is found by squaring the individual speeds of a set of molecules, finding the mean of the squares, and then taking the square root.

Thus, the thermal energy 'U_t' of the gas consisting of N atoms will have the sum average of these energies.

$$U_t = \tfrac{1}{2} Nm\langle V^2 \rangle = \tfrac{3}{2} NkT \qquad (1.6)$$

where, k is the Boltzmann constant which has value of 1.38×10^{-23} JK^{-1} and T is temperature in Kelvin.

Example: What is the root-mean-square(r.m.s) speed of helium molecules at 30°C? Take the mass of helium molecule as 6.68×10^{-27} kg.

$$T = 30 + 273 = 303 \text{ K}$$

$$\tfrac{1}{2} m\langle V^2 \rangle = \tfrac{3}{2} kT$$

$$\begin{aligned}
\langle V^2 \rangle &= 3/2 \, kT \div \tfrac{1}{2} m \\
&= 3/2 \, (1.38 \times 10^{-23})(303) \div \tfrac{1}{2} (6.68 \times 10^{-27}) \\
&= 6.27 \times 10^{-21} \div 3.34 \times 10^{-27} \\
&= 1877874.252
\end{aligned}$$

$$\begin{aligned}
\text{r.m.s} &= \sqrt{\langle V^2 \rangle} \\
&= \sqrt{1877874.252} \\
&= 1370 \text{ ms}^{-1}
\end{aligned}$$

Chemical Energy – The energy held in the covalent bonds between atoms in a molecule is called chemical energy. The food we eat combined with oxygen we breathe, store that energy as chemical which our body extract and convert into mechanical and thermal energy. One such reaction is part of our metabolism in which glucose burns with oxygen via this reaction.

$$C_6H_{12}O_6 + 6O_2 = 6CO_2 + 6H_2O$$

For one mole of glucose, this combustion releases 2.8 MJ of energy. One mole of glucose molecule is 180g (C = 12, H = 1, O = 16; (12x6 + 12x1 + 16x6 = 180)). Therefore, energy provide by one gram of glucose is 15.6kJ. Dietician normally work with calories as energy unit.

1 kilocalorie (kcal or Cal) is equal to 4.1858 kJ

Thus, one gram of glucose provides 3.73kcal.(Dietician normally take this as 4 kcal per gram). Molecules like glucose are referred to as carbohydrates. Carbohydrates yield less energy than hydrocarbon molecules. The reason is carbohydrates already have oxygen so they combine less with oxygen to release energy where hydrocarbon molecules have less or no oxygen thus their metabolism in our body yield more than twice (9 kcal per gram) the energy than carbohydrates. Fat are such molecules. The other mechanism which give our body energy are proteins and their energy content is the same as carbohydrates.

Energy from 4kcal. Protein = 16.7kJ, 9kcal. Fat = 37.7kJ and 4kcal. Carbohydrate = 16.7kJ

Example: A normal cheeseburger provides 367kcal. If this burger has 20g Fat, 27g Carbohydrates and 21g of Proteins. Calculate the calories provided by Fat, Carbohydrates and Proteins.

Amount of calories per gram = 367/68 = 5.40 kcalg^{-1}
calories provided by Fat = 5.40 x 20 = 108 kcal
calories provided by Carbohydrates = 5.40 x 27 = 145.8 kcal
calories provided by Proteins = 5.40 x 21 = 113.4 kcal

Example: Dietician recommend that the energy we need has to be blended which should be:
- 30% Fat
- 57% Carbohydrates
- 13% Protein

UK Department of health recommends a daily Calorie intake of 1940 calories for women and 2550 for men. What does an average men need in terms of fat, carbohydrates and protein.

Fat – 30% of 2550 = 765kcalories – at 9 kcalories per gram = 765 ÷ 9 = 85g
Carbohydrates – 57% of 2550 = 1453.5kcalories – at 4 kcalories per gram = 1453.5 ÷ 4 = 363.4g
Protein – 13% of 2550 = 331.5kcalories – at 4 kcalories per gram = 331.5 ÷ 4 = 82.88g

Sound Energy – is a form of energy produced by vibrating matter. Sound is sequence of pressure that propagates compressible medium such as sold, liquid and gas. During propagation sound energy can reflect, refract or attenuate. Close to a sound source, pressure and particle displacement are out of phase, but one wavelength from source they are in phase.
Travelling of sound is normally affected by these factors:

1. The speed of sound is affected by the density and temperature of the medium.
2. The viscosity of the medium also affect the energy of the sound waves. Waves carry energy which is proportional to the square of the amplitude of the wave. High viscosity will attenuate the sound energy.
3. Sound travels faster through solids than through liquids, and faster through liquids than through gases. Sound waves also travel faster through a warmer medium than through a cooler medium.

Sound waves are longitudinal waves, this means the direction of sound energy is the same as the oscillation. Sound energy is not measured in units of Joules but is measured in pressure and intensity using units such as Pascal and decibel. Sound pressure can be defined as the deviation of local pressure from ambient pressure caused by sound propagation. Another useful measure is sound pressure level. This is the threshold of human hearing at 1 kHz. Sound pressure level (SPL) is a logarithmic measure of the ratio of effective sound pressure to a threshold value. It is measured in decibel (dB) above a standard threshold level.

$$\textbf{SPL} = \textbf{20 log}_{10} \, (\, \textbf{P}_{rms} \div \textbf{P}_{t} \,) \, \textbf{dB} \qquad (1.7)$$

where P_{rms} is the sound pressure being measured and P_t is threshold of human hearing with a value of 20 µPa. A sound with twice the SPL is 6 dB louder.
- i.e. $20 \log_{10} (2) = 20(0.3) = 6$

Example: What is SPL for 1Pa ?

$$SPL = 20 \log_{10} \left(\frac{1}{20 \times 10^{-6}} \right) dB$$
$$= 20 \times 4.70$$
$$= 93.97 \, dB$$

Example: a) If sound A has 100 times the SPL of sound B, how much louder in dB is A than B?

$$dB = 20 \log_{10} 100 = 40 \, dB \text{ louder}$$

The sound energy produced by vibrating object is often referred to as sound energy flux, symbolised as **q**. This can be defined as the integral of the product of sound pressure and the velocity of the particles of the medium over a surface area. This can be written mathematical;

$$\textbf{q} = \int \textbf{pv. dA} \qquad (1.8)$$

When this equation is solved shows that the sound energy in a medium of density ρ with a velocity of propagation v, with sound energy flux through area A is:

$$E_s = (\, p^2 A / \rho v \,) \cos \theta \qquad (1.9)$$

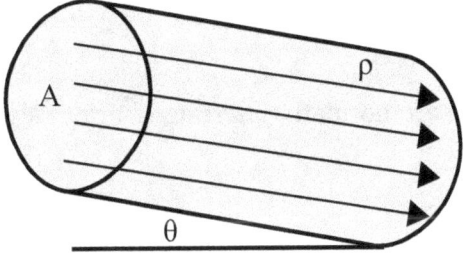

Figure 1.6 Sound flux through material with density ρ, cross-section A at angle θ to the horizontal.

Example: Calculate the energy of sound wave with 0.5 Pa pressure at standard condition through 1m² area of air
  a) parallel to the direction of sound wave.
  b) at an angle of 70°

At standard condition density of air is 1.2 kg m⁻³ and sound velocity 343 ms⁻¹. Since energy flux is in the same direction as wave so angle is zero.

a) $$E_s = (0.5^2 \times 1)/(1.2 \times 343) \cos 0°$$
$$= 0.61 \text{ mW}$$

b) $$E_s = 0.6 \cos 70°$$
$$= 0.21 \text{ mW}$$

Electrical Energy

Electrical Energy – Electrical energy refers to flow of charges along the conductor to create energy. Electrical energy is a secondary source of energy, which means that it is produced by converting other forms of energy. Electrical energy is a potential energy.

Electrical energy easily flow through metals as they are good conductors. Metals have lots of loose de-localised electrons which can be used to flow from one point to another. In order to flow these electrons we need to push them. For this we require force and force applied to a certain distance is known as work done which is energy and measured in Joules. So in order to push electron through electrical field we create a potential difference, and work need to be done on electron to move from lower energy to higher energy. Thus, potential difference can be defined as the amount of work required to move charge which is measured in coulomb.

$$\text{Potential difference, } V = \frac{W}{Q} \qquad (1.10)$$

The unit of potential energy is volt. One volt is equal to one joule per coulomb. So voltage is a measure of how much electrical energy is delivered for a certain amount of charge. This charge is carried by electrons and is measured in the unit of coulomb, and charge on one electron is equal to 1.6×10^{-19} coulomb. The current is defined as the flow of electrons and is measured in units of Ampere. One ampere is defined as when one coulomb of charge flowing past any given point in the conductor in one second. i.e.,

$$\text{Current, } I = \frac{Q}{t} \qquad (1.11)$$

In electrical circuit or appliances we are normally interested in rate of energy (work done), which is called Power. So

$$\text{Power, } P = \frac{\text{Work done}}{\text{time}} = \frac{W}{t} \qquad (1.12)$$

Substituting equation 1.10 and 1.11 into equation 1.12;

$$P = \frac{V \times Q}{\frac{Q}{I}} = VI \qquad (1.13)$$

Nuclear Energy – This usually means the energy which can be released by fusion or fission or naturally via radioactive decay.

Nuclear Fission: In this process, the nuclei of atoms are split, causing energy to be released. The element uranium is the main fuel used to undergo nuclear fission to produce energy. Uranium nuclei can be easily split by shooting neutrons at them. Once a uranium nucleus is split, multiple neutrons are released which are used to split other uranium nuclei. This phenomenon is known as a chain reaction.

Following equation will give an example of nuclear fission:

$$_{92}^{235}U + _{0}^{1}n \longrightarrow {}_{36}^{92}Kr + {}_{56}^{142}Ba + 2\,_{0}^{1}n + \text{Energy} \qquad (1.14)$$

When neutron strike a ^{235}U nucleus, it is at first absorbed by the nucleus. This creates ^{236}U which is unstable and this causes the atom to split or go fission. If we look closely at above equation then we have 236 nucleons on the left hand side and 236 nucleons on the right hand side of the equation. If we have not lost any mass then where does this energy has come from?

The mass of any atom is always more than the sum of the individual mass of its constituent particles called protons and neutrons. The extra mass results from the binding energy which keeps protons and neutrons together in the nucleus. You can make an analogy of butterscotch peanut bar, you can think of peanuts as protons and neutrons. To turn these peanuts into bar we need binding material. Now the final mass of the bar would be more than the mass of peanuts alone. If you split the peanut bar some of the binding material will be release as dust.

Likewise, when atom is split via fission process, some of the energy is released as radiation in the form of heat. According to Einstein's famous equation $E = mc^2$, energy and mass are equivalent, thus energy released is same as mass released. The amount of energy released in fission of ^{235}U is about 200 MeV. This is equal to 3.2×10^{-11} J of energy.

Nuclear Fusion: In nuclear fusion two lighter atomic nuclei (normally hydrogen) fuse together to form a heavier nucleus (helium) under high pressure and temperature. This process release huge amount of energy arising from the binding energy due to the strong nuclear force. Our sun radiates energy at the rate of 3.9×10^{26} Joule per second.

This fusion happens in several steps and different routes. We will look at just one route.

When two protons are fused together they make deuterium and release positron (anti-electron) and neutrino.

$$_{1}^{1}H + _{1}^{1}H \rightarrow _{1}^{2}H + _{1}^{0}e^{+1} + \nu$$

the next step is deuterium – deuterium fusion will produce helium-3

$$^2_1H + ^2_1H \rightarrow ^3_1He + ^1_1H + 4.03\,MeV$$

in the next step deuterium – helium-3 fusion will produce stable helium-4

$$^2_1H + ^3_1He \rightarrow ^4_2He + ^1_0n + energy$$

Lets calculate the amount of energy release in the final step. We need the following data:

Mass of deuterium nucleus = 3.3425 x 10^{-27} kg
Mass of helium-3 nucleus = 6.6425 x 10^{-27} kg
Mass of helium nucleus = 6.465 x 10^{-27} kg
Mass of neutron = 1.675 x 10^{-27} kg

Now total mass on the left hand side of the equation = 9.985 x 10^{-27} kg and,

total mass on the right hand side of the equation = 8.3215 x 10^{-27} kg

Mass difference = 1.6635 x 10^{-27} kg

Applying Einstein mass energy equation E = mc^2 to find the energy

E = (1.6635 x 10^{-27} kg)(3 x 10^8)^2
E = 1.497 x 10^{-10} J

to convert this energy to eV:

$$E = \frac{1.497 \times 10^{-27}}{1.6 \times 10^{-19}} = 935.72\,MeV$$

This is lot of energy.

Radiant Energy – Radiation is energy that moves through space from one object, the source, to another object where it is absorbed. Radiant energy is the energy of electromagnetic waves. Electromagnetic radiation is classified according to the frequency of its wave. The relationship of electromagnetic energy and frequency can be written as:

$$E = hf \qquad (1.15)$$

where E is the energy (J), h is Planck's constant = 6.63 x 10^{-34} Js, and f is frequency (Hz).

Example: Calculate the energy in the ultraviolet radiations with 7.50 x 10^{15} Hz frequency.

E = (6.63 x 10^{-34} Js) x (7.50 x 10^{15} Hz)
= 4.97 x 10^{-18} J

The Law of conservation of energy

The law of conservation of energy states that energy may neither be created nor destroyed but can only change from one form into another form of energy. Therefore the sum of all the energies in the system is a constant. For the idealised situation of train descending down the hill, the gravitational potential energy is converted into kinetic energy if we ignore the air resistance, friction and other energy transfers.

Example: a) The cyclist in figure 1.6 rode down the hill from the height of 15 m. Calculate the speed of the cyclist at the bottom of the hill on a straight road. Take $g = 9.8 ms^{-2}$ and mass of the cyclist plus bicycle as 85kg. Assume no energy loss.

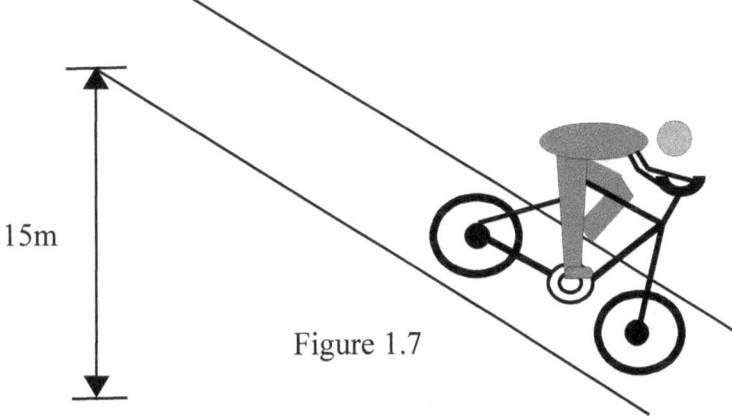

Figure 1.7

As no energy is lost, hence all the gravitational potential energy is converted into kinetic energy of the cyclist.

$$\tfrac{1}{2} mv^2 = mgh$$

solving for, v

$$\tfrac{1}{2} v^2 = gh$$
$$v^2 = 2gh$$

hence

$$v = \sqrt{(2gh)} = \sqrt{(2 \times 9.81 \times 15)}$$

$$= 17.15 ms^{-1}$$

b) Now cyclist provided 1300J of energy by paddling the cycle while coming down the hill and some energy was lost due to friction and air resistance. If arrived at the bottom of the hill with speed of $16 ms^{-1}$, how much energy was lost due to friction and air resistance?

$$\text{Total energy} = mgh + 1300J$$

$$= 85 \times 9.81 \times 15 + 1300$$
$$= 13.81 kJ$$

$$\text{Energy at the bottom of the hill} = \tfrac{1}{2} mv^2$$
$$= \tfrac{1}{2} \times 85 \times 16^2$$
$$= 10.88 kJ$$

Loss of energy = (13.81 − 10.88) kJ = 2.93kJ

Energy Conversion.

Energy in a system may be converted so that it resides in different forms. A car engine converts chemical energy of fuel into thermal energy which then rotate the crankshaft to move the car. Not all the energy is converted into useful energy. Some of the energy goes to waste. The ratio of useful energy to total input energy is called the efficiency of the energy converting system.

Efficiency (%) = (useful energy out ÷ total energy in) x 100. (1.16)

Energy conversion is a very useful process. For anything to happen there will always be an energy conversion. When you put the CD in a CD player, two obvious energies are converted. Namely, electrical energy gets converted into sound. But think of other energy conversions which happen inside the CD player to your speakers?

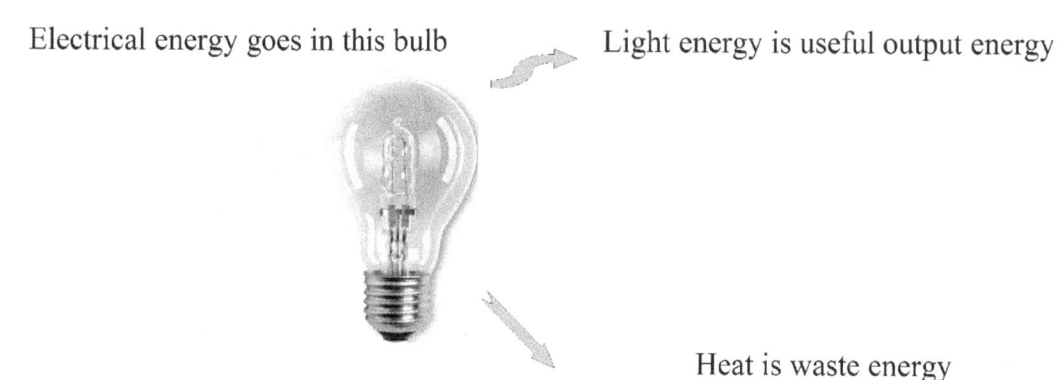

Figure 1.8

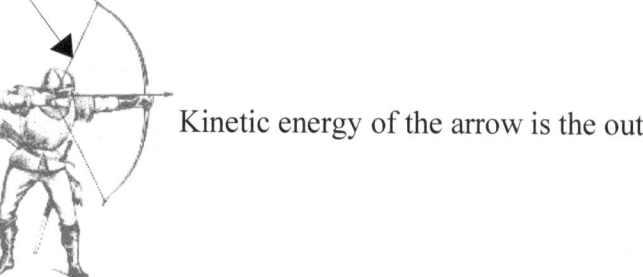

Figure 1.9

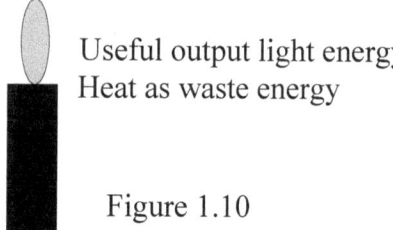

Figure 1.10

Energy conversion Flow Diagram

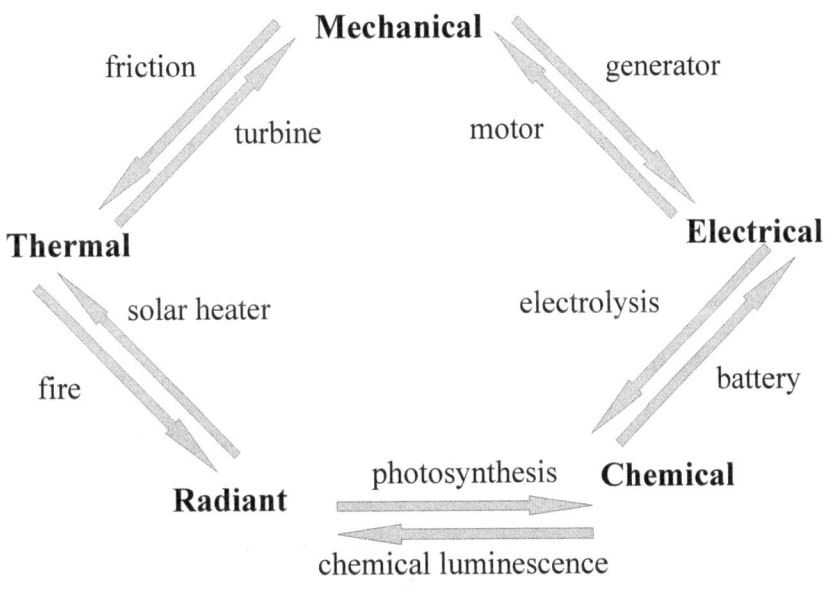

Figure 1.11

Sankey Energy diagram

Sankey diagrams are a useful representation of energy conversion. They are named after the Irishman Matthew Phineas Riall Sankey, who first used this type of diagram on the energy efficiency of a steam engine in 1898. In Sankey diagrams the width of the arrow is shown proportional to the energy flow.

Example: In normal incandescent lamp only 10% of the input electrical energy is converted into useful light energy and a rest is wasted as heat. Show this in Sankey diagram.

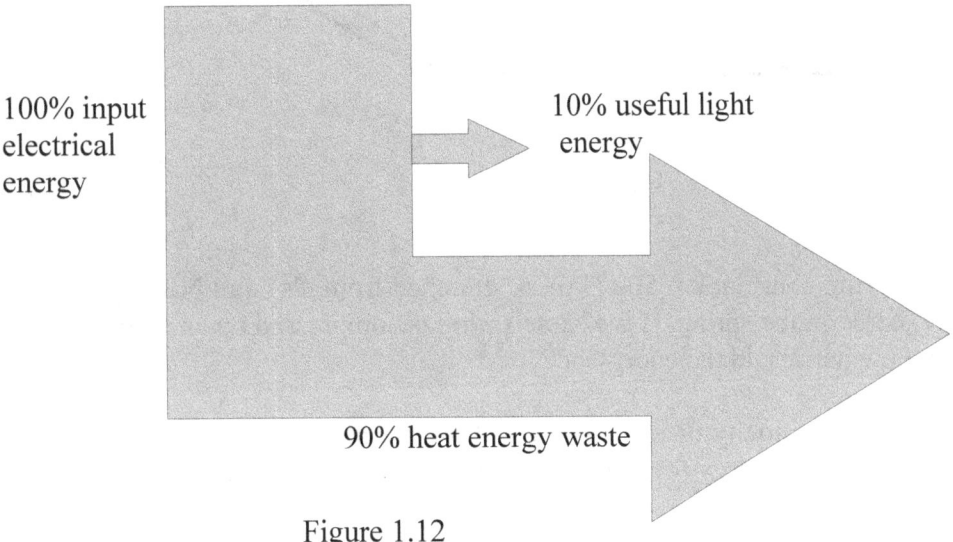

Figure 1.12

Exercise 1

Q1 – Calculate the kinetic energy of a car of mass 800 kg moving at a speed of 10 m/s.

Q2 – Calculate the mass of a car that has a kinetic energy of 32 800 J when moving at 8m/s.

Q3 – Calculate the speed of a lorry of mass 3000 kg that has 85 000 J of kinetic energy.

Q4 – Calculate the gravitational potential energy gained by a car of mass 1200 kg when it climbs up a hill of height 45 m. Take g = 10 ms^{-1}.

Q5 – Calculate the mass of a boy if he gains 2.4 kJ of gravitational potential energy when he climbs up stairs of height 4 m. Take g = 10 ms^{-1}.

Q6 – Calculate the weight of a car if it gains 60 kJ of gravitational potential energy when it climbs up a hill of height 5 m.

Q7 – Calculate the height 'h' of the hill climbed by the roller skater in figure 1.13. Assume no loss of energy due to friction or air resistance. Take g = 10 ms^{-1}.

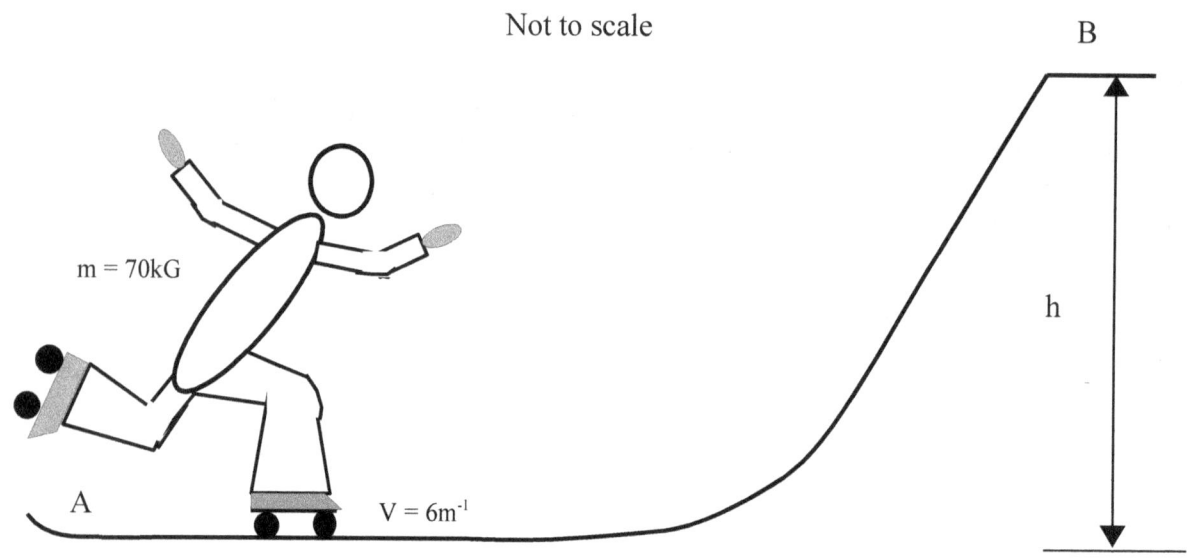

Figure 1.13

Q8 – A spring with spring constant k = 500 N/m, is stretched from it's equilibrium position by 1.25 m, find the energy stored in the spring. If the same spring is compressed by an amount of 1.5 m from the equilibrium, what would the energy be?

Q9 – Assume the average velocity of nitrogen is 850 meters per second. Using the kinetic energy formula, calculate the kinetic energy for nitrogen. Also calculate the temperature of nitrogen gas.

atomic mass nitrogen = 4.65 x 10^{-26} kg and Boltzmann constant =1.38 x 10^{-23} JK^{-1}

Q10 – One serving of bourbon biscuit of 13.5g provides 67 kcalories. If protein provides 3.48 Calories, carbohydrates provide 45.65 Calories fat provides 15.38 Calories and fibers provides 2.38 Calories. Calculate the amount of protein, carbohydrates, fat and fibrous quantity in grams.

I accept full responsibility for any error. I would welcome any suggestions as to how this book might yet be improved.

Q11 – Calculate the pressure level P_{rms} for sound with an SPL value of 120 dB.

Q12 – Calculate the energy of sound waves with 20 Pa pressure at standard condition through 2 m^2 area with an angle of 35^0 to the direction of the sound wave.

Q13 – 2.5 kW of power is consumed by electric heater connected across a potential difference of 230 V for two hours. Calculate the amount of charge passed through the heater wire.

Q14 – Calculate the frequency of the radiant light falling on solar cell when 3 x 10^{-16} J of energy is produced.

Q15 – Complete the energy conversion diagram shown in Figure 1.14. Write as many as you can.

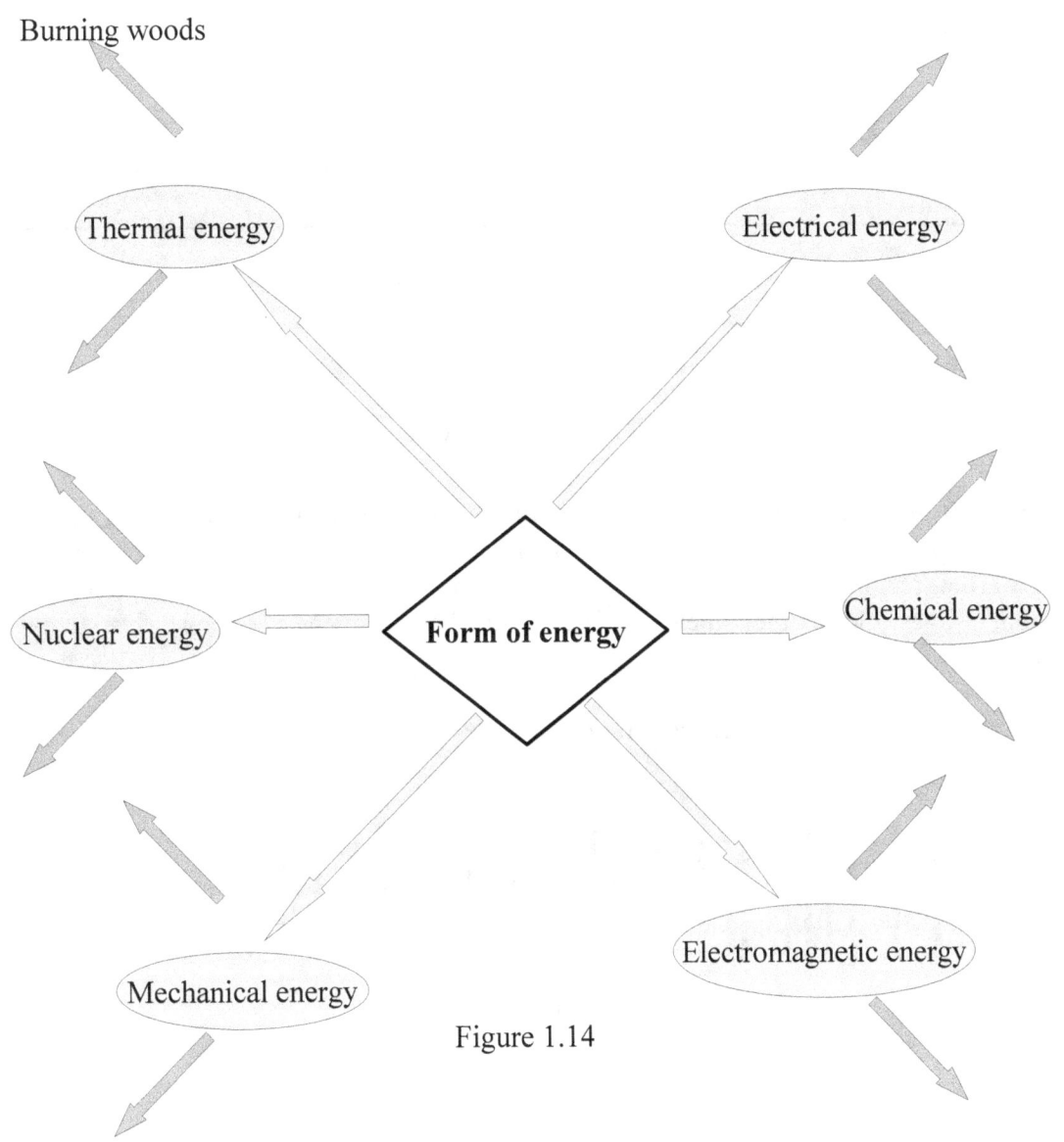

Figure 1.14

2. WAVES

A wave can be defined as a disturbance that travel through space and time. This wave travel transfer energy from one point to another. When wave travels they do not displace the particles of the medium. The important point about the wave is that they carry energy from one point to another point without moving the medium particles. There are two types of waves:

Transverse waves: In a transverse wave the particle displacement is perpendicular to the direction of wave propagation. Figure 2.1 shows that if the wave motion is from left to right that the particles of medium do not move along the wave but oscillate up and down about their individual equilibrium position. Light travel as transverse wave.

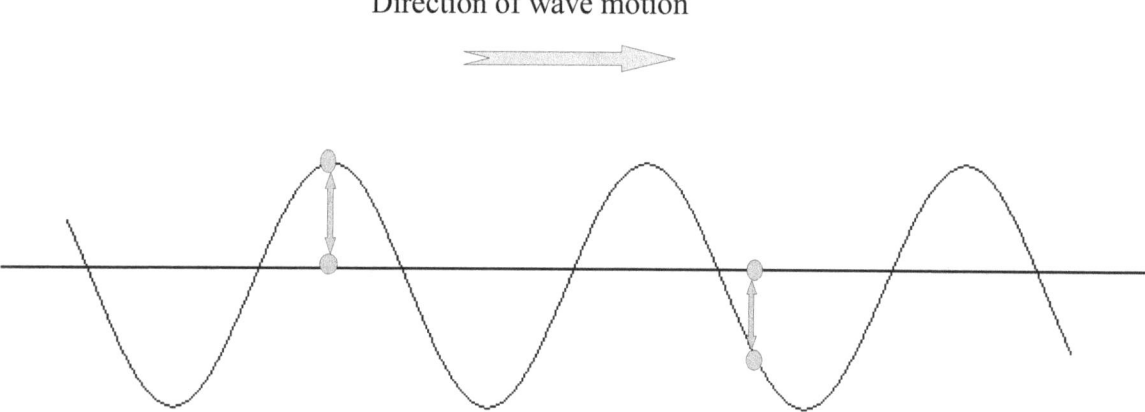

Figure 2.1- transverse waves

Longitudinal waves: In longitudinal waves the displacement of the medium is parallel to the propagation of the wave. Figure 2.2 shows that if the wave motion is from left to right then the particles of the medium also oscillate parallel to the direction of motion. Sound travels as longitudinal waves. Sometimes they are called mechanical waves because it's actual moves particles.

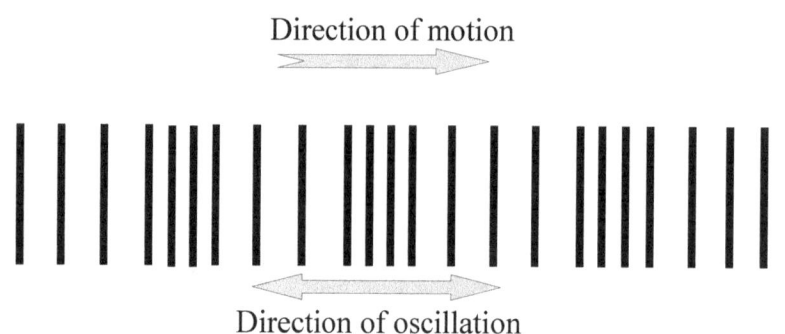

Figure 2.2 – longitudinal waves

Wave anatomy:

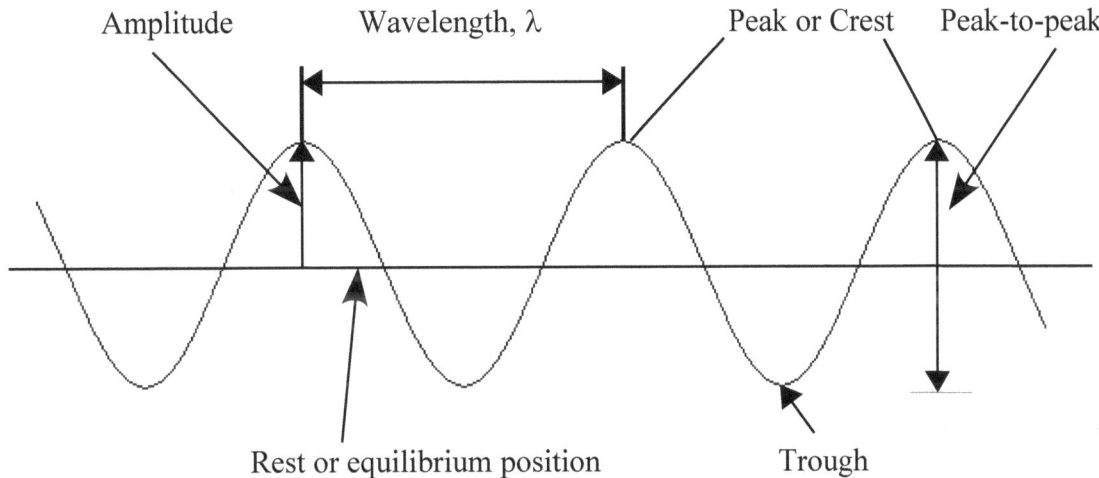

Figure 2.3 Anatomy of waves

Amplitude of a wave is the maximum displacement from the rest or equilibrium position.

Equilibrium position is the point of oscillation when the particle of the medium stops and about to change direction and the wave displacement is zero.

Crest is the highest part of the wave.

Trough is the lowest part of the wave.

Wavelength is the distance between two successive particles which are at exactly the same point in their paths and are moving in the same direction. Such pairs are said to be in phase.

Period is the duration of time to get from crest to crest or trough to trough at a given point. Also the time taken to complete one wavelength.

Frequency is the number of waves that pass a fixed place in a given amount of time.

Peak-to-peak is the distance between successive crest and trough.

Example: Give a name to each point on the wave below.

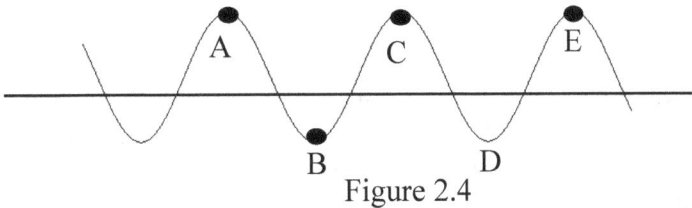

Figure 2.4

Point A represent Crest, point B represent Trough.
Point C to E or B to D shows the wavelength of this wave.

The Speed of Waves

From the above explanation of transverse wave it is clear that any point on a transverse wave moves up and down in a repeating pattern. The amount of time taken to return any point to its original position is called period. The number of periods taking place is called frequency of the wave. If period, T is measured in seconds then the unit of frequency is per second, s^{-1} or Hertz.

This can be written mathematically.

$$f = \frac{1}{T} \qquad (2.1)$$

Also, from the above explanation of wavelength show us that the shortest distance between peaks or trough is one wavelength, λ.

We know that speed is distance covered per unit time.

$$\text{Speed} = \frac{distance}{time}$$

In this case the distance covered is one wavelength per oscillation.

$$\text{Speed } v = \frac{\lambda}{T} \qquad \text{or} \quad v = f\lambda \qquad (2.2)$$

Example: Find the frequency and period of a transverse wave travelling at 350km/h with 20cm wavelength.

First we notice that speed and wavelength are not in appropriate units.
To change speed from km/hr to m/s we divide it by a factor of 3.6 and to change wavelength from cm to m we divide by 100.

Hence, $350 \div 3.6 = 97.22$m/s and wavelength is $20 \div 100 = 0.2$m

$$v = f\lambda \text{ rearranging } f = \frac{v}{\lambda} = \frac{97.22}{0.2} = 486.1 \text{Hz}$$

Now from 2.1
$$T = \frac{1}{486.1} \approx 2.06 \text{ms}$$

The speed of a wave is determined by the medium. One cannot increase the speed of a wave by increasing its wavelength. If we do this, the number of vibrations per second decreases and so the speed of a wave remain constant.

In general the speed of the wave can be thought of as being dependent on two properties of the medium - inertial property and elastic property. The inertial property of a material is the resistance to change in its state of motion or rest. Density of the medium is one of the examples of inertia. The greater the density of the material, less interaction between neighbouring atoms and this will result in slower wave speed. More dense material means it has more mass per unit volume, so high density material's molecule will be heavy and it require more energy to vibrate them. So sound waves will travel slower in dense medium.

Elastic property of a material is the resistance against any force which tries to deform the material. Therefore, elastic property keeps the shape of the material. When force is applied to a certain cross section area of the material is called stress. Metals have very high elasticity compared to rubber. At atomic level this property is due to atomic structure and interaction between them. This interaction can be viewed as springs attached between each atom and when they are subject to external force they will vibrate. If the spring is rigid the atoms will return to their rest position quicker and thus ready to move again quickly. This means they can vibrate at higher speed. Sound waves will travel faster in metal than liquid.

$$v_{solid} > v_{liquid} > v_{gas}$$

If sound waves pass through two materials with similar elastic properties but one's density is twice the other. Sound will travel faster in the less dense medium. We can write the above discussion mathematically:

$$v = \sqrt{\frac{elastic\ property}{inertial\ property}}$$

The quantity which influence the speed of sound in the material is due to the *Bulk modulus B* of the material which is a measure of how much material will compress under a given amount of external pressure. Thus,

$$v = \sqrt{\frac{B}{\rho}} \tag{2.3}$$

Example: If the bulk modulus (elastic property) of aluminium and gold are 68.9 GNm^{-2} and 220 GNm^{-2} respectively. But density of aluminium is 2700 kgm^{-3} whereas for gold density is 19300 kgm^{-3}. Allocate the following speeds of sound to it material: 5052 ms^{-1} and 3376 ms^{-1}.

The speed of sound through aluminium is 5052 ms^{-1} and through gold is 3376 ms^{-1}.

Another property which influences the speed of sound in the air is the temperature. The temperature will affect the elastic property of the air. When an air molecule occupies less volume their density will be higher. As molecules are squeezed closer together their bonding becomes stronger. As we have read above strong bonding means more rigid spring which will result in faster speed of sound. During the night air temperature is low hence air molecules are closer together and sound travels much faster.

However, at high temperature air molecules will have more energy and so they will vibrate faster as a result sound wave can travel faster. The speed of sound in normal air is about 343 ms^{-1} at 20 °C and at 1 atmospheric pressure. The general correction formula for air at different temperatures is given as:

$$v = 331.3\ \text{ms}^{-1} + 0.606\ T \tag{2.4}$$

where T is temperature in °C. This equation best work in the temperature range of 0 to 100 °C. The other property which may affect the speed is humidity. The humidity is just presence of water vapour in the air. This additional matter in the air will affect the mass density of the air. But this normally has a very minor influence and is normally ignored.

Speed of sound at 20°C = 331.3 + 0.606 × 20

= 331.3 + 12.12

= 343.42 ms^{-1}

How Energy is transported on waves.

As mentioned above that wave carry energy from one point to another. Waves are produced by the vibration of the material. This movement in one particle is transferred to the next particle which is actually passing of energy. This energy is passed from one molecule to the next.

The amount of energy carried by wave depends on the amplitude of the wave. High amplitude wave will carry more energy than low amplitude. The energy carried by a wave is proportional to the square of the amplitude of the wave.

$$\text{Energy} \, \alpha \, (\text{amplitude})^2 = E \, \alpha \, A^2 \qquad (2.5)$$

Example: A wave has an amplitude of 3m, if the amplitude is increased to 6m. What happens to the energy transported by such a wave.

As amplitude has doubled, so applying the above rule:

$$E \, \alpha \, (2A)^2 \, \alpha \, 4A^2 \qquad \text{so the energy will quadruple.}$$

Refraction of waves at plane boundaries.

You may have observed that when a pencil or straw is placed into a bowl of water it appears to bend at the water surface. This is due to bending of light (as you will see the light also travel as wave) rays as they cross the water to air boundary. When these rays reach the eye, the eye traces them back as straight lines. These rays will intersect at a higher position than where they have actually originated. This will make straw to appear higher and also water will appears shallower than it really is.

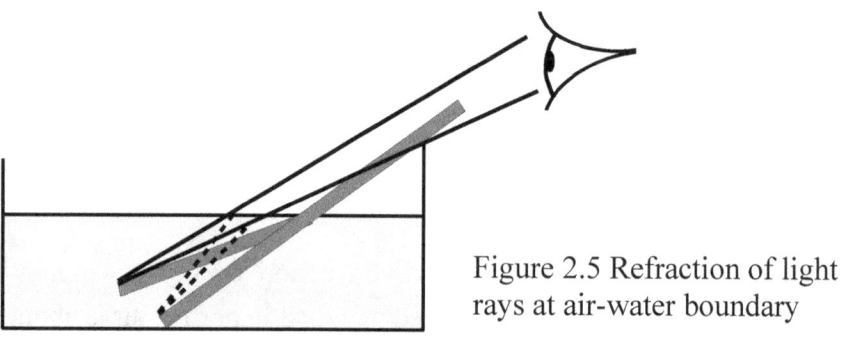

Figure 2.5 Refraction of light rays at air-water boundary

We have studied that speed of the wave depends on the elastic and inertial properties of the medium through which it travels. When wave change medium it also changes speed which then result in a change in the direction of the wave. Although refraction is normally applied to visible light, but it applies to sound and water waves.

When light rays enter from the air to denser medium such as glass, their wavelength becomes shorter. Also, when straight waves pass from deep to shallow water, their wavelength gets shorter. In both cases, frequency remained the same. Now, since the wave velocity, $v = f\lambda$, it means that the waves travels slower in denser medium (shallow water).

When the incident wave hit the boundary at an angle other than zero, the change in wavelength and velocity brings about a change in the direction of travel of the waves when the cross such boundary. This bending of the wave is called, *refraction*.

The Law of Refraction

The law of refraction was first accurately described by Arab scientist Ibn Sahl (c. 940 – 1000) in Baghdad in 984. But it was not until 1621 that Willebrord Snellius (1591 – 1626) found the exact relationship between the angle of incidence and refraction.
The law of refraction, which is generally known as **Snell's law**, governs the behaviour of light rays as they propagate across a boundary between two transparent medium with a different optical density. The Snell's law is stated as follows:

The ratio of the sine of the angle of incidence to the sine of the angle of refraction is a constant for a given pair of media.

Refraction involves the angle that the incident ray and the refracted ray make with the normal to the normal to the surface at the point of refraction. The **normal** is a construction line drawn perpendicular to the surface at the point of incidence. Consider a light ray incident on a plane interface between two transparent media, air and glass, labelled 1 and 2 in Figure 2.6 respectively.

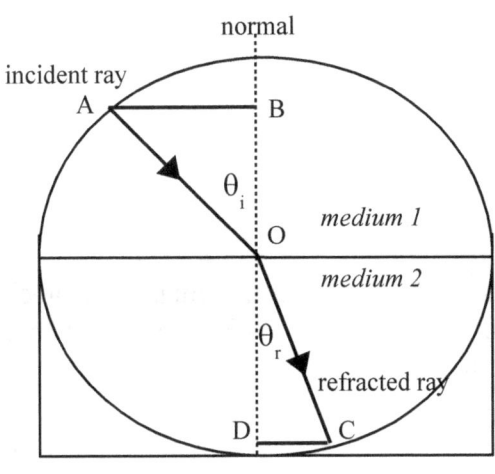

Figure 2.6

Referring to Figure 2.6, we see that

$$\sin i = \frac{AB}{OA}, \text{ and } \sin r = \frac{CD}{OC}$$

Now if we take the ratio of sin i to sin r, we get

therefore
$$\frac{\sin i}{\sin r} = \frac{AB}{OA} \div \frac{CD}{OC} = \frac{AB}{OA} \times \frac{OC}{CD}$$

since $\quad OA = OC \quad$ (radii of circle)

hence
$$\frac{\sin i}{\sin r} = \frac{AB}{CD}$$

Ratio of AB to CD is found as a constant for given mediums. The value of the constant for a ray passing from one medium to another is called the *refractive index* of the second medium with respect to the first; and is denoted by the symbol n.

If the refractive index of medium 1 is n_i and refractive index of medium 2 is n_r, then equation 2.6 can be written as:

$$n_i \sin i = n_r \sin r \qquad (2.6)$$

rearranging

$$\frac{\sin i}{\sin r} = \frac{n_r}{n_i} \qquad (2.7)$$

Refractive index of air is normally taken as one.

The index of refraction of some common materials are given below. Table 1.

Material	Refractive Index
Vacuum	1
Air	1.0003
Water	1.33
Salt	1.54
Crown Glass	1.52
Asphalt	1.635
Heavy Flit Glass	1.65
Diamond	2.42

Source : *CRC Handbook of Chemistry and Physics*

Example: In an experiment, a light ray is incident on o semicircular glass block from air (Fig. 2.7). Find the refractive index of glass when angle of incidence is 30 degree and angle of refraction is 19.2 degrees. Take refractive index of air as 1.

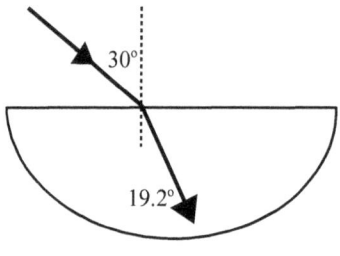

Figure 2.7

using equation 2.7

$$\frac{\sin 30^o}{\sin 19.2^o} = \frac{n_r}{1}$$

$$\frac{0.5}{0.329} = n_r = 1.52$$

Example: When light ray of light incident at an angle of 50° on the surface of the water. Determines the angle of refraction when refractive index of water is taken as 1.33.

rearranging equation 2.7

$$\sin r = \sin i \left(\frac{n_i}{n_r} \right)$$

$$\sin r = \sin 50^o \left(\frac{1}{1.33} \right)$$

$$\sin r = 0.766 \, (0.75)$$

$$\sin r = 0.575$$

$$r = \sin^{-1}(0.575) = 35.10^o$$

Example: Prove that θ_1 is equal to θ_2 as shown in the diagram in Figure 2.8

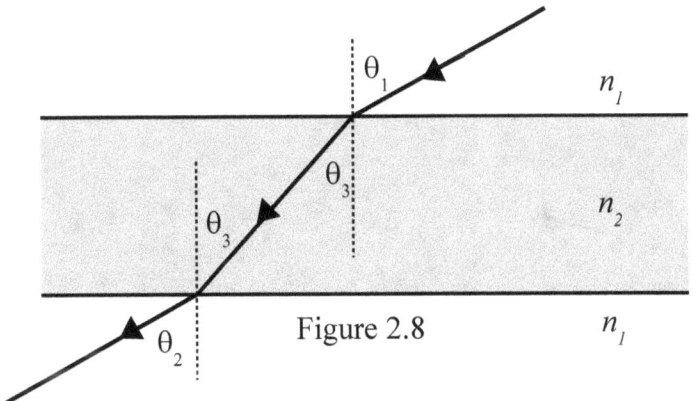

Figure 2.8

Using equation 2.7 and rearranging

$$\sin \theta_2 = \frac{n_2}{n_1} \sin \theta_3 \qquad \text{i}$$

$$\theta_2 = \sin^{-1} \left(\frac{n_2}{n_1} \sin \theta_3 \right) \qquad \text{ii}$$

likewise

$$\sin \theta_3 = \frac{n_1}{n_2} \sin \theta_1 \qquad \text{iii}$$

$$\theta_3 = \sin^{-1} \left(\frac{n_1}{n_2} \sin \theta_1 \right) \qquad \text{iv}$$

substituting for θ_3 (equation iv into equation ii)

$$\theta_2 = \sin^{-1} \left(\left(\frac{n_2}{n_1} \right) \sin \left(\sin^{-1} \left(\left(\frac{n_1}{n_2} \right) \sin \theta_1 \right) \right) \right) \qquad \text{v}$$

This simplifies to

$$\theta_2 = \sin^{-1} (\sin \theta_1) \qquad \text{vi}$$

hence

$$\theta_2 = \theta_1 \qquad \text{vii}$$

The speed of light and index of refraction.

We have studied that when light enters into the optically denser medium it slows down which is a result of shortening of the wavelength. They also move towards the normal. It is clear from Fig. 2.9 that the wavelength has changed from λ_i to λ_r. We also notice in the diagram that there are two right-angled triangles with angles θ_i and θ_r, and sides λ_i and λ_r respectively, together with common hypotenuse AB.

Therefore, the refractive index for waves passing from medium *i* to medium *r* is.

$$\frac{n_r}{n_i} = \frac{\frac{\lambda_i}{(AB)}}{\frac{\lambda_r}{(AB)}} = \frac{\lambda_i}{\lambda_r}$$

As frequency of the wave remains unaltered:
hence, using the wave velocity equation 2.2 $v = f\lambda$

$$v_i = f\lambda_i$$
$$v_r = f\lambda_r$$

therefore,
$$\frac{v(i)}{v(r)} = \frac{\lambda(i)}{\lambda(r)} = \frac{n_r}{n_i} \qquad (2.8)$$

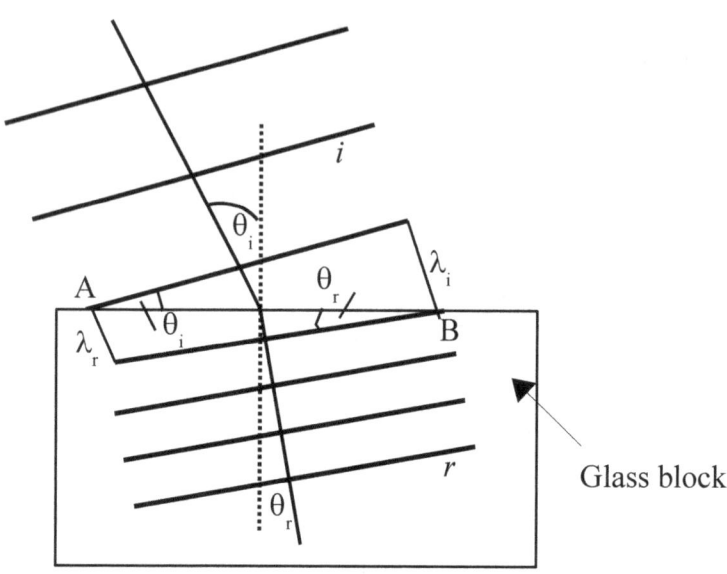

Figure 2.9 wavelength decrease when they enter from less to more dense medium and the result of this they move toward the normal

The amount by which light slows in a given material is described by the index of refraction, n. The index of refraction of a material is defined by the speed of light in vacuum c divided by the speed of light through the material v:

$$n = \frac{c}{v} \qquad (2.9)$$

Example: Light moving from an optical fibre to air changes wavelength from 419nm to 635nm. Take speed of light in the air as 3.00×10^8 ms^{-1}

a) What is the index of refraction of the fibre?
b) What is the speed of light in the fibre?
c) Show that the frequency of light does not change when it moves from optical fibre to air.

a) using equation 2.8
$$\frac{\lambda(fiber)}{\lambda(air)} = \frac{n(air)}{n(fiber)}$$

$$\frac{419\text{nm}}{635\text{nm}} = \frac{1}{n(fiber)}$$

$$n_f = 1.52$$

b) using equation 2.8

$$\frac{v_{fiber}}{v_{air}} = \frac{n_i}{n_f}$$

$$v_{fiber} = v_{air} \times \frac{n_i}{n_f}$$

$$v_{fiber} = 3 \times 10^8 \left(\frac{1}{1.52}\right)$$

$$= 1.97 \times 10^8 \text{ ms}^{-1}$$

c) frequency of light in optical fibre

$$f = \frac{v_{fiber}}{\lambda} = \frac{1.98 \times 10^8}{419 \times 10^{-9}} = 4.71 \times 10^{14} \text{ Hz}$$

frequency of light in the air

$$f = \frac{v_{air}}{\lambda} = \frac{3 \times 10^8}{635 \times 10^{-9}} = 4.72 \times 10^{14} \text{ Hz}$$

The Law of reflection

The law of reflection governs the reflection of waves off smooth surface. For light waves such surface could be glass mirror or polished metal. The law of reflection states that;

i- the incident ray, the normal, and the reflected ray are coplanar (in the same plane)
ii- the angle of incidence is equal to the angle of reflection (Fig. 2.10 angle i = angle r)

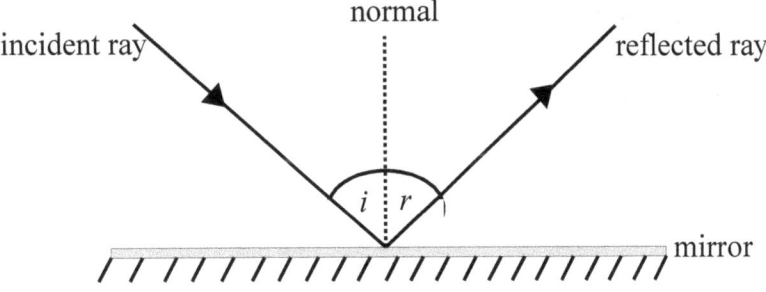

Figure 2.10 Reflection at the boundary of optical surface

Total internal reflection

When a wave hits a boundary of lesser index of refraction, it will be refracted through a larger angle than its angle of incidence. As we increase the angle of incidence the bigger the angle of refraction will become. We will reach to a certain angle of incidence when the refracted angle is exactly 90°. At this point the incidence angle is called **critical angle, c**.

If we increase this angle any further away from the normal, all the light wave will reflect back in the higher index medium. This phenomenon is called **total internal reflection**. The figure 2.11 clarify this.

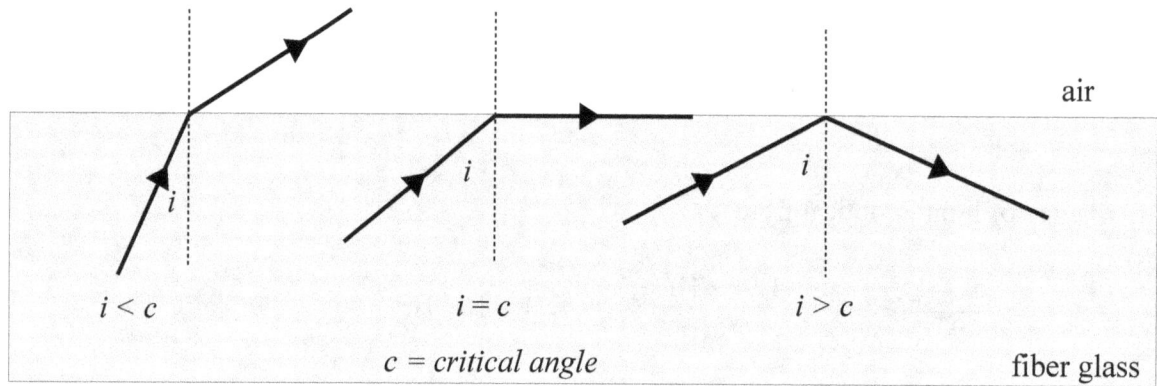

Figure 2.11 Critical angle

Using equation 2.7 we can derive the equation for critical angle. Note angle of incidence is now denoted with letter c.

$$\frac{\sin(c)}{\sin(r)} = \frac{n_a}{n_f}$$

At critical angle the refractive angle is at 90°.

therefore, $\sin r = \sin 90° = 1$

rearranging the equation for critical angle and substituting sin r.

$$c = \sin^{-1}(\sin r) = \sin^{-1}\frac{n_a}{n_f} \qquad (2.10)$$

Example: Find the critical angle for Crown glass – air boundary.

From table on page 22, refractive index for Crown glass is 1.52

Using equation 2.10

$$c = \sin^{-1}\left(\frac{1}{1.52}\right)$$
$$= 41.14°$$

Uses of total internal reflection (TIR)

Decorative cut glass and Diamond: The way the diamond is cut will use TIR. As light rays enters at the top of the diamond will reflect back and exit from the top which will make diamond spark brightly as shown in figure 2.12.

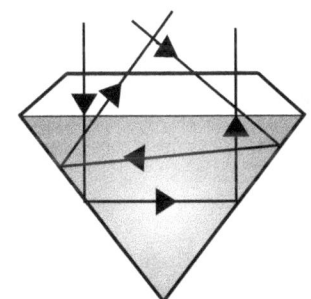

Figure 2.12 Sparkling diamond

Also in decorative tableware and chandeliers cut glass is used which contains heavy elements such as lead. This makes the critical angle smaller and when cut into facets it sparkle.

Telecommunications: Copper cables are used to send signals. Because copper has a certain resistance so the signals will get weaker after few kilometres and have to be re-amplified. This was acceptable for telephone conversation but as we entered into internet age which meant that vast amount of data needs to flow and copper cables are not able to cope with the demand.

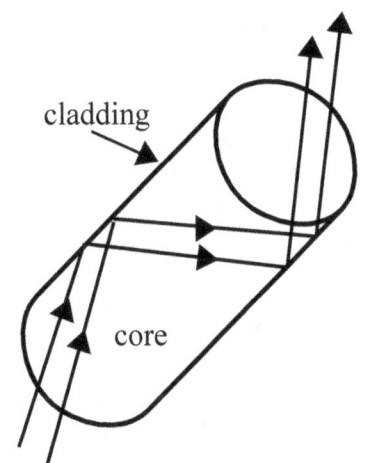

Figure 2.13 TIR in fiber optics

Fibre optics are very long and thin (human hair) strands of pure glass which are bundled together in optical cable. Optical fibre consists of following parts:

1- *Core* – thin glass centre where the light travels. 8 μm diameter.
2- *Cladding* – material that surround the core that reflects the light back into the core. 125 μm diameter.
3- *Buffer* – plastic coating which protects the fibre from damage. 250 μm diameter.
4- *Jacket* – Outer sheathings and armour to protect in conditions such as direct burial in trenches and insertion in paved streets. 400 μm diameter

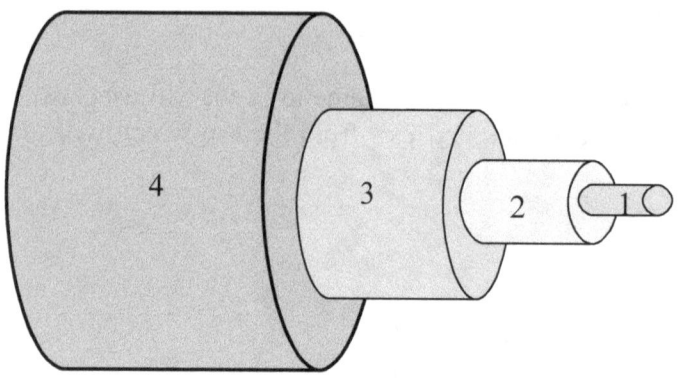

Figure 2.14 The structure of a typical single-mode fiber

Two types of optical fibre are used in telecommunications:

Single mode fibres – the core is made from one type of glass whose refractive index is a step above that of the outer cladding. The size of the core is normally about 9 microns. This allows for a high capacity to transmit signals as it can retain high fidelity of each signal over long distance. It also has a lower dispersion of the signal compared with multi-mode fibre. Signal attenuation is far less than multi-mode fibres.

Multi - mode fibre – the core is much larger than single mode fibre in the order of 62 microns and this allows a large number of modes. The refractive index of core gradually decreases further from the centre. The number of light reflections created as the light passes through the core increases, creating the ability for more data to pass through at a given time. Because of the high dispersion and attenuation rate with this type of fibre, the quality of the signal is degraded over long distances. They are typically used for short distance, data and audio/video applications in LANs. RF broadband signals.

Fibre-scope: A fibre-scope (also known as endoscope) is an instrument used to look inside the human body. Fibre-scope is made from bundles of fibre optics and allows smooth passage of light. At one end is a lens which is connected to an eyepiece for viewing. They use wide angle lenses at either end so they can capture light from a large area and then focusing into a smaller diameter of the entire bundle of cable. The other end can be viewed by a doctor or connected to the camera. The fibre-scope has two distinct functions. One the vision function and the second is illumination.

The vision function is made using optoelectronic system consists of:
a) a lens, b) image transmission, c) a display unit either eyepiece or video monitor.

The illumination function is achieved by using either a light generator, lighting fibre optic bundle.

A medical profession use of endoscopy can be for any of the following:

- investigation of symptoms, in the digestive system including abdominal pain, vomiting, nausea, problem with swallowing food or medicines and gastrointestinal bleeding.
- used in diagnosis, most commonly by performing a biopsy to check for conditions such as bleeding, anemia, inflammation and cancer.
- used in giving treatment, such as cauterization of a bleeding vessel, widening a narrow esophagus, clipping off a polyp or removing a foreign object.

Exercise 2

Q1 – A transverse wave is transporting energy from west to east. In which direction particles of the medium will move?

Q2 – War ships use sonar to detect submarines. Is sonar to be transverse wave or longitudinal wave?

Q3 - The illustration below shows a series of transverse waves. Label each part in the space provided.

a:.. b: ....
c: .... d: ....
e: .... f: ....

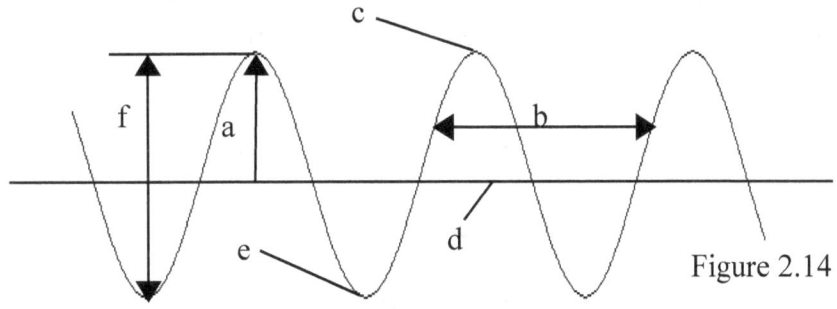

Figure 2.14

Q4 – Answer the following question with reference to Figure 2.15

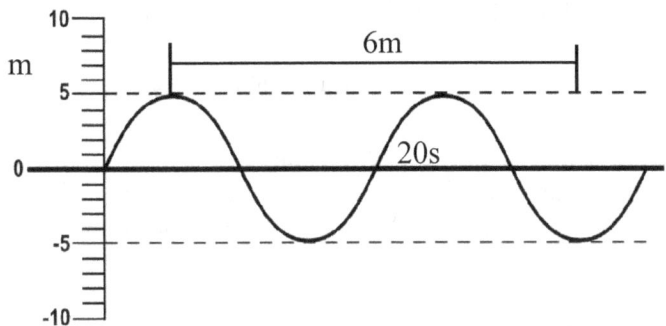

Figure 2.15

a) what is the amplitude?
b) what is the period?
c) what is the frequency?
d) what is the wavelength?
e) what is the peak- to – peak amplitude?
f) what is the speed of the wave?

Q5 – What happens to wavelength when frequency is increased?

Q6 – A wave carries a 50J of energy with an amplitude of 20cm. How much energy will be carried by this wave if the amplitude is increased to 40cm?

Q7 – A bird flaps its wing back and forth 75 times in each second. What are period and frequency of bird's flap?

Q8 – What is the frequency of the second hand on the clock?

Q9 – A student measured that the wavelength of a sound in the laboratory is 1.23m and the frequency is 270Hz. Calculate the speed of the sound in the laboratory.

Q10 – If now student doubles the frequency of the sound source to 540Hz. Calculate the new speed of sound?

Q11 – A scientist examining a clear solid determines that light travels at 1.95×10^8m/s through the material. Should that scientist take the solid to a jeweller, or should he grind it up on his dinner?

Q12 – A ray of light passes from kerosene to crown glass. The angle of incidence of the light is 45° and the relative refractive index from kerosene to crown glass is 1.08. Calculate the angle of refraction in the glass.

Q13 – Using figure 2.16, calculate the angle of refraction in the lower medium.

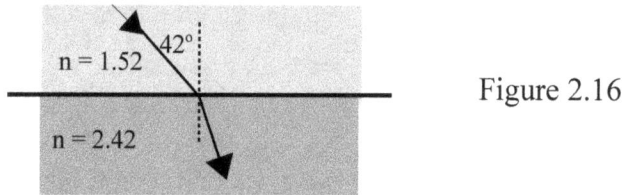

Figure 2.16

Q14 – What is the critical angle for the fibre optics with refractive index 1.44?

Q15 – What is the refractive index of a decorative glass with critical angle 38°?

Q16 – A beam of light travels from water into a piece of diamond in the shape of a triangle, as shown in diagram 2.17. Step-by-step, calculate the beam angles as shown in the diagram until it emerges from the diamond. Copy the diagram and draw the rays throughout and show where it will exit? Also answer the following question as well?

a) What is the speed of light inside the diamond?
b) What is the critical angle between diamond – air interference?

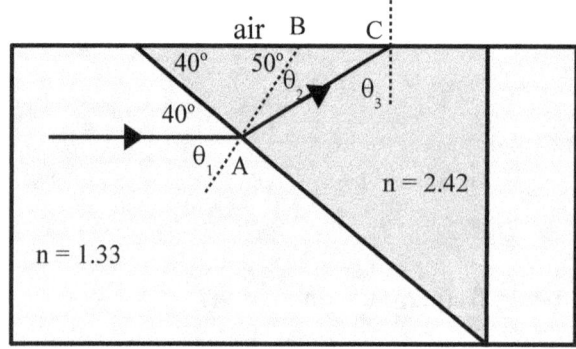

Figure 2.17

3 – Electromagnetic Spectrum

Nature of electromagnetic waves

Electromagnetic waves are either emitted as continuum emission by charged particle such as electron when they change their velocity or direction. While line emission result from atomic processes that only have very specific quantised energy. Continuum emission is not quantised as photon emits with continuous energy distribution.

Figure 3.1 Continuous electromagnetic spectrum

Figure 3.2 Line emission electromagnetic spectrum of iron

There are two main sources of electromagnetic radiations, thermal or non thermal.

Thermal emission: All matter with a temperature greater than absolute zero emits thermal radiations. The temperature of the material causes the atoms within the material to move around. For example if gas is heated the molecules will move faster and will collide with other molecules thus change their direction of motion. A change in direction is equivalent to acceleration and so these molecules will emit continuous radiations. If we place a metal rod on flame for few minutes and if we then remove it and place it one side. Although we may not be able to see but this rod is emitting electromagnetic radiation in the infra-red region. If we again place the rod in the flame for a longer period, it will start glowing red thus emitting higher energy radiations. Scientist call this black body radiations.

A black body is defined as a body that absorbs all radiation that falls on its surface. Once the object reaches an equilibrium temperature it will start re-radiating energy in characteristic spectrum. The peak wavelength of the spectrum depends on the temperature of the object. The diagram in figure 3.3 shows black body spectra of objects at three different temperatures. We can see that at higher temperature object emit more radiation at shorter wavelength.

The emission spectrum of black body was first fully explained by Max Planck (1858 – 1947). The radiation power (energy per unit time) from a black body is proportional to the fourth power of the absolute temperature.

$$P \propto T^4$$

This can be fully described by **Stefan – Boltzmann Law** as

$$P = \sigma A T^4 \qquad (3.1)$$

where,

P is power measured in Watts (W)
σ is the Stefan – Boltzmann constant 5.6703×10^{-8} Wm^{-2}K^{-4}
A surface area of the emitting body (m^2)
T absolute temperature in *Kelvin* (K)

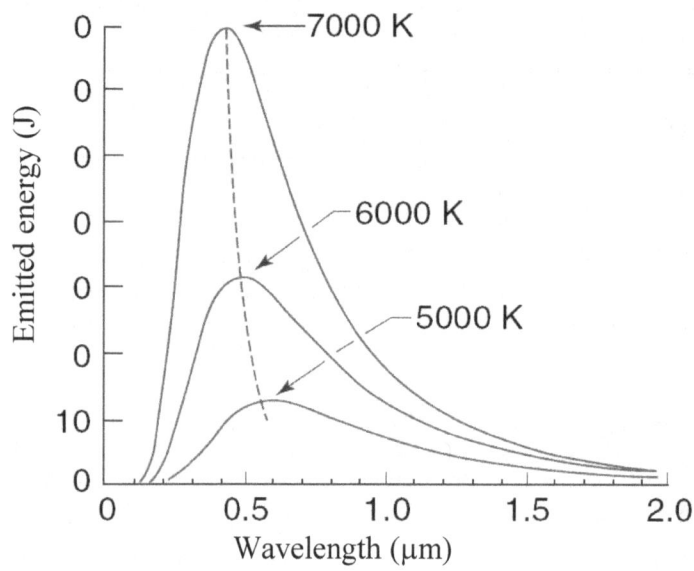

Figure 3.3 Blackbody radiations energy at various wavelengths.

Non – Thermal emission: Non – thermal emission does not behave like black body radiation. In fact, emission increases at longer wavelength which is opposite to what happens in black body radiations. Most common forms of non – thermal emission is synchrotron emission. These arise by the acceleration of charged particles (moving very close to the speed of light) in a strong magnetic field. The electromagnetic radiation is emitted in a narrow cone in the forward direction, at tangent to the particle's orbit.

Line emission of electromagnetic spectra.

By the early 1900 scientist observed that when gas at low pressure (so atoms do not experience many collisions) was given energy in the form of electric discharge, the gas atoms appear to be emitting electromagnetic radiations which were not continuous but discrete. This means that they only emit certain frequencies only. When such emission lines were superimposed on continuous spectrum, these spectral lines existed in the series in the different regions (infrared, ultraviolet etc.) of the electromagnetic spectrum. Figure 3.2. Each element has its own unique emission spectra.
When a scientist placed the same gas at low temperature and subjected to radiations they observed a similar line spectrum but this time rather than emitting certain colour the appeared as black lines. This is called absorption spectra.

The full understanding of this phenomenon wasn't understood until Niels Bohr (1885 - 1962) proposed the atomic model in 1915. He showed that inside the atom electrons orbit around the nucleus just like planets go around the sun. Just likes planets orbits are stable so are electron orbits.

But the similarity ends here. Electron orbits are called energy levels. The lowest energy level is called *ground state*. Next level is called *first excitation energy level* and one above the *second excitation energy level* and so on. Energy is *quantised*. This means only certain orbits are allowed with discrete energy and nothing else is allowed in between. Electrons can make transitions between the orbits i.e. energy levels by absorbing or emitting the energy difference between the orbits.

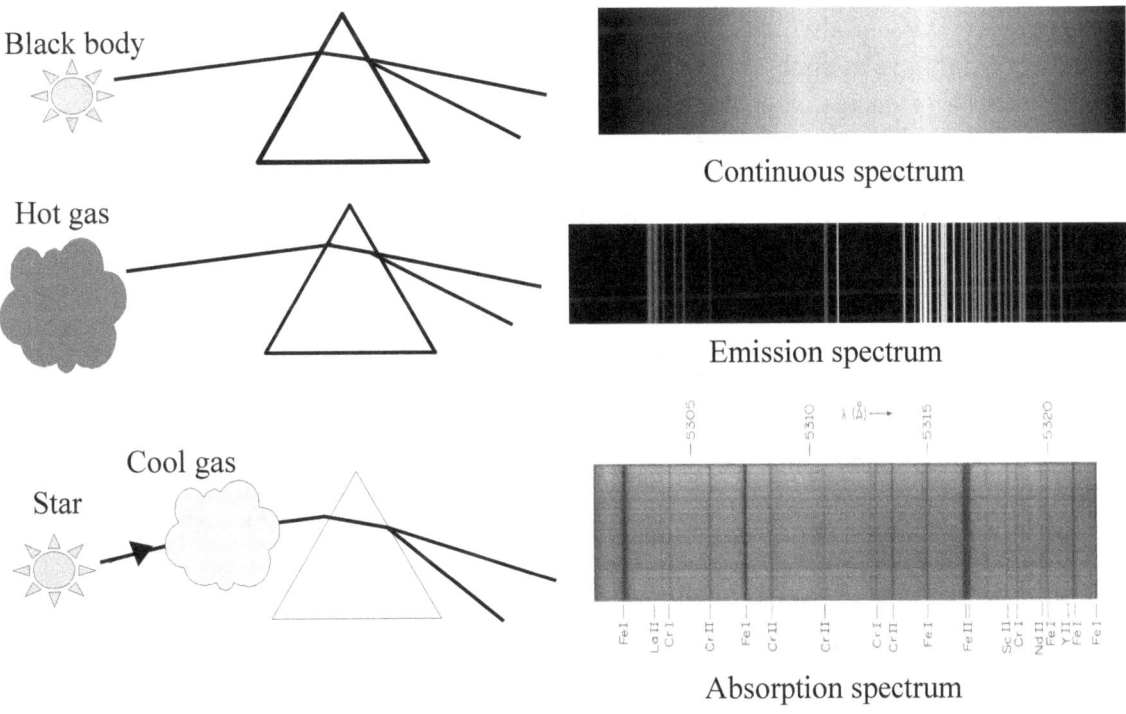

Figure 3.4 The origin of continous, emmission and absorptio spectrum

Now we can explain the emission. When the electrons in the atoms are excited by means of heating or collision with another electron or photon. This extra energy absorbed by the electron will push it to the next energy level as long as energy is equal to the difference between two energy levels. So far example the ground state energy of hydrogen electron is -13.6eV (negative sign signifies that it take that much energy to unbind the electron from the atom) and the first excitation energy level is -3.40 eV. Then electron needs minimum if 10.20eV energy to move there. Once the electron jump to a higher energy orbital it cannot stay there. It falls back down and by doing this it will lose energy in the form of electromagnetic radiation. This energy will be will be re-emitted in the form of a *photon*. This energy is quantised and come as a packet of quanta and that is why there is a discrete line on the emission spectrum for that unique transition. Figure 3.5

The fact that only some colour lines are emitted for a given element's emission spectra that means only certain frequencies are emitted. Plank has already give this relationship but nobody understood the actual meaning of this until Niels Bohr explained it. This equation is the corner stone of quantum mechanics.

$$E = hf \qquad (3.2)$$

Where E is energy in J, f frequency in Hz and h called Planck's constant.
h = 6.63 x 10^{-34} Js

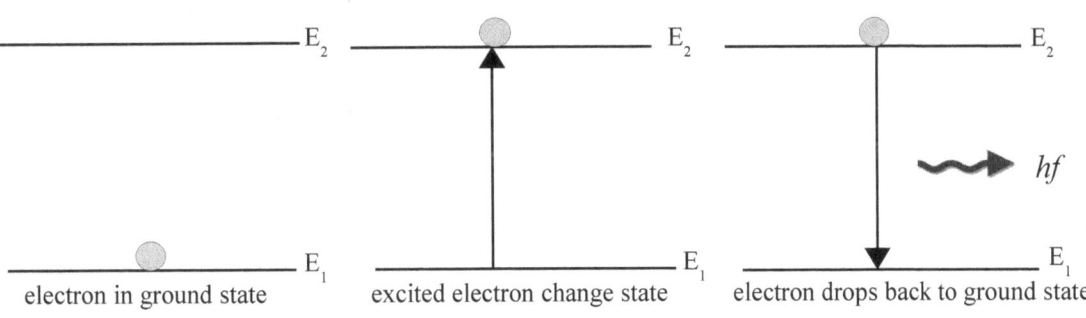

Figure 3.5

Because an electron bound to an atom can only have certain energies so electron can only absorb certain energies. We can also understand absorption spectra. We know hot object emits like black body so when looking at such object and in between is a cold gas then if the energy carried by photons match with energy gap of the intervening gas then it will absorb that energy and those photons will be missing and appears as dark lines in our spectrum.

Energy level diagram of hydrogen is shown in figure 3.6. You can notice that there are many possibilities available for electron to drop to ground energy level and these become more as electron goes to a higher energy level. In this diagram only transition from n = 4 to n = 1 are shown.

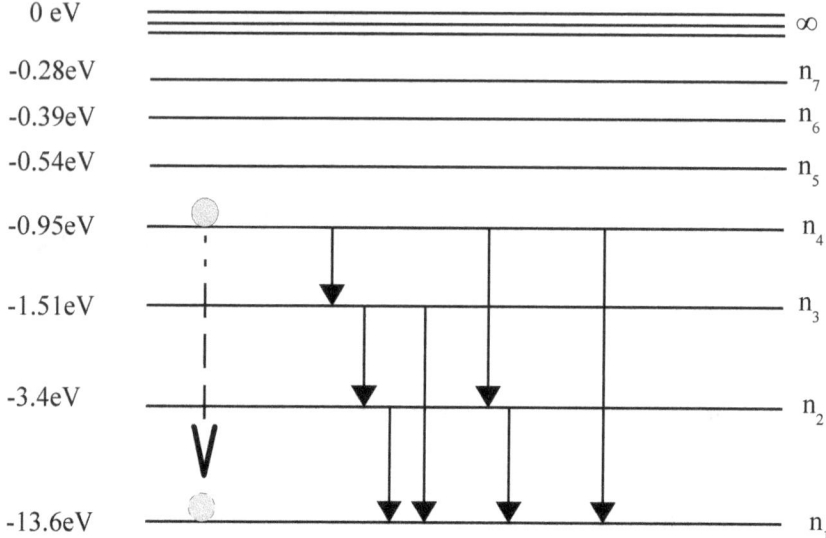

Figure 3.6 - hydrogen energy levels with transition shown for electron coming to ground state from n_4. Energy levels are not to scale.

Long before the discovery of hydrogen quantum model astronomers have been looking at hydrogen spectrum of stars as they are mostly made up of hydrogen. The very first person who observed these lines was Swiss scientist Johann Jakob Balmer (1825 – 1898) in 1885. He observed 4 lines with wavelengths of 410.2nm (violet), 434.1nm (blue), 486.1nm (cyan) and 656.3nm (red). As they are in visible region of the spectrum so Balmer was able to see them. Now we know there are 4 more lines in this series which are in the ultraviolet region and Balmer was unable to see them. This series is a transition of an electron from upper energy levels to second energy level, n_2

Each line has been designated by a letter 'H'. Line with the transition from 3 → 2 is called H_α and next from 4 → 2 is called H_β etc.

Another series was discovered by American Physicist Theodore Lyman (1874 – 1954) when he was studying the ultraviolet spectrum of electrically discharged hydrogen gas in 1906. All the wavelengths in Lyman series are in the ultraviolet region. This series is a transition of an electron from upper energy levels to ground energy level, n_1. Each line is designated with letters 'Ly'. So line with the transition from 2 → 1 is called Ly_α.

Paschen series was first observed by German physicist Friedrich Paschen (1865 – 1947) in 1908. This series is in the infra red region of the spectrum. This series is transitioning energy level 3, n_3. More series of hydrogen atom have been observed like Brackett series, Pfund series and Humphrey series.

Some hydrogen spectral lines for three series and their resulting wavelengths are shown in figure 3.7

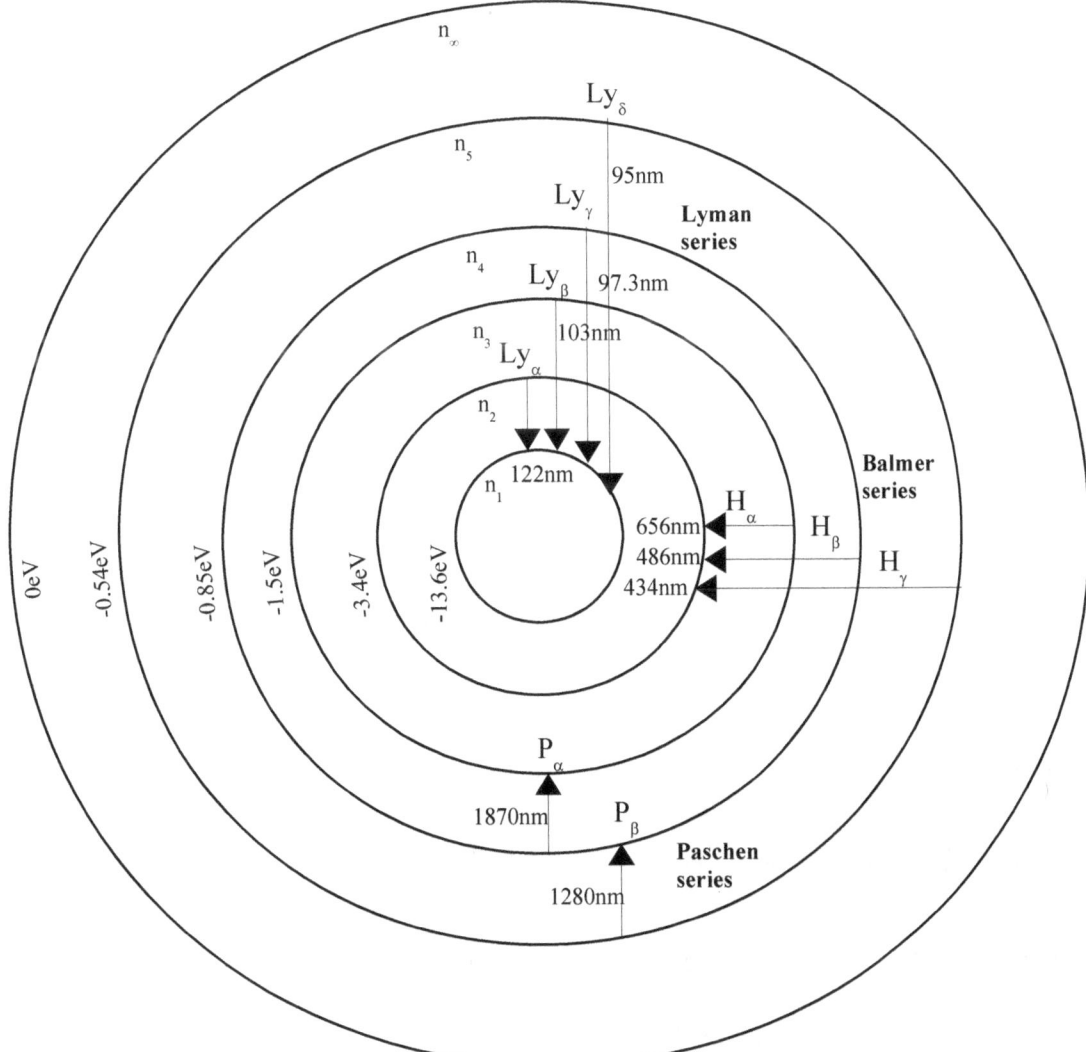

Figure 3.7 Energy levels are not to scale

You must have noticed that we have described energy levels with eV (electron volt). Normally we use joule for energy. Historically, scientists were accelerating particle in large electric potential and particle with charge q has energy qV after passing through the potential V. q is normally an integer value of the elementary charge (charge on the electron, e which has a value of 1.602×10^{-19}C). Thus one eV is equivalent to 1.602×10^{-19}J.

Equation 3.2 allows us to calculate the energy of photons, given their frequency. If wavelength is required then we can use equation 2.2 but this time velocity is the speed of light, c.

i.e.,
$$E = hf$$
and
$$c = \lambda f$$

therefore
$$E = \frac{hc}{\lambda}$$

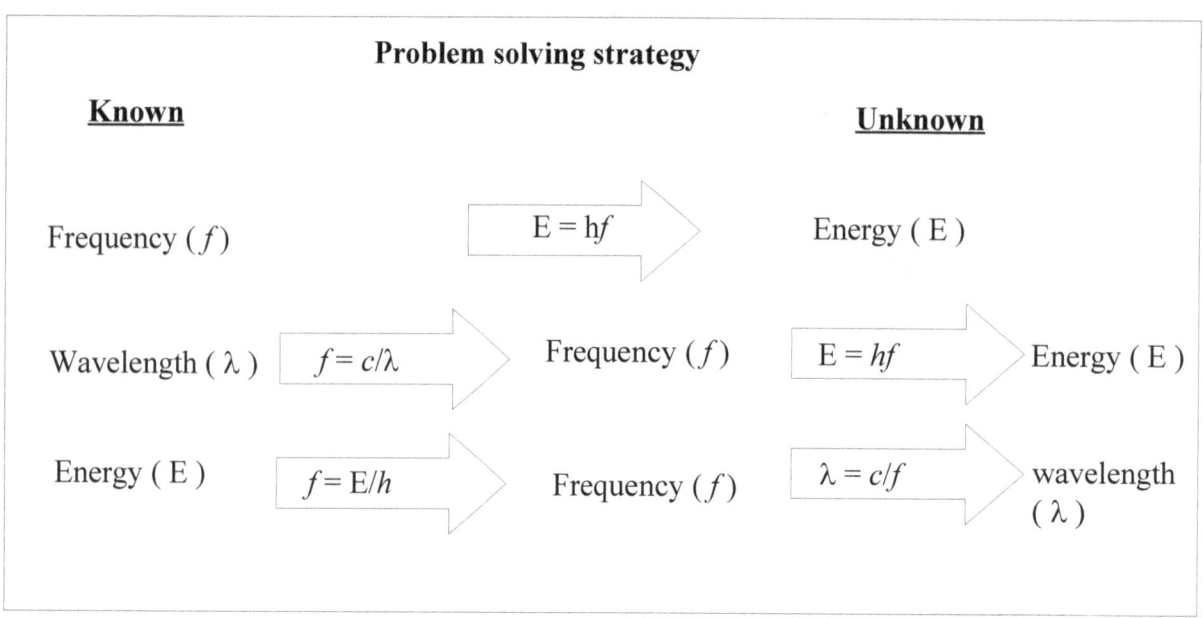

Example: Light with a wavelength of 525 nm is green. Calculate the energy of the photon both in Joule and electron volts.

Using the above box, we know the wavelength so using the middle strategy,

$$f = \frac{c}{\lambda} = \frac{3 \times 10^8}{525 \times 10^{-9}} = 5.71 \times 10^{14} \text{ Hz}$$

$$E = hf = (6.63 \times 10^{-34}) \times (5.71 \times 10^{14}) = 3.79 \times 10^{-19} \text{ J}$$

To convert Joule into eV we divide Joule by 1.602×10^{-19} C

$$\frac{3.79 \times 10^{-19}}{1.602 \times 10^{-19}} = 2.36 \text{ eV}$$

Example: Sun radiates 3.84×10^{26} energy per second. The diameter of the sun is 1.39×10^9 m. If sun radiates as black body, calculate the sun's surface temperature and compare it with the true surface temperature of 5778K.

Using equation 3.1, we need to calculate the sun surface area first. We consider sun a perfect sphere.

The surface area of sphere is $4\pi r^2 = 4\pi \left(\frac{1.39 \times 10^9}{2}\right)^2 = 6.07 \times 10^{18}$ m²

rearranging equation 3.1 for T,

$$T = \sqrt[4]{\frac{P}{\sigma A}} = \sqrt[4]{\frac{3.84 \times 10^{26}}{(5.6703 \times 10^{-8})(6.07 \times 10^{18})}} = 5775 \text{K}$$

Very close agreement.

Example: An X-ray tube used for diagnostic purpose has an operating voltage of 60kV. Calculate:
a) the maximum energy of of x-ray photon that can be produced.
b) the minimum wavelength of the x-ray photon.

a) we need to convert energy from eV into Joule to use equation 3.2

To convert eV into J, we multiply Joule by 1.602×10^{-19}C
$$(60 \times 10^3) \times (1.602 \times 10^{-19}) = 9.612 \times 10^{-15} \text{J}$$

This is the maximum energy that photon of x-ray can carry.

b) Using problem solving strategy box, the last technique is needed.

$$f = \frac{W}{Q} = \frac{9.612 \times 10^{-15}}{6.602 \times 10^{-34}} = 1.59 \times 10^{19} \text{Hz}$$

$$\lambda = \frac{c}{f} = \frac{3 \times 10^8}{1.59 \times 10^{19}} = 1.88 \times 10^{-11} \text{m}$$

Creating an electromagnetic wave.

We have studied waves in the last chapter. In this chapter, we have explained electromagnetic radiations are carried by particles called *photons*. But we have studied in the last chapter that light which is just small part of the electromagnetic spectrum travels as a wave. This is one of the mysteries of quantum physics that light is both wave and particle, and our observation makes the choice. Early scientist thought that electricity and magnetism are two different forces. Danish physicist Hans Christian Ørsted (1777 – 1851), made the surprising observation on 21 April 1820 that compass needle was deflected when a current in the nearby wire was switched on and off. Although he could not give any explanation of his observation apart from speculating that somehow electric wire is behaving like a magnet and later that turned out to be true. But many great scientists worked on this observation. Some names are worth mentioning like French physicist André Marie Ampére (1775 - 1836), British physicist Michael Faraday (1791 – 1867), German physicist Heinrich Hertz (1857 – 1894) and finally by Scottish mathematical physicist James Clarke Maxwell (1831 – 1879).

They all agreed on four pillars of electromagnetism as follow:

1- Like charges repel whereas unlike charges attract each other. The force of repulsion or attraction is directly proportional to charge on them and inverse proportional to the square of their distance.
2- Like magnetic poles repel whereas unlike pole attract each other just like an electric charge. But they always come in pair: every south pole is yoked to a north pole.
3- A current is seen to flow in a looped wire when it is moved away brought towards the magnet. The direction of current depends on the movement.
4- A current in a looped wire creates a circular magnetic field around the wire. The direction of the magnetic poles depends on the direction of flow of current.
Focus on these two facts:
   1. an oscillating electric field generates an oscillating magnetic field
   2. an oscillating magnetic field generates an oscillating electric field
Those two points are key to understanding electromagnetic waves.

An electromagnetic wave (such as a radio wave) propagates outwards from the source (an antenna, perhaps) at the speed of light. What this means in practice is that the source has created oscillating electric and magnetic fields, perpendicular to each other, that travel away from the source. The E and B fields, along with being perpendicular to each other, are perpendicular to the direction the wave travels, meaning that an electromagnetic wave is a transverse wave. The energy of the wave is stored in the electric and magnetic fields.

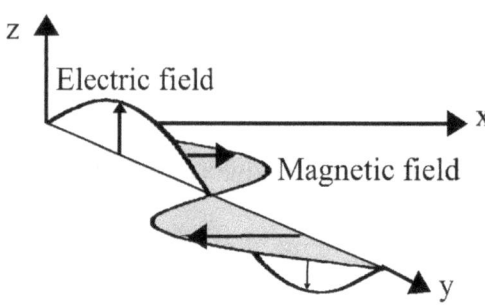

Figure 3.8 propagation of electromagnetic waves

Energy in an electromagnetic wave

The electromagnetic wave carries energy as they travel through empty space. The energy in electromagnetic wave is equally split into electric and magnetic field. In general, the energy per unit volume in an electric field is given by:

Electric field energy density $\eta_E = \frac{1}{2} \varepsilon_0 E^2$ (3.3)

Magnetic field energy density $\eta_B = \frac{1}{2} \frac{B^2}{\mu_0}$ (3.4)

where, ε_0 is a constant called permittivity of free space and has a value of 8.85×10^{-12} Fm^{-1}
μ_0 is also a constant called permeability of free space and has a value of $4\pi \times 10^{-7}$ Hm^{-1} or NA^{-2}

Regions of the Electromagnetic Spectrum (EM)

The electromagnetic spectrum is the distribution of EM radiations according to energy. There are seven distinct regions of the EM. A table 3.1 illustrates regions of the EM spectrum.

Table 3.1 – Regions of EM spectrum

Region	Wavelength (m)	Frequency (Hz)	Energy (eV)
Radio	$> 10^3 - 10^{-3}$	$10^5 - 10^{11}$	$10^{-10} - 10^{-4}$
Microwave	$10^{-3} - 25 \times 10^{-6}$	$10^{11} - 1.2 \times 10^{14}$	$10^{-4} - 10^{11}$
Infrared	$25 \times 10^{-6} - 750 \times 10^{-9}$	$1.2 \times 10^{13} - 4 \times 10^{14}$	$10^5 - 10^{11}$
Visible	$750 \times 10^{-9} - 400 \times 10^{-9}$	$4 \times 10^{14} - 7.5 \times 10^{14}$	$10^5 - 10^{11}$
Ultraviolet	$400 \times 10^{-9} - 1 \times 10^{-9}$	$7.5 \times 10^{15} - 3 \times 10^{17}$	$10^5 - 10^{11}$
X-Rays	$1 \times 10^{-9} - 1 \times 10^{-12}$	$3 \times 10^{17} - 3 \times 10^{20}$	$10^5 - 10^{11}$
Gamma Rays	$< 1 \times 10^{-12}$	10^{20}	$10^5 - 10^{11}$

Radio Waves: They have the lowest frequency and longest wavelength. There are five main classes of radio waves.

1- Long wave radio (10^3m – 10^4m) – AM radio (international broadcast)
2- Medium wave radio (10^2m – 10^3m) – AM radio (national broadcast)
3- Short wave radio (10^1m – 10^2m) – Amateur radio, police, ambulance
4- Very high frequency (10^0m – 10^1m) – FM radio
5- Ultra high frequency (10^1m – 10^{-2}m) – TV, radar, air traffic control

Heinrich Hertz produced the first time radio waves by using spark gap attached to an induced coil and he also used spark gaps with antenna wire as a receiver. Radio waves are emitted by all heavenly bodies. The atmosphere of the Earth is transparent to radio waves with wavelengths of a few millimetres to about twenty metres. So astronomers can pick up the radio signals on ground based radio telescopes. Since radio waves are much larger than the visible light so the radio telescopes are much larger than optical telescope. The world largest radio telescope is in Arecibo, Puerto Rico. It is 300 metres in diameter and is constructed in a natural bowl shaped depression. The most distant object was also discovered by radio telescope called quasar (quasi stellar radio source).

Figure 3.9 The Wireless Telephone" U S Patent Office in Washington, DC

Figure 3.10 World's largest single-aperture radio telescope at Arecibo telescope in Puerto Rico

Sources: Transmitters (using inductance and capacitance), sparks (e.g., from the brushes of unsuppressed motors).
Detection: Receivers containing inductance and capacitance which are set into resonance by the wave.
Uses: Medium range airborne communication. Radio astronomy. For navigation of ships and aircraft the radio range, radio compass, and radio time signals are widely used. Radio signals sent from global positioning satellites can also be used for a precise location. Radio frequency energy has been used in medical treatments, generally for minimally invasive surgeries and coagulation, including the treatment of sleep apnea. Magnetic resonance imaging uses radio frequency waves to generate images of the human body.

Microwaves: Microwaves are actually a portion of the radio band. They are distinguished from radio waves the way they are generated and detected. Just like radio waves they are also grouped.

There are 14 bands range from 1 to 170GHz. Only two are mentioned here.

    1. C-band – Medium length microwaves which can easily penetrate through clouds, smoke, dust and snow. So it's useful to study the Earth from a satellite. This band is also used by communication satellites.
    2. L-Band – Mainly used by Global Positioning System (GPS).

Microwaves are used in RADAR to detect planes. This radar altimeter in spacecraft beams microwaves at two different frequencies (13.6 and 5.3 GHz) at the sea surface and measures the time it takes the pulses to return to the spacecraft. Combining data from other instruments that calculate the spacecraft's precise altitude from the sea surface height within just a few centimetres! Microwaves are also used for cooking. These microwaves use frequency which water molecule absorbs and thus heat up the food.

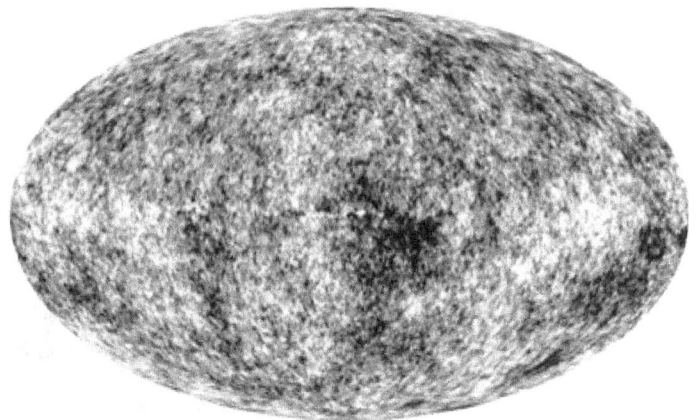

Figure 3.11 WMAP image of the CMB temperature anisotropy

Microwave Background Radiation (or Cosmic Background Radiation) is the primordial radiation field that fills the universe, having been created in the form of gamma rays at the time of the big bang. It has now cooled so that its temperature today is 2.73K and its peak wavelength is near 1.1mm (in the microwave portion of the EM spectrum). These waves have now been detected and have confirmed the Big Bang theory.

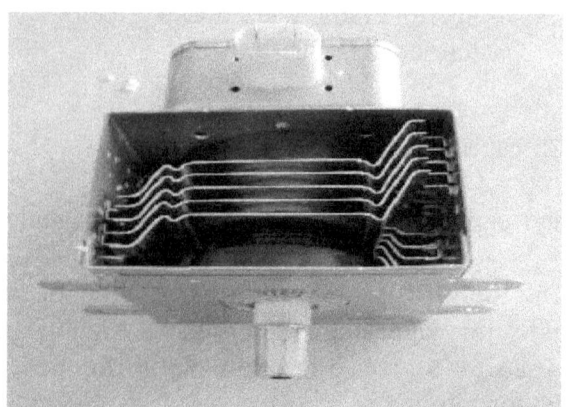

Figure 3.12 A magnetron with section removed.

Sources: Special electrical circuits (klystron, magnetron).

Detection: Resonance in similar special electrical circuits (also klystron, magnetron).
Uses: Greatly used in airborne communications, for example with mobile phones. Cooking. Remote sensing. Satellite link.

Infra – Red (IR): Infrared start from around 0.025mm and extend to the red edge of the visible spectrum at 0.74µm. Infra-red is also divided into five bands.

1- Near infra-red (0.75-1.4µm): They are commonly used in fibre optics communication. Night visions also use this band as image intensifiers are more sensitive to these wavelengths. They are also used to see the health of vegetation as healthy plants will reflect more near IR.
2- Short-wavelength infra-red (1.4-3µm): Also used in fibre optics for long distance telecommunications.
3- Mid-wavelength infrared (3-8µm): This is the band in which homing head of passive IR heat seeking missiles work. As the jet engine exhaust plume emits a signature infrared at this band.
4- Long-wavelength infrared (8-15µm): This band is also called *thermal infrared*. In this band sensors can obtain a complete passive picture of the outside world.
5- Far-infrared (15-1000µm): This band is for laser.

Infra-red were discovered by Sir Frederick William Herschel (1738 - 1822) in 1800 when he was measuring the difference in temperature between colours. He discovered even higher temperature beyond red colour. As we have already studied that as object heat up they start glowing through red to violet. But before they start glowing they emit lower end of the spectrum. Human body (all warm blooded animals) also emits IR. Our eyes are not sensitive to IR but we have made sensors and they can see these waves.

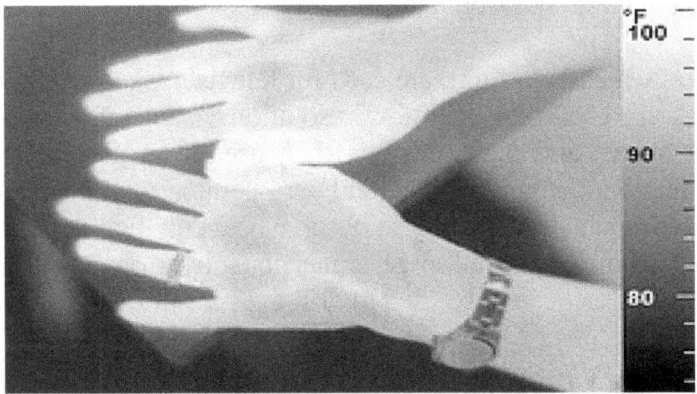

Figure 3.13 An IR image of a hand using thermographic camera.

Many heavenly bodies are too cool to emit visible radiations and they glow in infrared like interstellar gas, nebula, planets. The Cassini spacecraft has explored many such objects and that has made a major contribution in understanding the dynamic of the universe. Astronomer's are waiting for the launch of *the James Webb space telescope* scheduled to be launched in 2018. This IR telescope will find the first galaxies that were formed in the early universe. This will also see through dust clouds to see the new star formation.

Sources: All hot bodies.
Detection: Photographic plates, heat sensitive detectors (e.g., IR thermometers), thermopile.

Uses: Infra-red radiation is mainly *heat*. All moderately hot bodies emit infrared. This is often used to detect human beings by groups like the army to see the enemy, and the police trying to locate criminals who are trying to hide. Firefighters also use infrared detectors to locate the sources of fires which are naturally hotter than the blazes around them.

Infrared photography also enables pictures to be taken in the dark or in hazy conditions.

Infrared radiation is also used in optical fibres for long distance communication. Infrared is also used for point to point communications.

Infrared is also used for therapeutic purposes. Doctors use infrared lamps to treat skin diseases and relieve the pain of sore muscles. In these treatments, the infrared rays pass through the patient's skin and produce heat.

Scientific advances pertaining to infrared have allowed buildings to keep cool even in the summer. Gold transparent films on the windows of large office buildings reflect infrared and help to keep temperatures cool.

Visible Light: This is the only part of the spectrum that human eyes are sensitive. Our sun is dominate source of visible light waves. This region is split into seven colours. Longest wavelength starts with Red colour and finish at Violet colour and in between these are Orange, Yellow, Green, Cyan and Blue. The spectrum is continuous, with no clear boundaries between one colour and the next. Most light we interact with is in the form of white light, which consists of some are all seven colours. Each colour has its associated temperature. Our sun's colour is yellow because its surface temperature is 5778K. If our sun's surface temperature was 2500K, it would glow red like Betelgeuse. If on the other hand it's colour was 13000K, it would glow in blue colour like Bellatrix in Orion.

Sources: Very hot bodies (progressively red-hot, yellow-hot and then white-hot), discharge lamps (e.g., most bulbs), fluorescence.

Detection: Photographic plates, photocells, the human eye.

Uses: Cathode ray tubes, which emit light, are used for televisions, computer monitors and the like. LED and LCD displays are used for cheap low resolution visual information. Bulbs are used for lighting which human beings and other animals then use as an aid for (amongst other things) location resolving, navigation, communication, and peripheral/accessory movement (e.g., walking). Optical telescope on high mountains are busy to observe heavenly bodies. Lasers are used in Micro-surgical procedures such as making small, precise incisions, liver operations, and capillary surgery, causing little loss of blood. Lasers are also used to drill eyes in surgical needles.

Ultraviolet radiation (UV): Ultraviolet radiation can be split into the shorter wavelength far ultraviolet and the longer wavelength near ultraviolet (the boundary between the two being at approximately 200nm). The extreme ultraviolet range overlaps with the far ultraviolet at wavelengths of between 1 and 100nm). Our sun produces ultraviolet radiation at all wavelengths in the range. Scientists have divided them into three bands. UV-A, UV-B and UV-C. UV-C are most harmful to human life but thanks to our atmosphere which completely absorb these rays. 95% of UV-B is absorbed by the ozone layer. If we are exposed to these waves for prolong period can cause sunburn and also increase the risk of DNA and other cellular damage to human beings. The damage caused by UV-B include direct formation of *thymine dimers* or other *pyrimidine dimers*. These are formed from thymine or cytosine bases in DNA by photochemical reaction produced by ultraviolet exposure which can change the structure of the DNA. They are the primary cause of melanomas, type of skin cancer. UV-B in high intensity are harmful to the eyes and can cause photokeratitis or arc eye and may lead to cataract, pinguecula and pterygium formation.

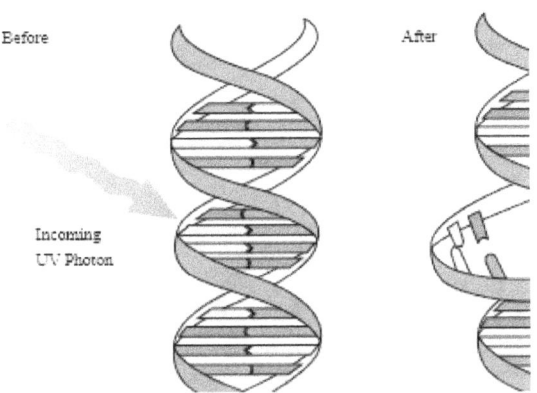

Figure 3.14 DNA Lesion-Thymine Dimer

UV-A is the most common type of UV radiation that we encounter. Ozone absorbs very little of this portion of the UV. We need UV-A for synthesis of vitamin D. UV-A contribute to skin cancer via indirect DNA damage like free radicals and reactive oxygen species such as *hydroxyl*, which in turn can damage DNA which consist mostly of single-strand breaks in DNA. On 13th April, 2011 the International Agency for Research on Cancer has classified all categories and wavelengths of ultraviolet radiation as a Group 1 carcinogen.

Astronomers have divided UV into three regions: the near UV, the far UV and the extreme UV. The near ultraviolet abbreviated NUV. *Photographic* plates detect NUV coming from the blue sky as violet.

Sources: Ultra hot bodies, mercury vapour lamp, electric arcs (sparks). Germicidal lamps, Excimer lasers and third harmonic Nd:YAG lasers.

Detection: Photographic plates, fluorescence of certain chemicals, photocells, photoelectric devices.

Uses: UV radiations are used in the treatment of skin conditions such as *psoriasis* and *vitiligo*. For this treatment 311nm wavelength is used. Suitable doses of Ultraviolet rays cause the body to produce vitamin D, and this is used by doctors to treat vitamin D deficiency and some skin disorders. UVB helps us when we hang washing outside to dry, as some of the bacteria present in the washing are inactivated by exposure to UVB. This helps to protect us from infection. The ability of UV to inactivate bacteria and viruses is used to produce UVC to sterilise surfaces of things such as medical equipment.

X-Rays: X-Rays were first discovered by German physicist Wilhelm Röntgen (1845 – 1923) on 8th November 1895, and who had named it X-Radiation to signify an unknown type of radiations. X-rays are divided into two bands. One with energy up to 10keV is called *soft X-rays* and from 10keV to 120keV are known as *hard X-rays*. Soft X-rays can hardly penetrate matter whereas hard X-rays can not only penetrate uncompressed gases and liquid but solid as well. They are used in radiography and crystallography. To distinguish X-rays from Gamma rays was normally done by some arbitrary wavelength. But now they are distinguished in their origin. X-ray are emitted by electrons outside the nucleus, while gamma rays are emitted by the nucleus.

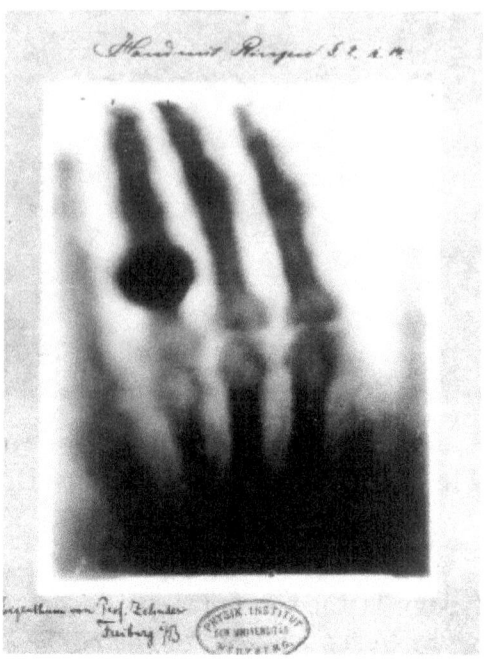

Figure 3.15 *Hand mit Ringen* (Hand with Rings): First ever image of x-ray taken by Wilhelm Röntgen on 22 December 1895 of his wife.

X-ray radiation is absorbed by the Ozone in the Earth's upper atmosphere in common with other high energy wavelengths of EM radiation. Sun's corona is very hot and radiate mostly in the X-ray region. Scientists knew that all very hot bodies will emit X-rays so the universe will be full of such bodies. X-rays can be observed from ground based telescopes so on 12th December 1970, Uhuru, the first X-Ray satellite was launched. More satellites have followed since i.e. the Einstein laboratory (1978) and ROSAT (1990).

Sources: The bombardment of targets of heavy atoms (typically tungsten) by fast moving electrons causes the energy levels in the target to change. When the target atoms' excited electrons drop back to their original level, they release fixed quanta of electromagnetic energy. (This is called the photoelectric effect). Heavenly bodies.

Detection: Photographic plates, photostimulable phosphors, fluorescence of certain chemicals (e.g., barium platnocyanide), Scintillator.

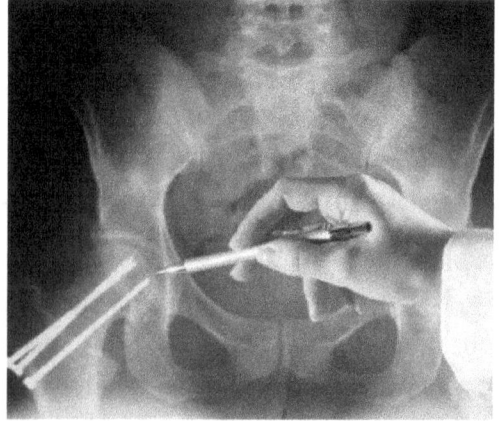 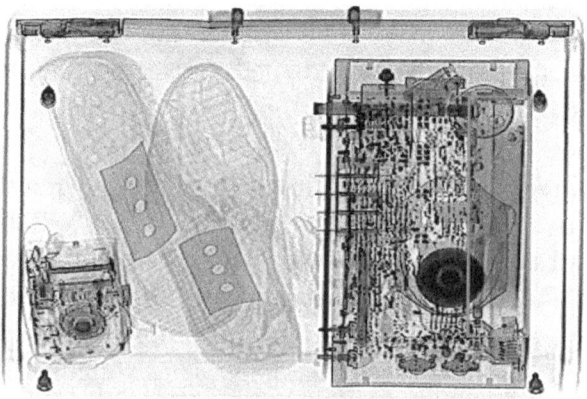

Figure 3.16 X-ray applications

Uses: The most well known use of X-rays is for medical scans. X-rays computer tomography, usually called, CAT Scan, uses X-rays to produce slices of specific area of the human body.

Another use of X-rays in the medical treatment is the use of short wavelength X-rays in a fashion similar to gamma rays for the killing of cancer cells.

Radiography is also used in industry for the examining of potentially damaged machinery to ascertain the cause of any difficulties, or to verify castings or welded joints prior to distribution.

X-ray are used to examine the work of Art. They also have been used to analyse the colour degradation in precious paintings like Van Gogh.

They are used at the airport to check passenger's luggage for security reasons.

Gamma Rays: Gamma rays were first discovered by French chemist and physicist Paul Ulrich Villard (1860 – 1934) in 1900 when he observed radiations emitted by radium during its gamma decay. They were named "gamma rays" by Ernest Rutherford (1871 – 1937) in 1903. They are the most energetic radiations in the EM spectrum. Gamma rays have the shortest wavelength and the highest frequency of all EM radiation. Only the very highest energies can reach the surface, the rest is absorbed by Ozone in the Earth's upper atmosphere. Gamma rays are produced when a radioactive atom (at higher energy) decay to more stable atom (at lower energy) it will shed the energy difference as gamma rays. These gamma rays have energies from few hundred keV up to 10MeV. Some heavenly bodies emit gamma rays with energy up to 10TeV. This amount of energy is not possible via radioactive decay. These gamma ray bursts (GRB) are coming from distant galaxies. Gamma rays of cosmic origin have also been observed from supernova explosions and black holes. These radiations were first detected in 1967 by Vela satellites, a series of satellites were launched by USA to keep an eye on USSR weapon tests.

Sources: Gamma rays are emitted from the nuclei of radioactive atoms during decay. Radioactive decay is spontaneous.
On Earth gamma rays originate largely from cosmic radiation. One high-activity source of gamma rays is cobalt-60 (^{60}Co).
In nuclear power station, gamma rays are produced as fission process takes place.

Detection: Bubble chambers, photographic plates (silver iodide emulsion darkens when hit by gamma radiation). Bubble chambers use a pressurized tank of H_2. High energy particles are sent through the Hydrogen, ionize it, and one can then follow the paths of the bubbles of H_2 produced using photographs.

Uses: Because gamma ray can kill living cells, they are used to kill cancer cells without having to go through difficult surgery.
Gamma radiations can kill microbes, so they are used to sterilise food so that it will remain fresh for a longer period.
Medical instruments are sterilized (after packing, to prevent contamination and to make handling easier) by brief exposure to gamma rays. This treatment kills any microbiological organisms which may be on the instruments, thus preventing unintentional spreading of disease. Gamma rays can be used to examine metallic castings or welds in oil pipelines for weak points. The rays pass through the metal and darken a photographic film at places opposite the weak points. In industry, gamma rays are used for detecting internal defects in metal castings and in welded structures.

Exercise 3

Q1 – In each of the following pairs, circle the form of radiation with the LONGER WAVELENGTH:

    a. yellow light **or** orange light

    b. infra-red **or** red light

    c. radio waves **or** microwaves

    d. violet light **or** UV radiation

Q2 – In each of the following pairs, circle the form of radiation with the GREATER FREQUENCY:

    a. green light **or** red light
    b. x-rays **or** gamma rays
    c. microwaves **or** radio waves
    d. UHF radio waves **or** FM radio waves

Q3 – In each of the following pairs, circle the form of radiation with the LOWER ENERGY

    a. x ray **or** orange light

    b. UV radiation **or** microwaves

    c. infra-red radiation **or** red light

    d. radio waves **or** UV radiation

Q4 – When free electrons collide with atoms in their *ground state*, the atoms can be excited or ionised.

    a. State what is meant by ground state.
    b. Explain the difference between excitation and ionisation.
    c. An atom can also become excited by the absorption of photons. Explain why only photons of certain frequencies cause excitation in a particular atom.
    d. The ionisation energy of hydrogen is 13.6eV. Calculate the minimum frequency necessary for a photon to cause the ionisation of a hydrogen atom. Give your answer to an appropriate number of significant figures.

AQA Unit 1 12 January 2012

Q5 – The lowest energy levels of a mercury atom are shown in Figure below. The diagram is not to scale.

```
                                                    0
        n =4  ─────────────────────────────── -0.26
                                                           x 10⁻¹⁸ J
        n =3  ─────────────────────────────── -0.59
        n =2  ─────────────────────────────── -0.88


        n =1  ─────────────────────────────── -2.18
```

    a. Calculate the frequency and wavelength of an emitted photon due to the transition level $n = 4$ to level $n = 3$.
    b. Draw an arrow on the Figure above to show a transition which emits a photon of a longer wavelength than that emitted in the transition from level $n = 4$ to level $n = 3$.

AQA unit 1 13 January 2010

Q6 – In an atom, a photon of radiation is emitted when an electron makes a transition from a higher energy level to a lower energy level as shown.

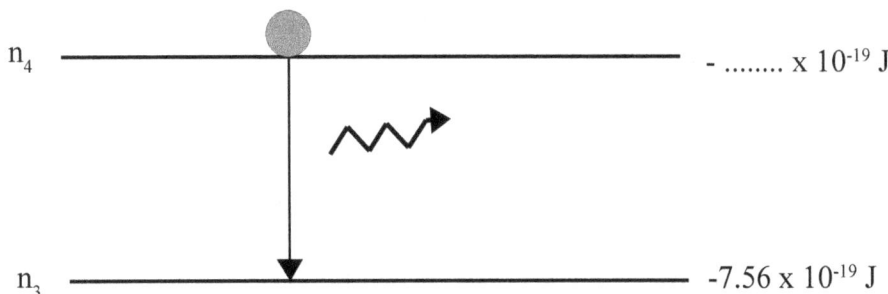

If the wavelength of the photon emitted is 680nm, calculate the energy of the energy level n_4.

Q7 – Gamma rays can travel faster than light, true or false?

Q8 - Which of the following has the highest frequency?
      Ultra violet   Microwaves   X-Rays   Radio waves

Q9 – What is the evidence that light is a wave?

Q10 – Calculate the energy of EM radiations with 760nm wavelength.

Q11 – Covert the energy in Q4 into eV.

Q12 – If a spherical steel ball of radius 20cm is heated to a temperature of 1200K. Calculate the energy emitted by steel ball per second.

Q13 – A radio station is broadcasting in FM band at 102MHz. The transmitter of the radio station has 120kW power. Calculate:

a) Wavelength of this FM band?
b) How much energy per photon is emitted by the antenna?
c) How many photons are transmitted per second?

Q14 – What is the wavelength of a photon that has a frequency of 2.10×10^{14} Hz? Give your answer in nm and determine what type of radiation this is.

Q15 – A hydrogen atom in its ground state absorbs light with a wavelength of 102.6 nm. Refer to figure 3.6 and calculate the energy level of the resulting excited state (n = ?).

Q16 – If you want to remove the hydrogen electron from its ground state. Calculate the wavelength of EM radiation required and also in where is this on the EM spectrum?

Q17 – Sodium vapour lamps are often used in residential street lighting. They give off a yellow light having a frequency of 5.09×10^{14}Hz. What is the wavelength of this light in nanometres?

Q18 – What is the energy of a yellow colour with average wavelength of 545nm?

Q19 – What is *waving* in electromagnetic radiations?

Q20 – Why can infra-red and microwave observations sometimes be conducted from Earth's surface, while X-ray and Gamma Ray measurements must be done from orbit?

Q21 – The ultraviolet rays that cause sunburn is called UV-B rays. They have a wavelength of approximately 300 nanometres. What is the frequency of a UV-B ray?

Q22 – Name three things that can produce gamma rays?

Q23 – What is the wavelength range for the FM radio band (88 MHz – 108 MHz)?

Q24 – Which two of the following are inversely proportional ?
a) energy and wavelength
b) wavelength and frequency
c) frequency and energy
d) Speed and energy

Q25 – Calculate the energy needed to remove the electron from the hydrogen in its state with -3.4 eV energy. If this is achieved by photoelectric effect what must be the wavelength of incoming photons.

Q26 – Capital FM broadcast music from its London studio at 95.8 MHz. How long would it take a particular song to travel from Earth (London) to Mars if the distance between the two planets is 8.00×10^7 km?

Q27 – What is the wavelength of the signals from Capital FM?

Q28 – List five characteristics which all electromagnetic radiations share.

Q29 – How electromagnetic waves are produced?

Q30 – Astronomers use ground based telescopes which are either optical or radio telescope. To observe in microwaves or infra-red part of the spectrum they build telescopes on the highest mountain like a UK Infra-red telescope on Mauna Kea, Hawaii (4205m above sea level), while to observe in X-ray and gamma ray they launch telescope in earth's orbit. Why?

Q31 – Calculate the total energy density of electromagnetic wave which has a magnetic field strength of 647mT and electric field strength of 2×10^8 N/C.

4 – Optics

Lenses have been in use for centuries. The earliest known lens was made as early as 700BC for Assyrian on the upper Tigris in northern Mesopotamia, present day northern Iraq. Some statues from ancient Egypt are fixed with lenses dated around this time, their purpose was ornamental. In ancient India, the philosophical school of *Samkhya* and *Vaisheshika* developed the theory of light around 6 BC. Greek mathematician Euclid wrote a book *optics* first ever book on the topic. Al-Kindi (c. 801- 873) was an earliest optical scientist in the Islamic world. Ibn Sahl (c. 940-1000) in Baghdad wrote a treatise *On burning Mirrors and lenses* in which he has tried to explain how curved mirrors and lenses bend and focus light. Ibn al-Hytham (965-1040) was the first scientist who correctly said that the vision occurred because of light rays entering the eye. In medieval Europe English bishop Robert Grosseteste (1175 – 1253), Sir Franciscan, Roger Bacon (1214 – 1294) did lots of work on optics.

A *lens* is an optical device which is made of transparent material such as plastic or glass and passes the light through and also refract light and as a result converge or diverge beam of light.

Types of Lenses: Lenses are classified by the curvature of the two optical surfaces.
- A convex lens is thicker at the centre than at the edge. They are called converging lens.
- A concave lens is thinner at the centre than at the edge. They are called diverging lens.

Features of Lenses
- The principal axis of a lens is the line joining the centre of curvature of its surface.
- The principal plane of a lens is the line passing vertically through the middle of the lens.
- The principal focus of a lens is that point on the principal axis to which all rays originally parallel and close to the axis converge, or from which they diverge, after passing through the lens.
- The focal length is the distance between principal focus and principal plane.
- The centre of curvature is twice the distance of the principal focus to the surface of the lens.

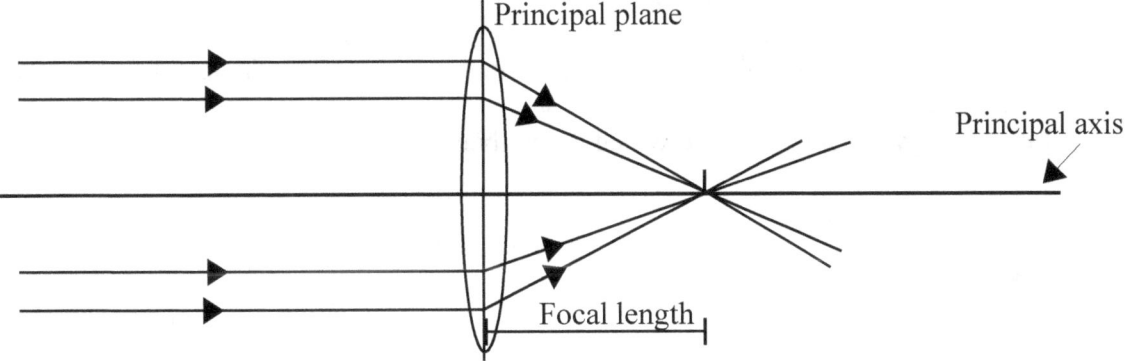

Figure 4.1 convex lens

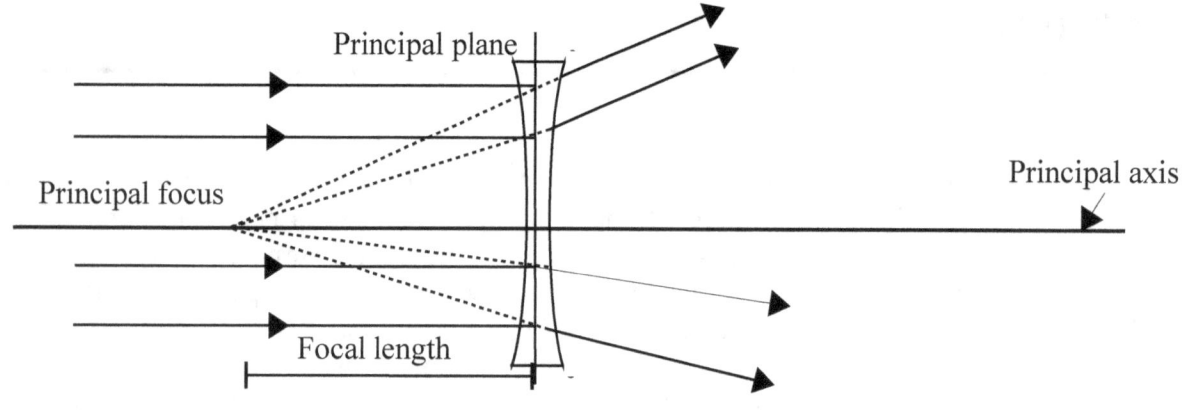

Figure 4.2 concave lens

Construction of ray diagram and formation of images by convex lens.

Since light may pass through a lens in either direction, there will be two principal foci equidistant from the principal plane, one on either side of the lens. These are denoted the symbol *f* on the image side 'v' and *f'* on the object side 'u'.

Three particular rays are used in geometrical construction to locate the image formed by a convex lens:

i. Rays parallel to principal axis which pass through the principal focus (*f* & *f'*) after refraction through the lens.

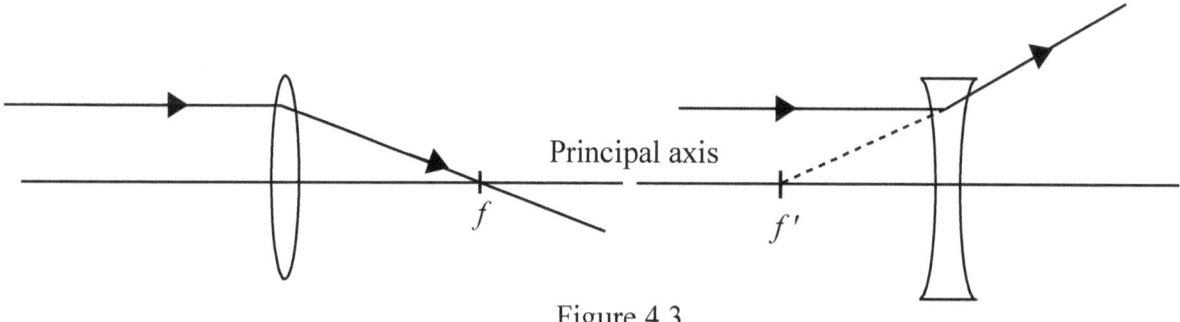

Figure 4.3

ii. Rays through the principal focus which emerge parallel to the principle axis after refraction through the lens.
iii. Rays through the optical centre which do not bend.

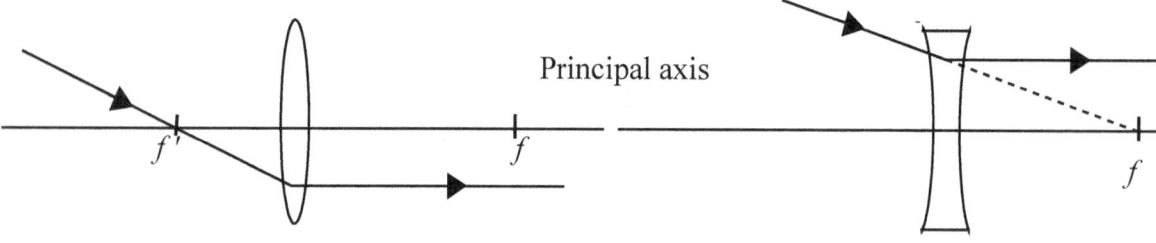

Figure 4.4

50

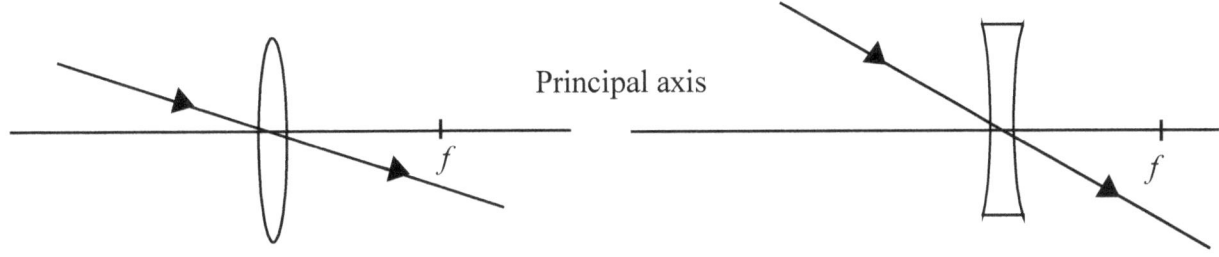

Figure 4.5

If the rays originating at an object point actually converge on an image point, so that they could be received on a screen, the image is called real. If the rays do not actually converge but appear to come from the image point, the image is called virtual.

Two of these rays only are sufficient to locate an image, and which particular pair is chosen is merely a matter of convenience.

Object at infinity

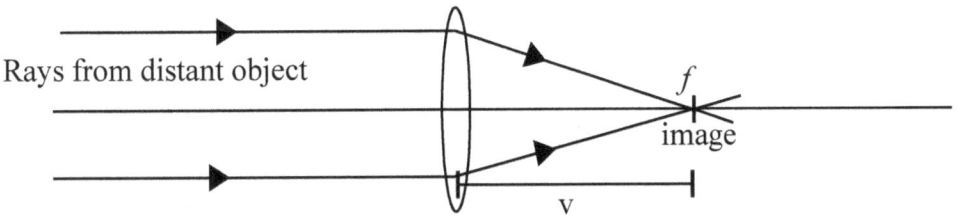

Figure 4.6

Object distance, u	Image distance, v	Position of the image	Nature of the image	Applications
u = ∞	v = f	Formed at focal point	Real, inverted, point size	Converging sun rays at a point to burn

Object at f'

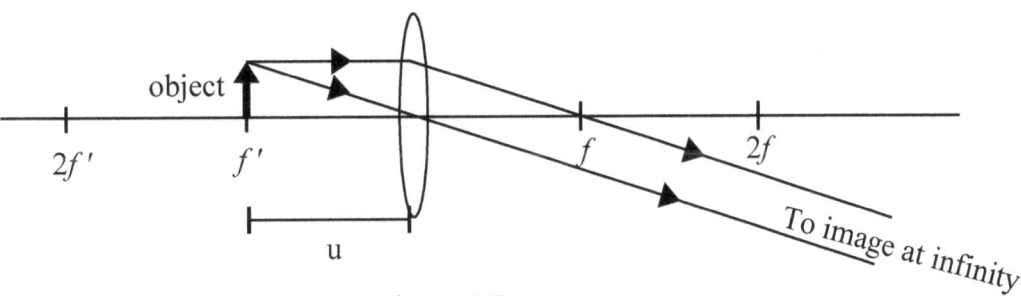

Figure 4.7

Object distance, u	Image distance, v	Position of the image	Nature of the image	Applications
$u = f'$	$v = \infty$	image at infinity	Real	To produce a parallel beam of light. Spot light

Object between f' and $2f'$

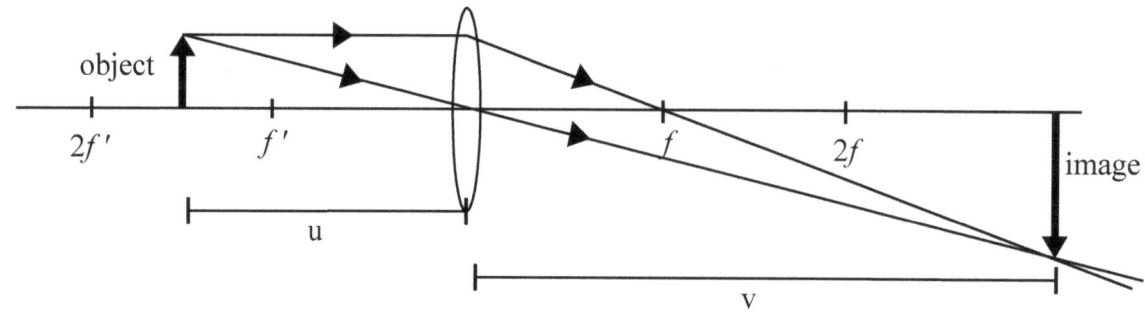

Figure 4.8

Object distance, u	Image distance, v	Position of the image	Nature of the image	Applications
$f' < u < 2f'$	$v > 2f$	Formed beyond $2f$	inverted real magnified	Projector photograph enlarger

Object at $2f'$

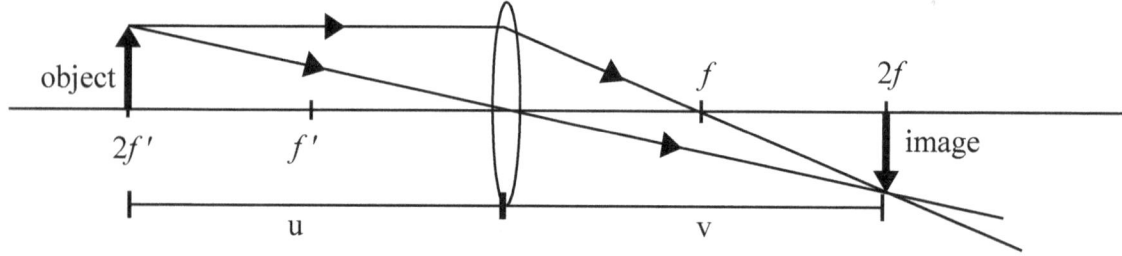

Figure 4.9

Object distance, u	Image distance, v	Position of the image	Nature of the image	Applications
$u = 2f'$	$v = 2f$	Formed at $2f$	inverted real same size	Photocopier

Object beyond $2f'$

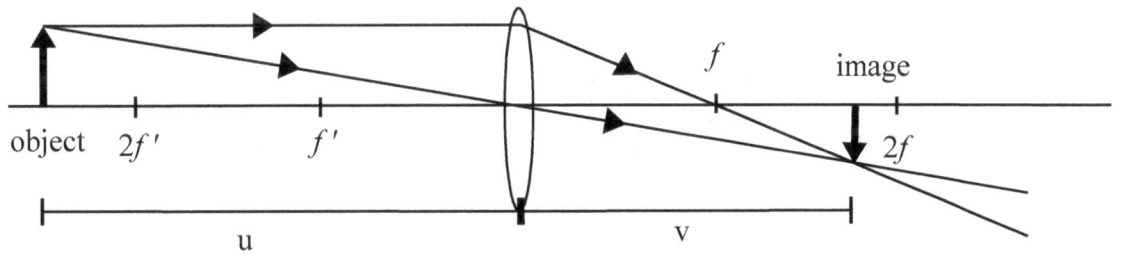

Figure 4.10

Object distance, u	Image distance, v	Position of the image	Nature of the image	Applications
$u > 2f'$	$f < v < f'$	Formed between f and $2f$	inverted real diminished	Camera Eye

Object between lens and f'

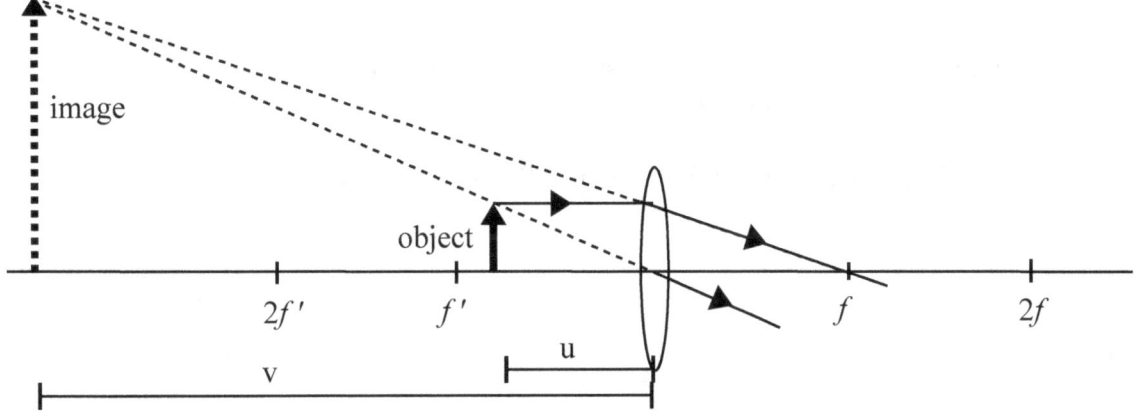

Figure 4.11

Object distance, u	Image distance, v	Position of the image	Nature of the image	Applications
$u < f'$	$v > 2f'$	Image behind the object same side of the lens	upright virtual magnified	Magnifying glass

Construction of ray diagram and formation of images by a concave lens.

To construct ray diagram for concave lens is easy. Because it does not matter where the object is, the image will always be virtual, upright and diminished. Only one diagram is needed to show the construction of ray diagram.

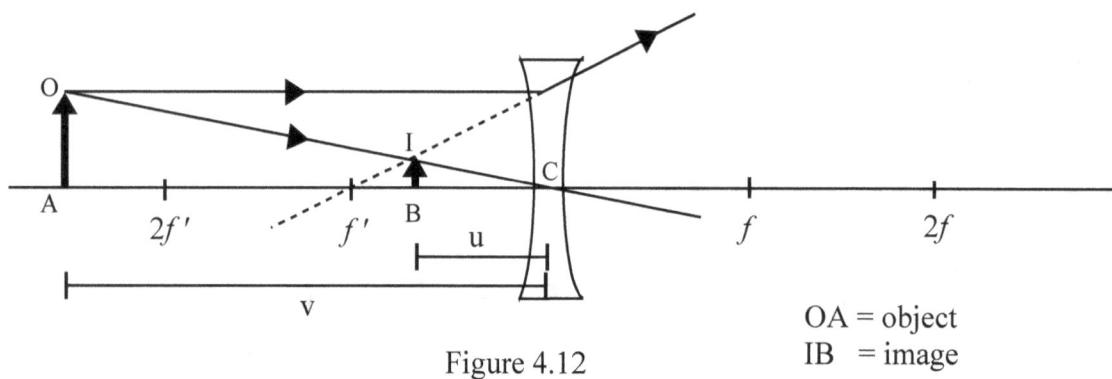

Figure 4.12

OA = object
IB = image

Magnification

We have seen from the above ray diagrams that the image produced by a lens varies according to the position of the object. The *linear magnification* is thus the ratio of the height of the image (h_i) to the height of the object (h_o), and is usually denoted by the letter *m*.

$$m = \frac{h_i}{h_o} \qquad (4.1)$$

Referring to figure 4.11, it will be found easy to prove that the angles of triangle CIB are respectively equal to the angles of triangle COA.

Hence the triangles CIB and COA are similar, and it follows that

$$\frac{IB}{OA} = \frac{IC}{AC}$$

now,   IB = h_i and OA = h_o

also,   IC = image distance, v
and
        OC = object distance, u

therefore, $$m = \frac{h_i}{h_0} = \frac{v}{u} \qquad (4.2)$$

Lens equation

By identifying two sets of similar triangles, we will be able to derive the thin lens equation. In similar triangles the ratio of their vertical sides is equal to the ratio of their horizontal sides.

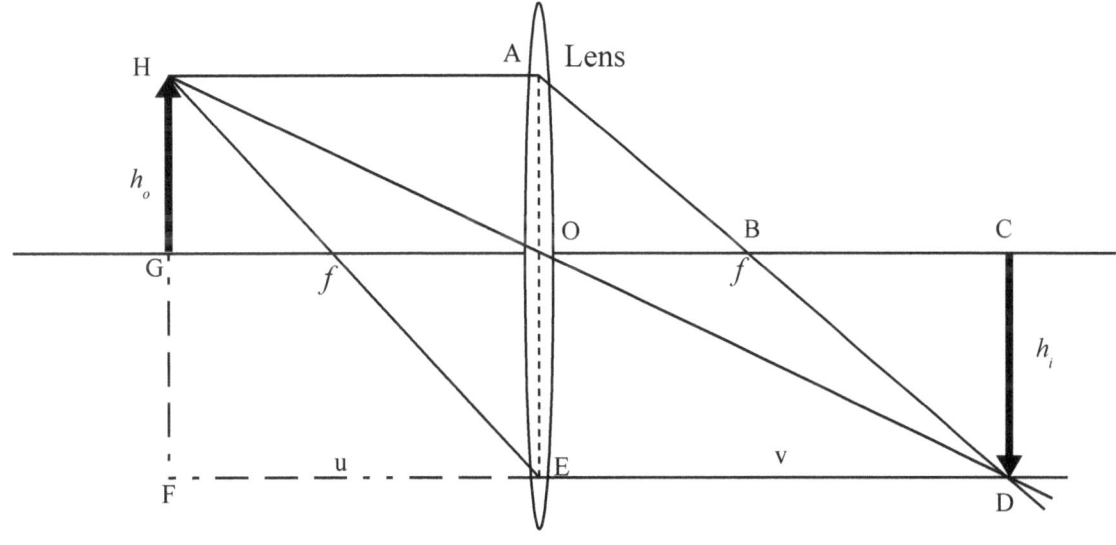

Figure 4.13

The triangle ABO and ADE are similar, so the ratio of their horizontal sides, is equal to the ratio of their vertical heights.

hence,
$$\frac{v}{f} = \frac{h_o + h_i}{h_o} \qquad \underline{i}$$

Also, triangle KOG and DOC are similar, so the ratio of their horizontal sides, is equal to the ratio of their vertical heights.

Therefore,
$$\frac{v}{u} = \frac{h_i}{h_o} \implies h_i = v \frac{h_o}{u} \qquad \underline{ii}$$

plugging $\underline{ii}$ into $\underline{i}$
$$\frac{v}{f} = \frac{v \times \frac{h_0}{u} + h_0}{h_0}$$

divide top and bottom by h_o
$$\frac{v}{f} = \frac{v}{u} + 1$$

divide both sides by v, we get lens equation

$$\frac{1}{f} = \frac{1}{u} + \frac{1}{v} \qquad (4.3)$$

Sign conventions
When using the lens formula correct sign have to be applied to object distances, image distances and focal lengths, according to convention chosen.
- f is +ve if the lens is a convex (converging) lens
- f is -ve if the lens is a concave (diverging) lens
- v is + if the image is real and located on the opposite side of the lens
- v is -ve if the image is virtual and located on the object's side of the lens
- h_i is + if the image is an upright and virtual.
- h_i is -ve if the image is inverted and real

Example: An object is placed (a) 25cm, (b) 5cm from a convex lens of focal length of 10cm. Find the nature, position and magnification of the image in each case.

a) u = 25cm (real object)
   f = 10cm (convex lens)

a) Using lens equation,
$$\frac{1}{10} = \frac{1}{25} + \frac{1}{v}$$

thus,
$$\frac{1}{v} = \frac{1}{10} - \frac{1}{25} = \frac{5-2}{50} = \frac{3}{50}$$

therefore,
$$v = \frac{50}{3} = 16.67 \text{cm}$$

$$m = \frac{v}{u} = \frac{16.67}{25} = 0.67$$

A real image is formed 16.67cm from lens on the side opposite to object, of magnification 0.67 or smaller than the object.

b) u = 5cm (real object)

$$\frac{1}{10} = \frac{1}{5} + \frac{1}{v}$$

$$\frac{1}{v} = \frac{1}{10} - \frac{1}{5} = \frac{1-2}{10} = \frac{-1}{10}$$

$$v = -10 \text{ cm (in front of the lens)}$$
$$m = \frac{10}{5} = 2$$

A virtual image is formed 10cm from the lens on same side as object, of magnification 2.
A real image is formed 16.67cm from lens on side opposite to object, of magnification 0.67 or smaller than the object.

$$m = \frac{v}{u} = \frac{1.96}{24 \times 10^{-3}} = 80$$

$$v = 80u$$

Also, we know that $\qquad v + u = 3\text{m}$

$$80u + u = 3\text{m}$$
$$81u = 3\text{m}$$
$$u = \frac{3}{81} = 37.0\text{mm}$$

b) The thin lens equation then gives

$$\frac{1}{f} = \frac{1}{u} + \frac{1}{v}$$

$$\frac{1}{f} = \frac{1}{u} + \frac{1}{80u}$$

$$= \frac{80}{80u} + \frac{1}{80u}$$

$$\frac{1}{f} = \frac{81}{80u}$$

$$f = \frac{80}{81} \times (37\text{mm}) = 36.54\text{mm}$$

Example: An object of 5cm height is placed at a distance of 60cm from a convex lens having a focal length of 20cm. Determine the image distance and image size.

$$\frac{1}{20} = \frac{1}{60} + \frac{1}{v}$$

$$\frac{1}{v} = \frac{1}{20} - \frac{1}{60} = \frac{6-2}{120} = \frac{4}{120}$$

thus
$$v = \frac{120}{4} = 30\text{cm}$$

$$m = \frac{v}{u} = \frac{30}{60} = 0.5$$

$$h_i = m \times h_o = 0.5 \times 5\text{cm} = 2.5\text{cm}$$

Power of the lens: The function of the lens is to converge or diverge incoming light rays. How much is converged or diverged depends on the power of the lens. The power is simply measured as the reciprocal of the focal length.

$$P = \frac{1}{f} \qquad (4.4)$$

The unit of lens power is *dioptres* denoted by the letter D. This is equal to m^{-1}.

Example: A diverging lens has the power of 6 ⅔ D and object is placed 60 cm from the centre of the lens. Locate and describe the image formed by the lens.

We have to be careful solving the problem. Since the lens is diverging so the rays will focus on the object side of the lens and according to sign convention focal length is -ve.

From equation 4.4 $$f = -\frac{1}{\left(\frac{20}{3}\right)} = -\frac{3}{20} \text{ m} = -\frac{300}{20} \text{ cm} = -15 \text{cm}$$

Using lens equation
$$\frac{1}{-15} = \frac{1}{60} + \frac{1}{v}$$

$$\frac{1}{v} = \frac{1}{-15} - \frac{1}{60}$$

$$\frac{1}{v} = \frac{-4-1}{60} = \frac{-5}{60}$$

therefore, $$v = \frac{-60}{5} = -12 \text{cm}$$

Since the image distance is negative, therefore the image is virtual. The nature is virtual, erect and diminished.

Lensmaker's equation:

The focal length of a lens *in air* can be calculated from the lensmaker's equation which is used to determine whether a lens will behave as a converging or diverging lens based on the curvature of its faces and the relative indices of the lens material n_1 and the surrounding medium n_2.

$$P = \frac{1}{f} = \left(\frac{n_1}{n_2} - 1\right)\left(\frac{1}{R_1} - \frac{1}{R_2}\right) \qquad 4.5$$

where,
- n_1 = Refractive Index of Lens Material
- n_2 = Refractive Index of Ambient Medium
- R_1 = Radius of Curvature of the First Surface
- R_2 = Radius of Curvature of the Second Surface

If this equation yields a negative value for *P*, then the lens is diverging; a positive *P* means that the lens is converging.

Compound Lenses:

A *compound lens* is a collection of lenses of different shapes and each lens has different refractive indices, arranged one after the other with a common axis. If the lenses of focal lengths f_1 and f_2 are "thin", the combined focal length *f* of the lenses is given by

$$\frac{1}{f} = \frac{1}{f_1} + \frac{1}{f_2} \ldots \qquad 4.6$$

If two thin lenses are separated in air by some distance *d*, the focal length for the combined system is given by

$$\frac{1}{f} = \frac{1}{f_1} + \frac{1}{f_2} - \frac{d}{f_1 f_2} \qquad 4.7$$

Exercise 4

Q1 – What are the nature and location of an image formed by the converging lens if an object is placed 15 cm from the centre of the lens if the focal length is 10 cm?
What are magnification and the height of the image, if the height of the object is 4 cm?

Q2 – A diverging lens has the power of 2.5 dioptres. If an object which its height is 3 cm is placed 60 cm in front of the lens, then determine :

a) The focal length of the diverging lens
b) The distance of the image from the lens
c) The magnification of the image
d) The height of image
e) Sketch rays diagram to scale
f) The nature of the image

Q3 – An object is placed 1.5cm in front of a converging lens having a focal length of 0.5cm. A second converging lens, with a focal length of 1cm, is placed 1cm behind the first lens as shown in Figure 4.13

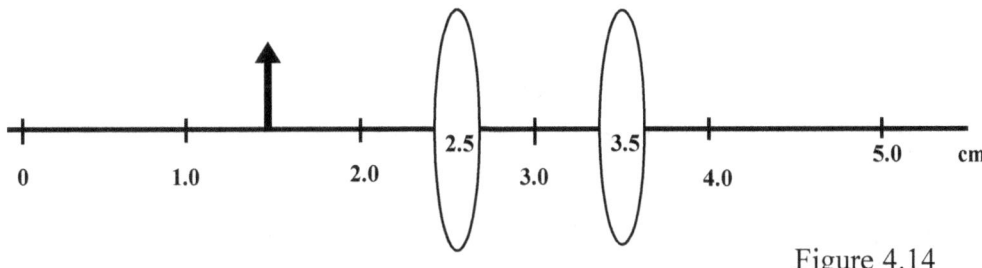

Figure 4.14

a) Where is the first image being formed?
b) What is the magnification of the image formed?
c) The first image will now serve as the object of the second lens. How far it is located in front of the second lens?
d) Where is the image formed by the second lens?
e) What is the magnification of this second image?
f) What is the final magnification of this final image of this lens system?
g) What is the nature of the final image?

Q4 – An object is 10cm in height and is located at a distance of 25cm from a convex lens. The focal length is 8cm. What is the image location (v) and the height (h_i) of the image.

Q5 – An object of height 10 cm is viewed through a diverging lens of focal length of 10 cm. Determine the location and height of the image of the object is held away from the lens at 35cm. Verify your calculations using crude ray diagrams.

Q6 – Converging lenses are_____at the centre and_____at the edge, whereas Diverging lenses are_____at the centre and_____at the edge.

Q7 – Determine image distance, height, and magnification of a 20 cm tall object placed 60 cm from a converging lens having a focal length of 20 cm.

Q8 – A system of two converging lenses needs to be designed as shown in figure 4.14, where the image must be formed between the two lenses at 20 cm from the second lens. Calculate the position of the object in front of the primary lens. Also, calculate the magnification.

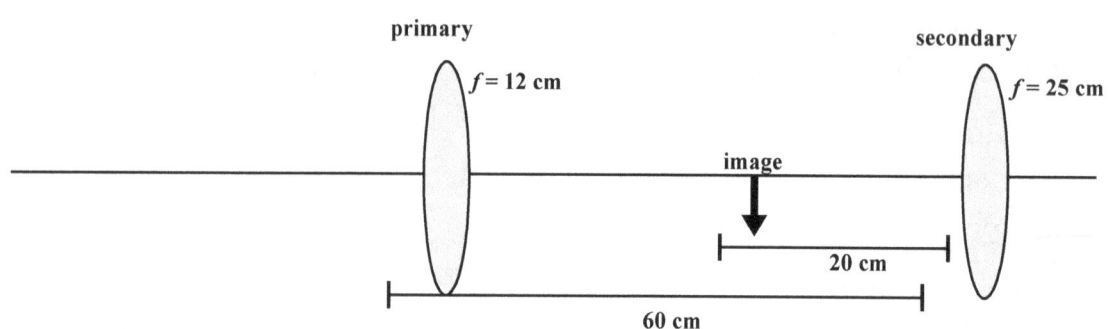

Figure 4.15

Q9 – An object is located 60 cm from a converging lens of focal length 45 cm. A second converging lens is placed 210 cm from the first lens. A final image with height 70 cm is formed on a screen 195 cm from the second lens.

Draw this the ray diagram of the system and calculate the focal length of the second lens. Calculate both magnifications. Also find the height of the original image.

Q10 – A divergent lens with focal length of f. Find the location and nature of the image, when

a) object at $3f$
b) object at f
c) object at $½f$

Q11 – An object is placed 54 cm in front of a converging lens with a focal point of 36 cm.

a) Calculate the distance of the image from the lens.
b) Calculate the magnification of the lens
c) What is the nature of the image.

Q12 – Write four characteristics of concave lens image.

Q13 – Find the power and focal length of the following thin lenses when they are placed in close combination: f_1=80cm, $f_2 = -50$ cm, $f_3 = -40$ cm.

Q14 – Find the power of the following lenses *in air*, where the index of refraction of the lens A is $n_A = 1.50$ and lens B is $n_B = 1.55$

5 – Radioactivity

Radioactivity was discovered by French physicist Henri Becquerel (1852 – 1908) in 1896 when he was studying fluorescence material. Fluorescence can be produced in a number of different chemicals simply by exposing them to sunlight. Becquerel placed some crystals on top of a photographic plate which he had wrapped in black paper and then placed them in sunlight so that crystals fluoresced. Afterwards he developed the photographic plate to see if it has been exposed. He had many failures but eventually he did get darkened plate when using a uranium salt. He was even more surprised to see exposure when he did not even expose the crystal to sunlight.
Two scientists took interest in Becquerel work. They were Marie Curie (1867 – 1934) and her husband Pierre Curie (1859 – 1906). They found that ray from uranium caused the ionisation of air and were able to measure the intensity of the radiation.

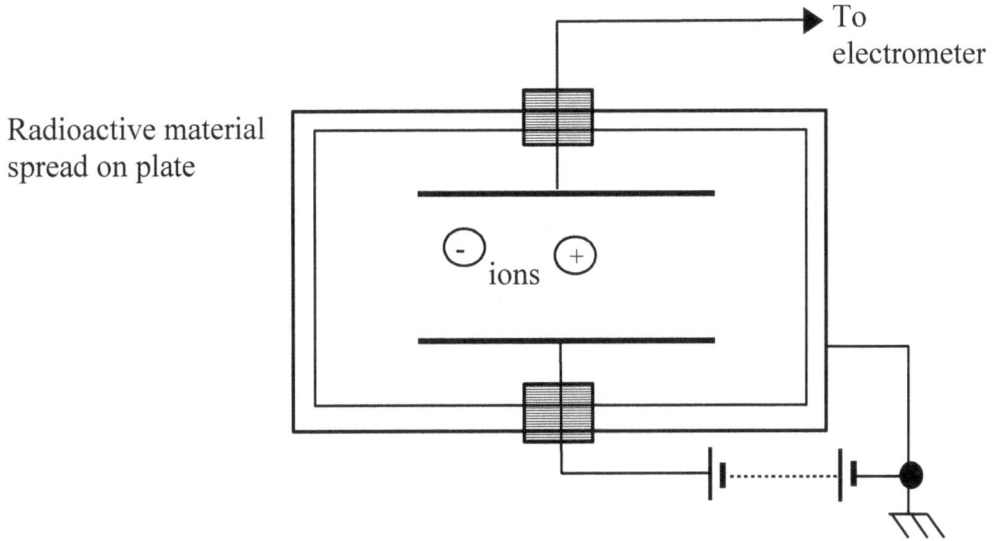

Figure 5.1 Intensity measurement by an ionisation chamber

They used the ionisation chamber which consisted of two insulated metal plates inside an earthed metal chamber. The lower plate was raised to a higher potential while upper plate was connected to an electrometer. When the radioactive substance was spread on the lower plate the rays it gave off caused the ionisation of the air inside the chamber, thus forming positive and negative ions. These ions were attracted towards the appropriate plate by the electric field between two plates. A tiny current flowed and the magnitude of ionisation current was used to measure the radioactivity of the sample. They used many other chemicals like *polonium* and *radium*.

A radioactivity decay happens when atom with one type of nucleus, called a *radionuclide,* is transformed to an atom with a nucleus containing different numbers of protons or neutrons, or nucleus with a different energy state. Either of the new atom is called the *daughter nuclide.* Radioactive decay is a stochastic process which means it is impossible to predict when a particular atom will decay.

The nucleus is composed of protons and neutrons. We know that two protons will repel each other via electromagnetic force. So what keeps the nucleus intact?

There exists another force in the nucleus which binds protons and neutrons together and is called the *strong nuclear force*. At atomic scale it is about 100 times stronger than the electromagnetic force. The actual strong force acts within the nuclear particles like protons and neutrons. Each of these particles is made up of three *quarks* and these quarks are held together with strong force. Strong force does not diminish in strength with increasing distance till the size of the nucleons. Outside the nucleons (proton and neutron) the residual effect of the strong force is called the nuclear force. Its this residual force which glues protons and neutrons. This force however, diminishes rapidly with distance. So as the nucleus gets bigger with increasing number of nucleons, it becomes less stable when atomic number becomes larger than 82 with element lead.

The ratio of neutrons to protons in stable atomic nucleus range from 1 to 1 at the lower end of the scale of atomic number to 1.6 to 1 at the upper end. Any ratio falls outside of this stability band will represent an unstable nucleus and will be radioactive. See Figure 5.2

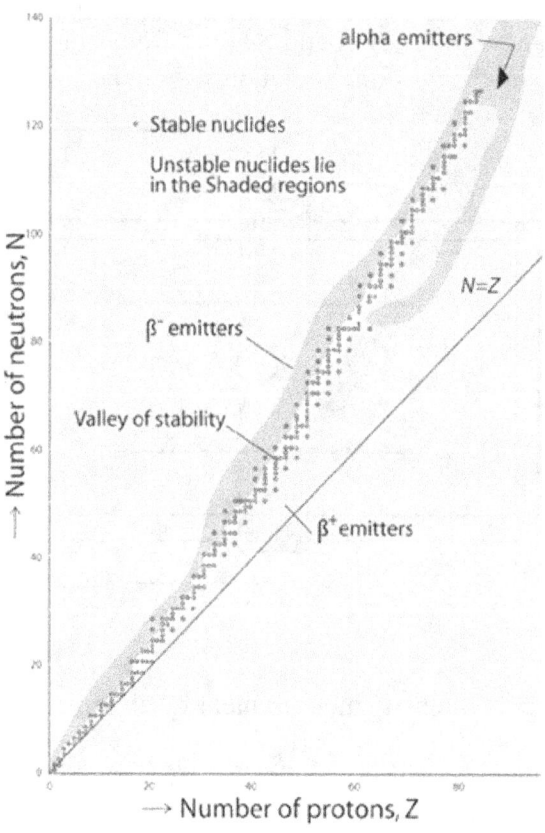

Figure 5.2 plot of the number of neutron verses number of protons inside the nucleus.

Looking closely at the graph in figure 5.2 we can see few general rules:

- the stable nuclide lies in a narrow band of neutron-to-proton ratio.
- Nuclei above the valley of stability are neutron rich. To become stable they will have to decrease their ratio by decreasing neutron numbers and increasing proton numbers by emitting β^-.
- Nuclei below the valley of stability are proton rich. To become stable they will have to increase their ratio by decreasing proton numbers and increasing neutron numbers by emitting β^+.

Alpha Decay

When nucleus has the number of neutrons significantly greater than a proton, there will be excessive electromagnetic repulsion. To reduce this repulsion nucleus eject alpha particle. Alpha particle consists of two protons and two neutrons bound together in a particle identical to helium nucleus. The symbol for alpha particle is α^{2+}, He^{2+} and ^4_2He. The ejected alpha particle will typically have a kinetic energy of 5MeV and speed about 5% of the speed of light (15 x10^6 ms^{-1}). You may wonder where this energy comes from. There is a difference in mass between the original nucleus and the sum of the mass of the alpha particle plus the resulting nucleus. This mass deficit is converted into energy according to Einstein equation $E = mc^2$. This energy is then carried by the alpha particle and recoiling of parent nucleus.

Because of their relatively large mass and +2 electric charge, they are highly ionising form of particle radiation. This means they will interact with other atoms and thus will lose their energy, so they will be stopped within few centimetres of air. They have low penetration distance. Alpha particle can be stopped by a sheet of paper.

Alpha particles are normally emitted by large radioactive nuclei like uranium, thorium and radium etc., As we have seen that alpha decay occurs due to the electromagnetic repulsion so this process must have a minimum size nucleus that can eject alpha particle. The smallest atomic nuclei which has been observed to eject alpha particle is tellurium **Te** with atomic number 52. Figure 5.3 shows the decay of uranium via alpha particle into thorium.

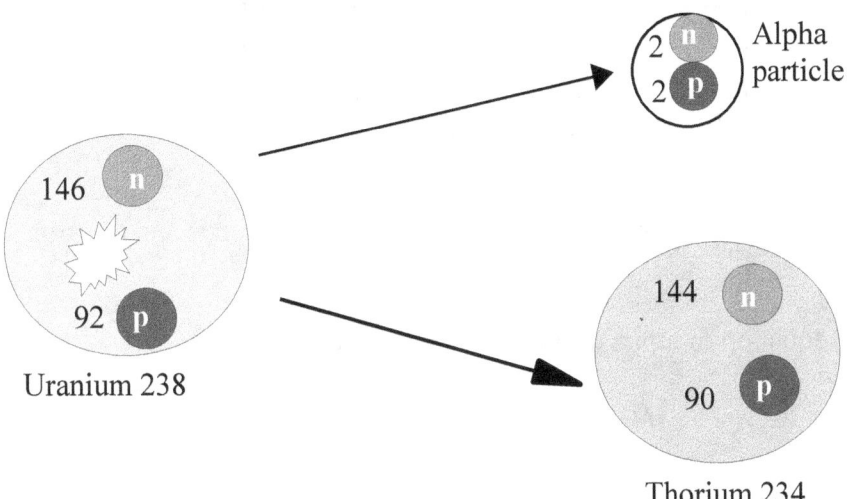

Figure 5.3 Radioactive decay of uranium nucleus.

This can be written as nuclear equation. You can see that right hand and left hand sides are balanced.

$$^{238}_{93}U \rightarrow ^{234}_{90}Th + ^4_2\alpha$$

This radioactive process is mono-energetic. But the isotope of radium $^{226}_{88}Ra$ some time decay into alpha particle and radium nucleus is still left excited as not all the energy has been taken by alpha particle as kinetic energy. This energy is emitted as gamma radiations. See figure 5.4

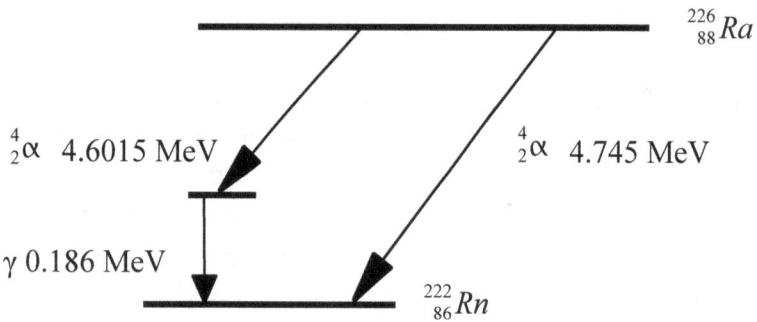

Figure 5.4 Radon nucleus modes of alpha decay

This can also be written as nuclear equation.

$$^{226}_{88}Ra \rightarrow\, ^{222}_{86}Rn + \alpha^{4}_{2} + \gamma$$

Table below will illustrate what change occurs in radioactive nucleus.

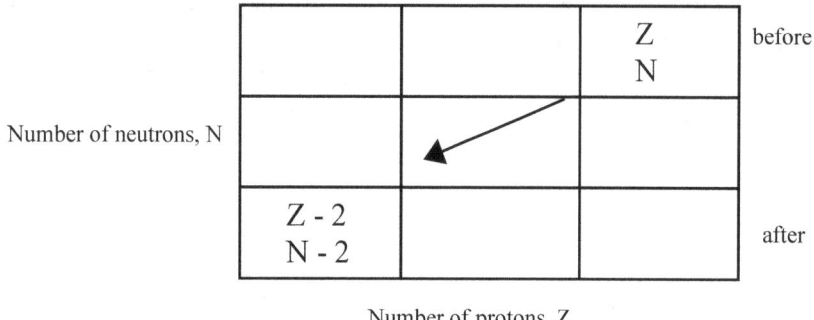

Table – 1 After alpha emission original nuclei has 2 fewer protons and neutrons

So the general nuclear equation of alpha decay can be written as:

$$^{A}_{Z}N \rightarrow\, ^{A-4}_{Z-2}X +\, ^{4}_{2}He$$

Biological effects

Alpha particles are given off by uranium-238 and radium-226 and other radioactive nuclei. These are present in nearly all the rocks, soil and water. Their danger to humans become great when rocks or soils are disturbed during soil quarry and mining.

Generally, alpha particles are not dangerous to human body as we have seen that their penetrating power is low so they will be absorbed by the skin or even the air around the body. But they can be hazardous if they are inhaled (known to cause lung cancer as alpha emitting polonium-210 is present in tobacco smoke), injected (absorbed into the blood stream) or ingested (swallowed). They can damage chromosome up to 1000 times greater than by an equivalent amount of beta or gamma radiation dose. In 2006 Russian dissident and ex-FSB and KGB agent Alexander V. Litvinenko was killed in London by mixing Polonium-210, alpha emitter in his tea.

Water contaminated with uranium can cause kidney damage, as alpha particles harms the renal system as it move through the body.

Uses

Alpha particles are more readily used in smoke detector, which uses Americium-241. Smoke detectors have two conducting plates. Alpha particles emitted by Americium-241 will *ionise* the air between the small gap as a result a tiny current will flow between the rods. In the event of fire, smoke particles will enter the air gap between the rods and hence reduce the current flow which will result in sounding the alarm.

Radium-233 or radium- 226 (alpha emitters) is injected directly into tumorous tissue to irradiate and kill cancer cells. This process is called, unsealed source radiotherapy. Since they are not very penetrating, they pose less threat to damage healthy tissues.

Polonium-210 is used to eliminate static electricity. The positive double charge on alpha particle will attract free electrons, therefore reducing the potential for static electricity. Alpha source radio nuclei are used in thermo-electric generator to power satellites and spacecrafts. Plutonium-238 is used as a source to produce alpha particles which result in heat which is converted into electricity. Same source is used in pacemaker batteries.

Example: Determine how much energy is released when Uranium-238 decays to Thorium- 234 via alpha decay.

The equation of the reaction is this $\quad ^{238}_{92}U \rightarrow ^{4}_{2}\alpha + ^{234}_{90}Th$

It is possible to look at their mass from a data book.

So in terms of mass we get $\quad 238.0508U \rightarrow 4.0026U + 234.0436U$

where U is called atomic mass and is equal to 1.66×10^{-27} kg

Now adding the mass on the right-hand side give us 238.0462U which is smaller than the left-hand side.

So 238.0508U – 238.0462U = 0.0046U mass is missing after the decay.

Using $E = mc^2 = (0.0046 \times (1.66 \times 10^{-27})) \times (3 \times 10^8)^2 = 6.83 \times 10^{-13}$ J

This energy is normally quoted in eV, hence $\quad \dfrac{6.83 \times 10^{-13}}{1.6 \times 10^{-19}} = 4.30 \times 10^6$ eV = 4.30 MeV

Beta Decay

In 1899 Ernest Rutherford separated radioactive emission into two types: alpha and beta. He also found that the charge on beta particle is negative. In 1900 Henri Becquerel measured the charged to mass ratio of beta particle using Sir Joseph John Thomson (1856 - 1940) method of finding the charge to mass ratio of the electron. Becquerel found that the ratio is exactly the same as J. J. Thomson found in electron, so he inferred from that the beta particles are actually fast moving electrons. In 1901 Rutherford and English radio chemist Frederick Soddy (1877 – 1956) explained that radioactivity is due to the transmutation of elements. Later Soddy and Kazimierz Fajans independently proposed their radioactive displacement law also known as Fajans and Soddy law. The law describes which element and isotope is produced during the particular type of radioactive decay:

- In alpha decay, an element is produced with an atomic number less by 2 and mass number less by 4 of that of the parent nucleus.
- In beta decay, the mass number remains unchanged while the atomic number number becomes greater by 1 than that of the parent nuclei.

This law is illustrated in figure 5.5. The decay via β^+ is explained later.

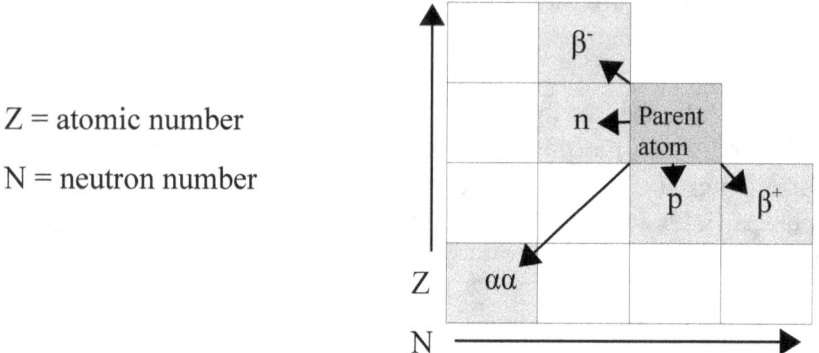

Figure 5.5 Fajans and Soddy displacement law

In beta decay mass number, A, does not change but the atomic number, Z change in the nucleus. Nucleons with the same mass number but different atomic number are called *isobaric nuclide*. For example ^{40}S, ^{40}Cl, ^{40}Ar are isobaric nuclide. On the other hand if two nuclide have the same neutron number, N, but different proton number (atomic number) Z, then these nuclide's is called *isotonic nuclide*. For example, S-36, Cl-37, Ar-38 all have 20 neutrons.

Beta minus decay occurs in neutron rich nucleus. This happens when one of the neutron is changed into a proton. To conserve charge electron is emitted. Cs-137 decays by emitting a beta particle.

$$^{137}_{55}Cs \rightarrow ^{137}_{56}Ba + ^{0}_{-1}\beta$$

The cause of beta emission is an excess number of neutrons in the atomic nucleus. When there are significantly more neutrons than protons in a nucleus, the neutrons degenerate into protons and electrons, which are ejected from the nucleus at high speeds. This increases the atomic number of the atom and also increases its stability, an example of natural atomic alchemy which gives rise to a new type of atom.

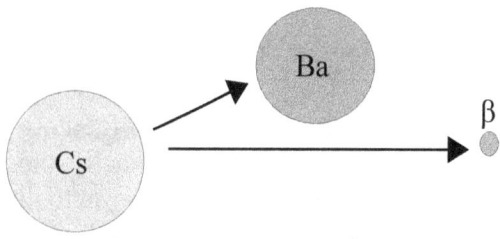

Figure 5.6 Cs-137 nucleus transmuted to Ba-137 by ejecting β particle

In 1911 German radio chemist Otto Hahn (1879 – 1968) and Austrian physicist Lise Meitner (1878 – 1968) showed that the energy of beta particle emitted had a continuous rather than discrete energy spectrum. Such spectrum means some energy is lost in the process which contradict well established law of conservation of energy.

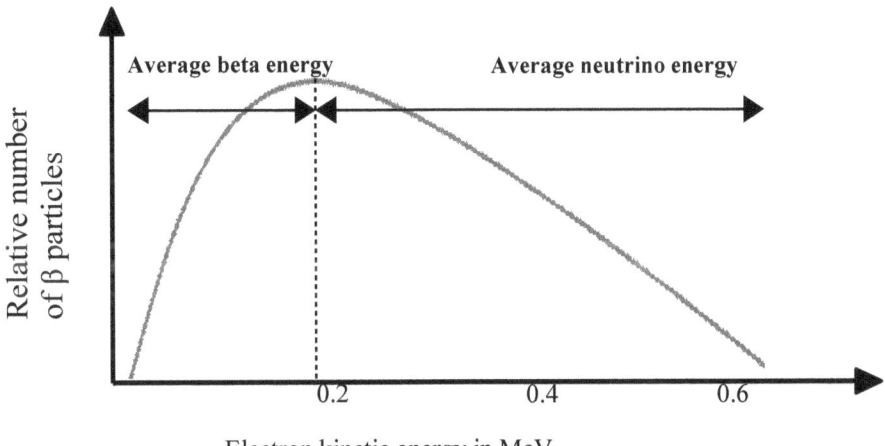

Figure 5.7 Energy spectrum of beta particle

This matter was concluded by Austrian physicist Wolfgang Ernest Pauli (1900 – 1958) when he wrote his famous letter dated 4th December 1930, to the physics institute of the Federal Institute of Technology, Zurich. Pauli suggested that another light "neutron" is also emitted with beta particle which carry that missing energy and also count for conservation of angular momentum. In 1931 Italian physicist Enrico Fermi (1901 – 1954) named this particle "neutrino" and in 1934 he published a full account of neutrino production with beta decay is now called Fermi's Golden Rule.

Protons and neutrons are not fundamental particles because they are made up of by constituent particles called quarks. There are six quarks *Up, Down, Charm, Strange, Top* and *Bottom*. Proton is made with two Up and one Down quark, whereas neutron is made with two Down and one Up quark.

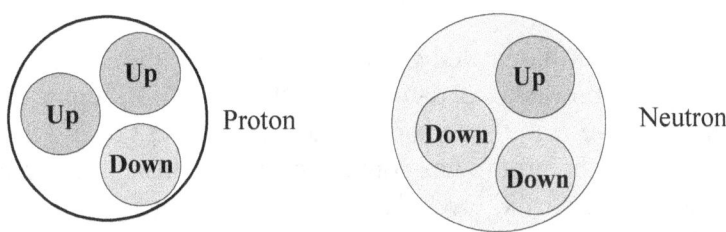

Figure 5.8 Quark model of nucleons

According to Quantum theory inside the nucleus *Up* quark can change into *Down* quark and vice versa via the weak force by exchange of $W^{\pm}$. So a neutron can turn into a proton and a proton can turn into a neutron. Now when neutron turn into proton this means a positive charge has been added to the nucleus. This violates the law of conservation of charge. As neutron is heavier than proton an excess mass is converted into energy which then converts back into mass of electron and anti neutrino. Neutrino are regarded as mass less particles and they take their share of energy as kinetic energy. This is illustrated in Feynman's diagram in figure 5.9, which shows that *down* quark in neutron change into *up* quark and becomes a proton by the exchange of W^- boson which immediately decay into an electron and electron anti-neutrino. Bar on the particle symbol signify that the particle is anti particle.

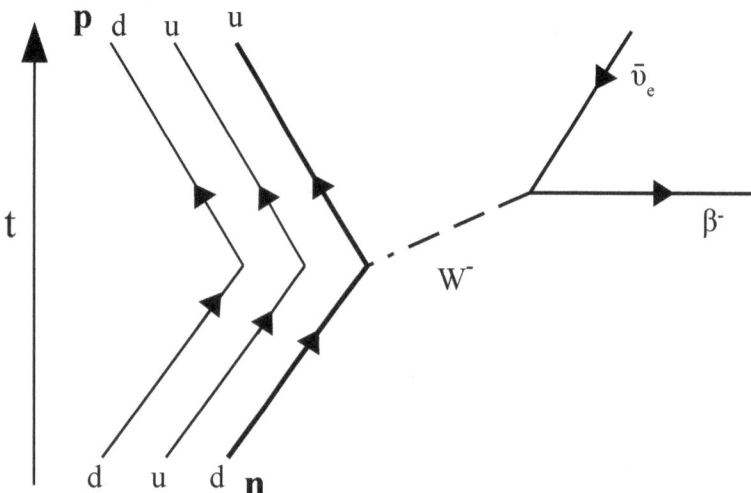

Figure 5.9 Feynman diagram of Beta decay

$$^{14}_{6}C \rightarrow ^{14}_{7}N + ^{0}_{-1}\beta + \bar{v}_e$$

French physicist Jean Frédéric Joliot-Curie (1900 – 1958) and his wife Iréne Joliot-Curie (1897 – 1956) both won the Noble prize in 1935 for their discovery of artificial radioactivity. They bombarded aluminium with alpha particle and produced a very short lived radioactive isotope of phosphorus.

$$^{27}_{13}Al + ^{4}_{2}He \rightarrow ^{30}_{15}P + ^{1}_{0}n$$

They first time produced radioactive isotope of phosphorus which does not exist in nature. This isotope with a half-life of 2.498(4) minutes, decay into silicon-30.

But something odd happened, the ejected particle had exactly the same mass as an electron but had positive charge and also produced a daughter nucleus with atomic number one less, which means a loss of proton. We have mentioned above that proton is less heavy than neutrons. But from the above decay it appears that the proton has converted into neutrons thus created mass (or energy) which is not possible according to the law of conservation of energy. In 1928, English physicist Paul Adrian Maurice Dirac (1902 – 1984) published a paper proposing that the electron can have both a positive and negative charge and in 1932 American physicist Carl David Anderson (1905 – 1991) discovered the positive electron known as positron.

So it was quickly established that phosphorus is emitting a positron. The decay equation of such process is this. Notice that this time electron neutrino is produced.

$$^{30}_{15}P \rightarrow ^{30}_{14}Si + ^{0}_{+1}\beta + v_e$$

In beta plus decay one of the *Up* quark in the proton is transmuted into *Down* quark via exchange of W^+ intermediate vector boson. The Feynman diagram is shown in figure 5.10.

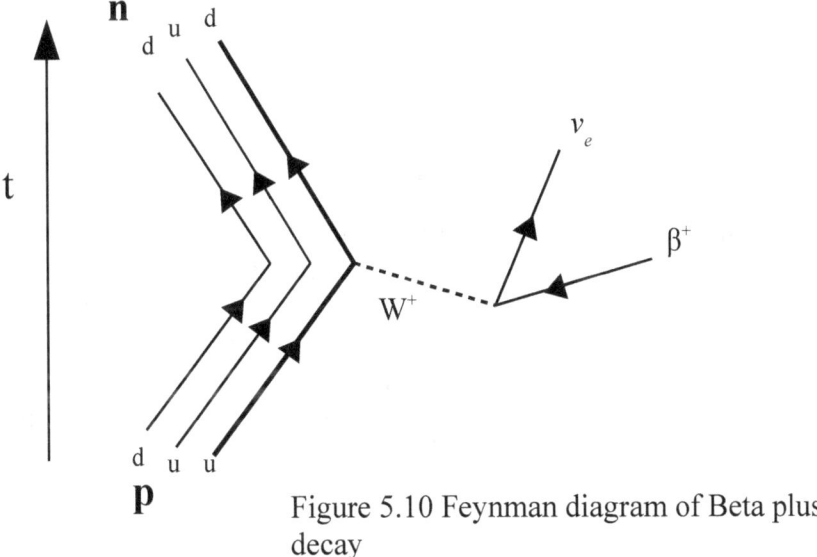

Figure 5.10 Feynman diagram of Beta plus decay

In proton rich nuclei where the energy difference between initial and final state is less than $2m_ec^2$ (1.022MeV), positron decay is not possible then nucleus will attract an electron from the first orbit or K-shell and will emit a neutrino. This mode of decay is called electron capture or K-capture. For example, rubidium-83 can only decay into krypton-83 via electron capture. This electron is captured by protons in the nucleus, forming a neutron and ejecting neutrino.

$$^1_1P + ^0_{-1}e \rightarrow ^1_0n + \nu_e$$

also,

$$^{22}_{11}Na + ^0_{-1}e \rightarrow ^{22}_{10}Ne + \nu_e$$

Biological effects

Beta radiation can cause both acute and chronic health effects. Contact with a strong beta source from which acute exposure could occur. Chronic effects are much more common, which result from low level exposure over a long period of time. The main risk is cancer. When beta emitters are taken internally they can cause tissue damage thus increasing cancer risk. Some beta emitters are distributed throughout the body, whereas others accumulate in specific organs and cause chronic exposure. Strontium-90 can affect bone and teeth and Iodine-131 concentration can cause thyroid cancer.

Uses

Beta particles are used in quality control to check the thickness of the paper. A certain amount of beta particles are allowed to be absorbed and if this number goes up or down the machine is adjusted accordingly to maintain the thickness to desired level.

Beta-lights are used where a permanent illumination without power source is required. A phosphorous coated glass sealed tube is filled with beta emitter tritium gas. The beta decay in tritium causes the phosphorous layer to fluoresce.

Higher doses of ionising radiation can kill cells; so beta particles are used to kill cancer cells.

Measurements of the amounts of radioisotopes and their decay products in rocks can be used to calculate the age of materials such as rocks, fossils and archaeological specimens. The proportion of the radioisotope carbon-14 to its stable decay product carbon-12 can be used to date organic materials.

Beta particles are used in brachytherapy source to treat benign eye growths that threaten vision. Beta brachytherapy sources are also used in the treatment of blood vessels following angioplasty.

Beta positive is used in diagnostic purpose. When positrons come close to normal electrons in the body, the reaction is one of particle-particle annihilation. The mass of the positron and electron will be converted into gamma rays. Positron tracer is given to the patient and where imaging is required, this annihilation will produce gamma rays. This scanning method is called PET which stands for positron emission tomography. This is very useful for cancer tumour imaging, to determine the places in the brain in which epilepsy occurs. PET scanning is also used in coronary artery disease.

Gamma Decay

A French physicist Paul Ulrich Villard (1860 – 1934) discovered gamma ray in 1900 while studying the radiation emanating from radium. He did not suggest any specific name for these radiations and so in 1903 Rutherford gave them the name gamma. Gamma rays (symbol γ) are high energy (above 100 keV), high frequency (10 exa Hertz) and very short wavelength (10 picometers) electromagnetic radiations.

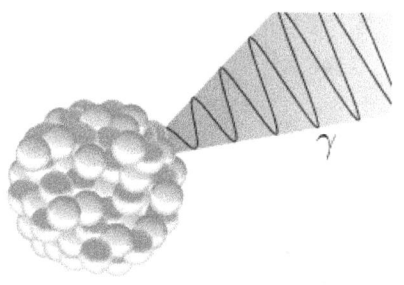

Figure 5.11 Emission of gamma ray from the nucleus.

One mechanism of the origin of this radiation is a redistribution of electric charge within a nucleus. Quantum mechanically there are many different possible ways in which the neutrons and protons can be arranged within a nucleus. When a nucleus transit from one such configuration to another and in terms of energy this transit normally happen from high to low energy, which is more stable state. The difference of energy is transmitted as gamma radiations. In such transition neither the mass number nor the atomic number is changed.

$$^{40}_{18}Ar^{\varnothing} \rightarrow {}^{40}_{18}Ar + \gamma$$

$Ar^{\varnothing}$ represents excited state.

Other common mode of gamma ray production is alongside other radioactive decay like alpha and beta. When a nucleus either emits alpha or beta particles, daughter nuclei is usually left in an excited state. This nucleus is then moved to lower energy by emitting gamma radiations.

$$^{137}_{55}Cs \rightarrow ^{137}_{56}Ba^{\varnothing} + ^{0}_{-1}\beta + \bar{v}_e$$

$$^{137}_{56}Ba^{\varnothing} \rightarrow ^{137}_{56}Ba + \gamma$$

Some nucleus are in excited metastable state and this state is also called *nuclear isomer*. When nucleus decay from such mode by emitting gamma rays and reach a lower and more stable energy state is called isomeric transition.

Gamma rays have the lowest ionisation power. It ionises matter via three processes: photoelectric effect, pair production and Compton scattering. It travels at the speed of light and they can only be stopped by material with high atomic number and high density. 1cm thick lead can reduce the intensity of gamma rays by half whereas 6 cm of concrete to reduce the intensity by 50%.

Sometime a radioactive nucleus decays but does not become stable and after some time (determined by its half life) it decays again and that again may not be stable state and it will keep decaying until the nuclei become stable. This is called a *decay chain*. Natural decay chain of Uranium-238 is as follow:

$^{238}_{92}U$ decays via α with half-life of 4.5 billion years to $^{234}_{90}Th$ (Thorium)

$^{234}_{90}Th$ decays via β with half-life of 24 days to $^{234}_{91}Pa$ (Protactinium)

$^{234}_{91}Pa$ decays via β with half-life of 1.2 minutes to $^{234}_{92}U$ (Uranium)

$^{234}_{92}U$ decays via α with half-life of 240,000 years to $^{230}_{90}Th$ (Thorium)

$^{230}_{90}Th$ decays via α with half-life of 77,000 years to $^{226}_{88}Ra$ (Radium)

$^{226}_{88}Ra$ decays via α with half-life of 1600 years to $^{222}_{86}Rn$ (Radon)

$^{222}_{86}Rn$ decays via α with half-life of 3.8 days to $^{218}_{84}Po$ (Polonium)

$^{218}_{84}Po$ decays via α with half-life of 3.1 minutes to $^{214}_{82}Pb$ (Lead)

$^{214}_{82}Pb$ decays via β with half-life of 27 minutes to $^{214}_{83}Bi$ (Bismuth)

$^{214}_{83}Bi$ decays via β with half-life of 20 minutes to $^{214}_{84}Po$ (Polonium)

$^{218}_{84}Po$ decays via α with half-life of 160 μs to $^{210}_{82}Pb$ (Lead)

$^{210}_{82}Pb$ decays via β with half-life of 22 years to $^{210}_{83}Bi$ (Bismuth)

$^{210}_{83}Bi$ decays via β with half-life of 5 days to $^{210}_{84}Po$ (Polonium)

$^{210}_{84}Po$ decays via α with half-life of 140 days to $^{206}_{82}Pb$ (Lead) which is a stable nuclei.

Biological effects

Gamma rays are more penetrating therefore causing damage throughout the body. They have sufficient energy to break bonds in the genetic material, cells and other biological molecules. The exposure to radiation can either be acute, in which radiation energy is absorbed over a few hours or days, or it can be chronic so the body is exposed over a longer period like months and years. If an individual is exposed to a small dose of gamma radiations or small dose over a long period, this will develop a case of mild radiation poisoning. In such exposure fast-growing cell like skin, hair and Gastrointestinal tract cell will be affected most. On average 10% of invasive cancer are due to radiations. Gamma radiation can strike the chromosome and break it which will result in an abnormal number of chromosomes. If radiation damage a cell in such a way that tumour suppressor genes are damaged it becomes partly functional which then may develop into cancer. Three stages are involved in the development of cancer by ionising radiations, an adaptation that favour tumour formation, morphological change to the cell and acquiring cellular immortality.

Uses

Gamma radiations are used to kill bacteria and other micro-organisms, sterilise medical devices. The sterilisation process use cobalt-60 radiations to kill micro-organisms. Gamma radiations are used to eliminate organisms from pharmaceuticals such as ointments and solutions.

Gamma radiations are used by a custom control agency to check trucks without unloading them.

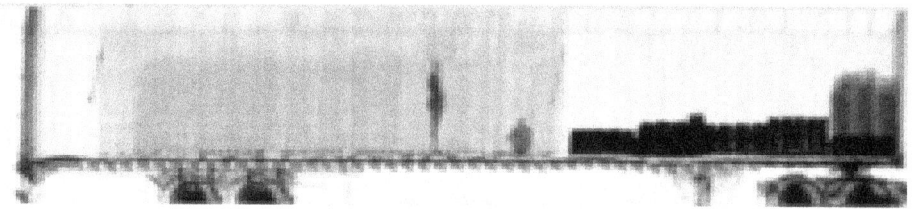

Figure 5.12 Scanning truck using gamma radiations

Gamma radiations are used to kill cancer cells by directly damaging their DNA or create charged particles (free radicals) within the cells that can in turn damage the DNA. The radiations that can be targeted from outside the body, is called external beam radiation therapy. This can also be done by placing radioactive material in the body, called brachytherapy.

Gamma radiations are used for diagnostic purpose in nuclear medicine in medical imaging. A patient will be injected with a small dose of radioactive tracer like Technetium-99. They will concentrate around the body part to be investigated, the detector outside the body will detect the radiations and with the help of the computer processing power will build a image of the organ and be able to see the what is wrong with the organ.

They are also used to check leaking pipes. To locate a leak in the underground pipe. A small amount of gamma emitting radioactive substance is added into the liquid that flow in the pipe. A GM detector is moved on the ground along the buried pipe. Radioactive substance will come out where there is a leak and so the reading on the detector will read high and leak point is detected.

Properties of semi-precious stones can be changed (changing white topaz to blue topaz) by using a process called gamma-induced molecular change.

Detecting radioactivity

A Geiger-Müller counter is a particle detector that is used to measure the ionisation radiations. The instrument consists of two main components: Geiger-Müller tube (G-M tube) and the processing unit with display. Initially developed by German physicist Hans Geiger (1882 – 1945) in 1908 which was further developed by one of his student Walter Müller (1905 – 1979) in 1928. The tube contains an inert gas at very low pressure. The tube contains one electrode serving as *anode* at the centre of the tube while the casing usually made of metal will serve as the *cathode*. Between these electrode a several hundred volt of electric potential is created. When ionising radiation enters into the tube via window they will knock off electron from the gas molecules. This will produce positive ions and electrons. See Figure 5.13.

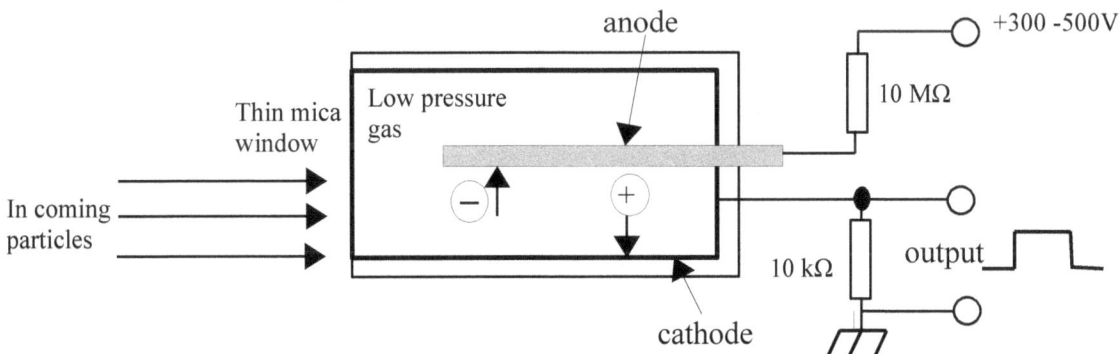

Figure 5.13 G-M tube and counter.

The strong electric field between the electrode will accelerate positive ions towards the cathode and electrons toward the anode. The electrons gain sufficient energy to further ionise the gas thus creating an avalanche. Electrons which are attracted by anode will flow in the resistor as tiny but measurable current. This then further processed by electronic circuits and displayed.

How to distinguish the radioactive source

1- If you have a material which is radioactive. How would you know whether it emits alpha, beta or gamma radiations. One way is to put source in front of the G-M tube and leave it for a few minutes until you get steady reading. Then place paper between the source and the G-M tube. If the count rate goes down this means the source is an alpha emitter. If nothing happens after placing the paper then the source is not alpha. Now place thin lead plate and if you observe fall on the counter then you have a beta source. To see if the source is gamma put a 4mm thick lead plate between the source and the G-M tube. Gamma rays will be stopped by this lead.

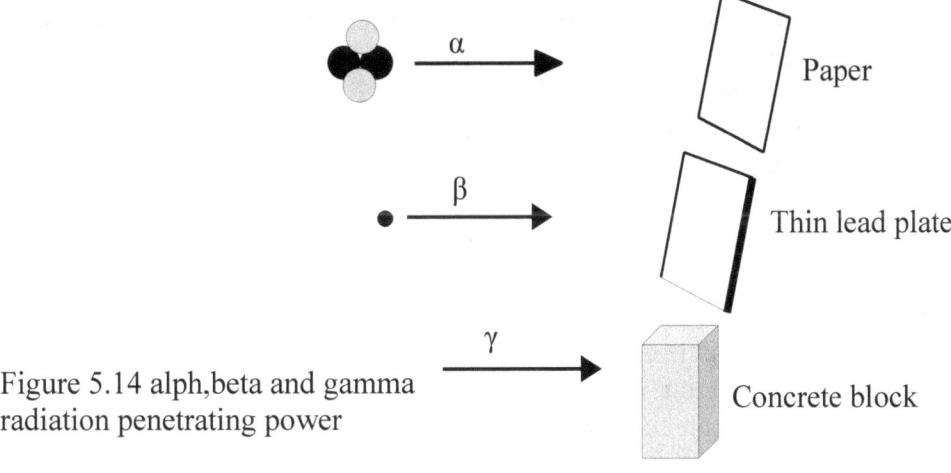

Figure 5.14 alph,beta and gamma radiation penetrating power

2- A moving charged particles in a magnetic field will experience force and their path is deflected. The direction of their motion can be explained using Fleming's left hand rule. This rule states that when thumb, first and second finger of left hand is placed at right angle to each other then, if first finger pointing in the direction of the magnetic field and second finger points in the opposite direction to the motion of negative charge then the thumb will point in the direction of the force on the charge. So the particle will deflect its direction of motion in the direction of the force and hence will move in a curve.

So if we direct the radioactive source toward the magnetic field and alpha and beta particle will deflect in the opposite direction to each other and using Fleming's left hand rule we can identify the source. Gamma radiation will not deflect and keep moving in a straight line because it does not have a charge. See Figure 5.15.

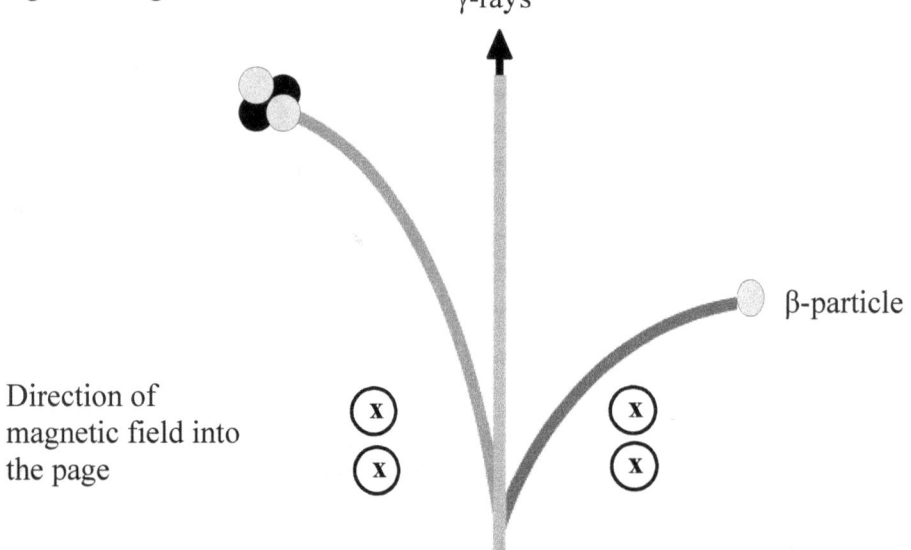

Figure 5.15 alpha, beta particles and gamma rays in uniform magnetic field.

3- Moving charged particles in a electric field will experience force and their path is deflected. If we direct the radioactive source in such a way that particles pass through electric field. Positive alpha particle are deflected towards the negative plate and the negative beta particle are deflected towards the positive plate. Beta particle will deflect more than the alpha particle due to their lower mass.

Law of Radioactive Decay

Radioactive decay is random and spontaneous act of transformation of one element into another. What this means is that it is impossible to know when a particular atom in radioactive material will decay. The probability that a particular radioactive nuclei will decay does not change with time. All the nuclei have equal probability of decaying. The number of nuclei decaying per second is called *activity* of the sample material. What is random for individual nuclei become a law for a large number of nuclei. The law of radioactivity states that:

"the activity of the radioactive nuclei always go down with time".

Let say there are N numbers of radioactive nuclei at time, t. As time passes we know that some nuclei will decay. It's obvious that this is proportional to the number of radioactive nuclei, N at a time. This means that if there are more nuclei there will be more decayed nuclei and longer the period again more decayed nuclei. This can be written in differential form as:

$$-dN \, \alpha \, N \cdot dt$$

the minus sign in the above equation signifies that the original number (N) of particle decreases. Changing proportionality into equality by introducing a constant of proportionality, λ then the above equation can be written as:

$$-dN = \lambda N \cdot dt$$

dividing both sides by N,

$$\frac{-dN}{N} = \lambda \cdot dt$$

If we integrate the above equation over a period of time from $t = 0$ to any time later, then the number of nuclei at time $t = 0$ are represented by N_0, and N at a later time.

Thus,

$$-\int_{N_0}^{N} \frac{dN}{N} = \lambda \int_{0}^{t} dt$$

integrating both sides and multiplying by (-1)

$$\ln[N]_{N_0}^{N} = -\lambda t$$

$$\ln[N] - \ln[N_0] = -\lambda t$$

$$\ln\left(\frac{N}{N_0}\right) = -\lambda t$$

taking natural log (ln) of both side

$$\frac{N}{N_0} = e^{-\lambda t}$$

or

$$N = N_0 \, e^{-\lambda t} \qquad (5.1)$$

This equation governs the radioactive decay law. This is an exponential equation. This means that the radioactive nuclei decays exponentially.

Any quantity will be subjected to exponential decay if it decreases at the rate which is proportional to its value. The constant λ is called the decay constant with unit s^{-1}.

If we plot a graph of the number of nuclei remained against time we will have exponential graph like the one shown in Figure 5.16. You can see that initial number of nuclei decrease fast and eventually they slow down. It is not easy to grasp this idea until you become familiar with the concept of half-life. Half-life *is the time required for the decaying quantity to fall to one half of its initial value and is denoted by symbol $t_{1/2}$.*

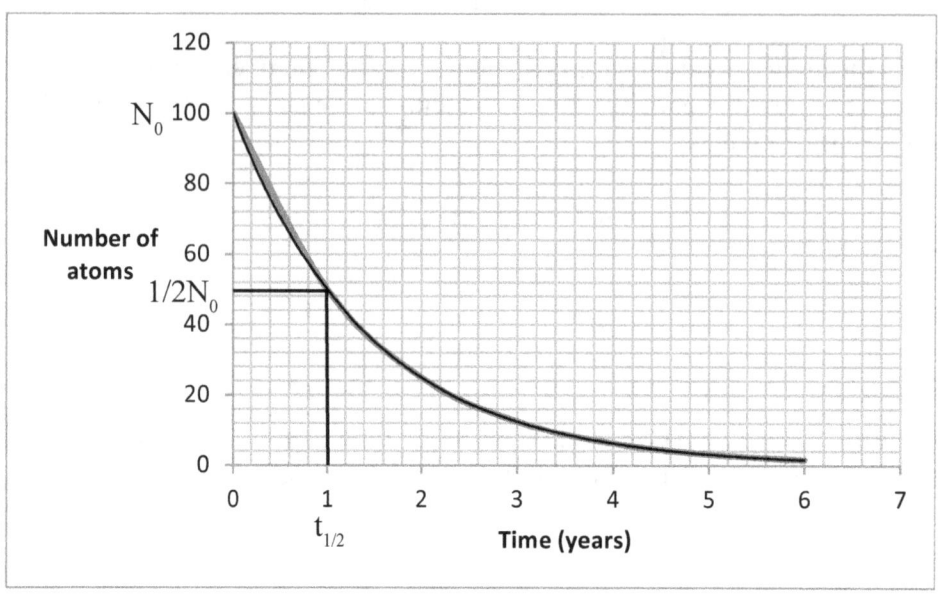

Figure 5.16 Exponential decay curve

Half – Life

From the definition of half life we can write this in an algebraic expression.

$$N = \frac{N_0}{2}$$

when $\quad t = t_{1/2}$

substituting these in equation 5.1

$$\frac{N_0}{2} = N_0 \, e^{-\lambda t_{1/2}}$$

cancelling N_0 from both sides

$$\frac{1}{2} = e^{-\lambda t_{1/2}}$$

Taking natural logarithm of both sides

$$\ln \tfrac{1}{2} = -\lambda \, t_{1/2}$$

$$-0.693 = -\lambda \, t_{1/2} \quad (\,'-'\text{ sign cancels on both sides})$$

rearranging $\quad t_{1/2} = \dfrac{0.693}{\lambda} \quad\quad (5.2)$

This equation links half-life with a decay constant.

Radioactive Decay
- independent of the age of the nucleus
- unaffected by heat
- pressure,
- magnetic and electric influences
- in fact, all external forces.

Radiocarbon dating

The radiocarbon dating method was developed by a team of scientists led by American Professor Willard Frank Libby (1908 – 1980) in 1949. All living organisms continually exchange CO_2 with the atmosphere. The rate of production of C-14 in the atmosphere is constant. Cosmic rays contains high energy protons interact with nuclei in the upper atmosphere and produce neutrons. These neutrons in turn bombard nitrogen. This produces radioactive carbon-14. This carbon isotope combines with oxygen and produce carbon dioxide.

$$^1_0n + {}^{14}_7N \rightarrow {}^{14}_6C + {}^1_1P$$

This C-14 reacts with the oxygen in the atmosphere and produces CO_2.

The ratio of C-14 to stable C-12 is fixed in our atmosphere and this ratio is:

$$\frac{{}^{14}_6C}{{}^{12}_6C} \approx 10^{-12}$$

Plants need carbon dioxide during photosynthesis and so the take C-14 at the above ratio. After the plants die or consumed by humans or animals this intake of C-14 stops. As C-14 is radioactive so the amount of C-14 will decrease exponentially from the plant or body which consumed the plants.

$$^{14}_6C \rightarrow {}^{14}_7N + {}^0_{-1}\beta + \bar{v}_0$$

If one compares the remaining amount of C-14 in the sample to that expected in the atmosphere one can determine the age of the sample. The half-life of C-14 is 5730 years.

Example: Find the activity rate of 5g sample of ${}^{243}_{95}Am$ with a half-life of 7370 years. Avogadro's number: 6.023×10^{23}.

The activity rate is $\frac{dN}{dt}$

From equation 5.1 $\quad N = N_0\, e^{-\lambda t}$

We need to calculate N and λ.

Now, $\quad \lambda = \dfrac{0.693}{t_{1/2}} = \dfrac{0.693}{7370} = 9.40 \times 10^{-5}$ year^{-1}

We need to find the number of seconds in a year $= 365.25 \times 24 \times 60 \times 60$

One year = 31557600 seconds

hence, $$\lambda = \frac{9.40 \times 10^{-5}}{31557600} = 2.98 \times 10^{-12} \text{ s}^{-1}$$

To calculate N, we use the following equation.

$$N = \frac{(Avogadro's\ number) \times (mass)}{mass\ number} = \frac{(6.023 \times 10^{23}) \times (5)}{243}$$

$N = 1.24 \times 10^{22}$ atoms

now, $$\frac{dN}{dt} = -\lambda N$$

$$\frac{dN}{dt} = -(2.98 \times 10^{-12}) \times (1.24 \times 10^{22})$$
$$= 3.69 \times 10^{10} \text{ atoms per second}$$

Example: The half-life of Zn-71 is 2.4 minutes. If one had 250 g at the beginning, how many grams would be left after 12 minutes has elapsed?

$$\frac{12}{2.4} = 5\ half\ lives$$

½ x ½ x ½ x ½ x ½ = 1/32 = 0.03125   or $(1/2)^5 = 0.03125$

0.03125 x 250 ≈ 7.8g remained

Example: After 24 days, 5 milligrams of an original radioactive material of 250 milligram sample remains. What is the half-life of the sample?

$N_0 = 250 \times 10^{-3}$g, $N = 5 \times 10^{-3}$g, t = 24 days = 24 x 24 x 60 x 60 = 2073600 seconds

rearranging equation 5.1

$$\frac{N}{N_0} = e^{-\lambda t}$$

taking natural logarithmic of both sides

$$\ln\left(\frac{N}{N_0}\right) = -\lambda t$$

rearranging for λ

$$\lambda = \frac{\ln\left(\frac{N}{N_0}\right)}{t} = \frac{\ln\left(\frac{5 \times 10^{-3}}{250 \times 10^{-3}}\right)}{2073600} = 1.89 \times 10^{-6} \text{ s}^{-1}$$

$$t_{1/2} = \frac{0.693}{\lambda} = \frac{0.693}{1.89 \times 10^{-6}} = 367330 \text{ s or 4.25 days}$$

Exercise 5

Q1- What are isotopes?
A. atoms that have the same number of neutrons
B. atoms of the same element with different masses
C. atoms of different elements with same masses
D. atoms that have the same number of valence electrons

Q2- An isotope of Bismuth has the nuclear symbol $^{207}_{83}Bi$. How many protons and neutrons are in the nucleus of this atom?
A. 83 protons and 83 neutrons
B. 83 protons and 212 neutrons
C. 83 protons and 124 neutrons
D. 124 protons and 83 neutrons

Q3- Which types of radiation will a thin piece of aluminium foil block?
I. alpha radiation
II. beta radiation
III. gamma radiation
A. I only
B. II only
C. I and II only
D. I, II and III

Q4- Polonium decays according to the following partial nuclear equation:
$$^{211}_{84}Po \rightarrow {}^{207}_{82}Pb + ...$$
What is missing from the equation?
A. an alpha particle
B. a beta particle
C. gamma radiation
D. an electron

Q5- An isotope of an element undergoes beta decay to form neptunium-237. Which of the following is the parent isotope?
A. $^{237}_{94}Pu$
B. $^{238}_{94}Pu$
C. $^{237}_{92}U$
D. $^{238}_{92}U$

Q6- Carbon-11, which has a half-life of 20.4 min, is useful for medical imaging. The whole medical test takes four half-lives to complete. When the test is complete, what percentage of carbon-11 will remain?
A. 3.125%
B. 6.25%
C. 12.5%
D. 50%

Q7- Write nuclear decay equation for the alpha decay of radon-198

Q8- Write nuclear decay equation for the Positron emission from silicon-26.

Q9- Bombardment of U-238 with C-12 produces an isotope of element 98 and 4 identical particles. Write the balanced nuclear equation.

Q10- In a radioactivity experiment, the count rate of a radioactive substance decreased from 1000 counts per second. The half-life of the substance was 30 minutes.
   i) What was the count rate after one hour?

   ii) How long would it take for the count rate of the substance to decrease to 25 counts per second?

Q11- Radioactive carbon found in living wood has a half-life of 5730 years. In a test:

   i) A sample of living wood was found to give a count rate of 512 counts per minute.

   ii) A sample of ancient wood of the same mass was found to give a count rate of 32 counts per minute.

Calculate the age of the sample of ancient wood.

Q12- What is the minimum mass of ^{99m}Tc that can have an activity rate of 3.5×10^8 ? Assume the half-life is 6 hours and that Avogadro's Number is 6.023×10^{23}.

Q13- Fluorine-21 has a half life of approximately 5 seconds. What fraction of the original nuclei would remain after 1 minute?

Q14- A scientist begins with a 64g sample of Radium – 226. What is the half - life of Radium – 226 in years if $\lambda = 0.00043$ s^{-1}. Round off to the nearest year.

Q15- In the table below the activity rate of two elements are given. Plot a graph of activity against time for both elements on the same graph. From your graph estimate the half-life of each element.

A_1 (dN/dt) s^{-1}	Element 1 time (s)	A_2 (dN/dt) s^{-1}	Element 2 time (s)
10.48	0	17.50	0
7.54	1	12.50	1
5.49	2	11.05	2
4.02	3	8.95	3
2.74	4	8.40	4
2.02	5	6.85	5
1.50	6	6.20	6
1.09	7	5.15	7
0.68	8	3.75	8
0.57	9	2.45	9
0.37	10	2.00	10
0.31	11	1.20	11
0.19	12	0.70	12
0.15	13	0.55	13

Q16 – Radioactive cobalt-60 (Co) is used to irradiate and sterilise surgical equipment. Many instruments can be irradiated simultaneously.

(a) Explain briefly which type of radiation (α, β, or γ) would be most appropriate to irradiate metallic surgical instruments in this way.

(b) The half life of cobalt-60 is 5.3 years.
(i) Calculate the decay constant of cobalt-60.
(ii) Calculate the activity of 1 mg of cobalt-60.

(iii) Calculate the time taken (in years) for the activity of a cobalt-60 source to drop from 4.16×10^{10} Bq to 1.04×10^{10} Bq.

Q17 – A uranium nucleus $^{238}_{92}U$ undergoes a series of radioactive decays before it attains a final stable state which is a nucleus of lead (Pb). The succession of particles emitted during its decay is listed below in the order in which they occur.

alpha, beta, beta, alpha, alpha, alpha, alpha, alpha, beta, beta,
alpha, beta, beta, alpha

Find the nucleon number and the proton number of the final stable Pb nucleus.

Q18 – The half-life of bismuth-210 for β-particle emission is 5.0 days. Find the percentage loss of activity in a sample after 15 hours.

Q19 – The nuclide $^{203}_{83}Bi$ is an alpha particle emitter. An initial measurement of the activity of a sample of this isotope gives a count rate of 1500 counts s^{-1}. After an interval of 24 hours the count rate falls to 200 counts s^{-1}. Both these readings have been corrected for background radiation.

(i) Show that the decay constant of $^{203}_{83}Bi$ is about 1.86×10^{-5} s^{-1}.
(ii) Calculate the half life of this nuclide.
(iii) Calculate the number N of nuclei in the sample when the count rate was 1500 s^{-1}.

Q20 – How long will it take the 50g of I-131 (half-life 8 days) to decay to approximately 1/100 its original mass?

6- Radiopharmaceuticals (Nuclear Medicine)

Father of radiopharmaceutical is Hungarian radio chemist George Charles de Hevesy (1885 – 1966). He used radioactive isotopes in studying the metabolic processes of plants and animal, by tracing chemicals in the body by replacing part of stable isotope with a small trace of radioactive isotope.

Radiopharmaceuticals are radioactive agents used to treat or diagnose certain diseases. They can be given to patients in many ways. For example, by mouth, injection or placed into the bladder or eyes. These drugs have two components, a radioactive tracer and a carrier molecule which delivers radioactive isotope to the area to be treated or examined. The physical characteristic of each radioactive nuclei is different and each nuclei is selected for a particular intended clinical use. Most radiopharmaceuticals incorporate a radioactive tracer atom into a large pharmaceutical active molecule, which will localised in the body, after which the radionuclide tracer can easily be detected using the gamma camera or other suitable imaging device.

Table 6.1 lists some common medical problems and their suitable radioactive nuclei.

Table 6.1

Medical problem	Suitable Radiopharmaceutical
Abscess and infection	Gallium(67) Citrate , Oxyquinoline, In-111
Blood vessel disease	Sodium pertechnetate ^{m99}Tc
Blood vessel disease of brain	Ammonia N-13, Iofetamine I 123, ^{m99}Tc Bicisate, Xe-133
Bone diseases	Sodium Fluoride(18), ^{m99}Tc Pyrophosphate and Oxidronate
Bone Marrow diseases	^{m99}Tc Albumin and Sulfur Colloid, Sodium Chromate(51)
Brain Tumour	Fludeoxyglucose F-18, Iofetamine I 123, ^{m99}Tc Exametazime, ^{m99}Tc Gluceptate, ^{m99}Tc Pentetate
Cancer	Gallium(67) Citrate, Fludeoxyglucose F-18, Indium In 111 Pentetreotide, Methionine C11, ^{m99}Tc Arcitumomab, ^{m99}Tc Nofetumomab Merpentan, Radioiodinated Iobenguane, Sodium Fluoride(18)
Heart diseases	Ammonia N-13, Thallous Chloride Tl 201, Fludeoxyglucose F-18, S ^{m99}Tc Teboroxime and Tetrofosimin
Kidney diseases	Iodohippurate Sodium I 131, Iothalamate Sodium I 125, ^{m99}Tc Gluceptate and Metiatide and Succimer
Liver diseases	Fludeoxyglucose F-18, Ammonia N-13, ^{m99}Tc Disofenin, ^{m99}Tc Lidofenin and Mebrofenin and Sulfur Colloid
Lung diseases	Krypton -81m, Xenon-127 and 133, ^{m99}Tc Pentetate
Red blood cell diseases	Sodium Chromate - 51
Thyroid diseases	Fludeoxyglucose F-18, Sodium iodide I-123&131,

All Radioisotopes emitting gamma radiations are used for diagnostic purpose and those which are beta emitters are used therapeutically. Some emit both hence they are used for both purposes. Technetium ^{m99}Tc is most widely used radionuclide because of its ideal short half-life and mono energy gamma emission. In metabolic therapy a radio element is administered to patients associated with a molecule which will preferentially focus on the tumour tissue in order to destroy it. Tumour-specific targeting helps increase the effectiveness of treatments while minimising harmful effects to healthy tissue. The most recent developments in the use of Rhenium-188 relate to cancer therapy, palliative care for bone pain caused by spreading of breast and prostate cancer, the treatment of synovitis and cardiology. Yttrium-90 effectively fights certain lymphomas (so called non-Hodgkin's), cancer type that originates in the lymphocytes (immune system).

Production of Radiopharmaceutical

About one third of the world supply, and most of the North America's supply of radionuclide are produced at the Chalk River Laboratories, Ontario, Canada. Another one third of the supply, and most of the European demand is produced at the Petten nuclear reactor in the Netherlands. Radionuclide are produced in nuclear reactors, cyclotrons or in radionuclide generators.

Production of a radiopharmaceutical involves two processes:
- The production of the radionuclide.
- The preparation and packaging of the complete radiopharmaceutical

Radionuclide are radioactive isotopes of elements (atomic number less than bismuth) that also have one or more stable isotopes. These are divided into four classes.

- Neutron-deficient radionuclide will decay by positron emission or electron capture. *Cyclotrons* produce such radionuclide by bombarding stable nuclei with high energy charged particles.
- Proton-deficient radionuclide are produced in *nuclear reactors*. They are produced from fission products
- Neutrons produced by the fission of uranium are used to produce radionuclide by bombarding stable material in the reactor with these neutrons is called *neutron activation*. Almost all radionuclide produced by this method decays via beta minus.
- Technetium-99 is very widely used in nuclear medicines but due to its short half-life of around 6 hours makes it impractical to store. This is overcome by parent Molybdenum-99 which ha relatively long half-life of 67 hours and continually produce Technetium-99. In a *radionuclide generator* daughter is easily separated.

Radioactive products should be stored, processed, packaged and controlled in dedicated and self-contained facilities. The equipment used should be reserved exclusively for radiopharmaceuticals. The facility has to be organized in such a way as to minimize the risk of cross contamination and mix-up.

CYCLOTRON:

A cyclotron is a type of charge particle accelerator in which particles accelerate outwards from the centre along a spiral path. The cyclotron was invented by Hungarian American physicist Leó Szilárd (1898 – 1964) and first manufactured and operated by American physicist Ernest Orlando Lawrence (1901 -1958) in 1932. Cyclotron accelerates charged particles by applying a high frequency alternating voltage between two D shaped electrodes, See figure 6.1

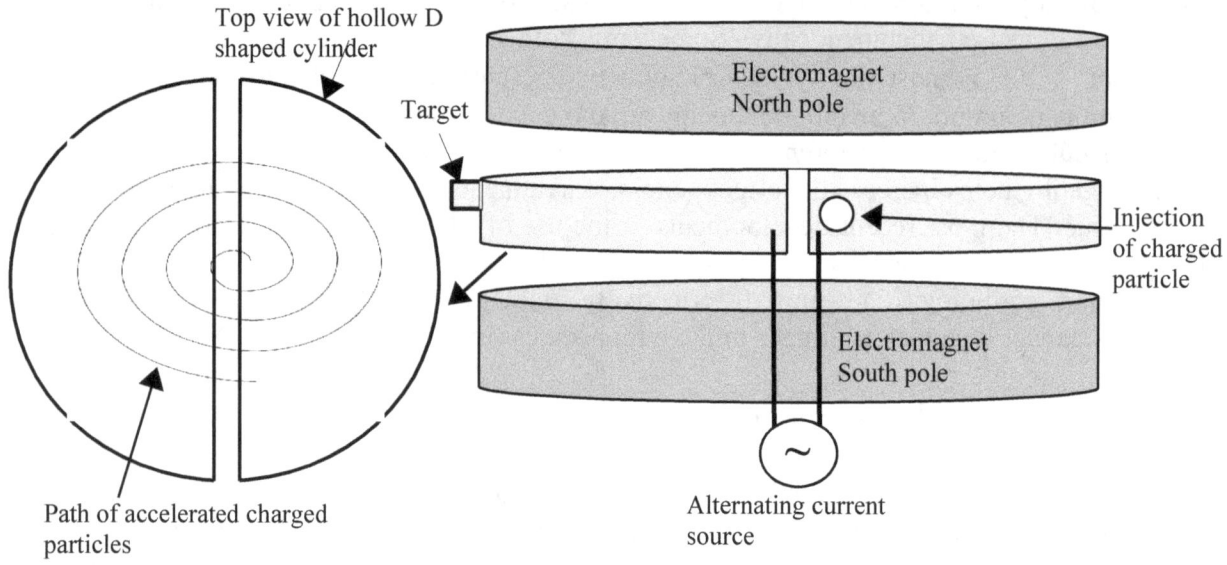

Figure 6.1 Cyclotron

The cyclotron is comprised of a vacuum contained within semi-circular electrodes, shaped 'D'. The magnetic field is applied perpendicular to the cyclotron. Each D is separated by a narrow gap and they are connected to high frequency voltage source. You have seen earlier that according to Fleming left hand rule that a charged particle in magnetic field will feel a force perpendicular to the direction of the motion which will move charged particle to follow a curved path and so the direction of the motion is toward the centre of the curve. The force on charged particle in a magnetic field can be represented by equation 6.1

$$F = q.v.B \qquad (6.1)$$

where q is charge in Coulombs(C), v is velocity in ms^{-1} and B is the magnetic field in Tesla (T)

Also in circular motion force on mass 'm' with velocity 'v' is given by equation 6.2

$$F = \frac{mv^2}{r} \text{ where r is radius} \qquad (6.2)$$

equating 6.1 and 6.2

$$q.v.B = \frac{mv^2}{r}$$

rearranging in terms of r

$$r = \frac{mv}{qB} \qquad (6.3)$$

Since q, B and m are constant for a given particle, so as v increases so the radius will also increase.

When charged particle travels through the first 'D', at that instant the polarity on 'D' is reversed and this will make second 'D' to attract the particle and the first 'D' will repel, the particle will accelerate. The particle now travelling through the second 'D' in a circular arc with a larger radius as explained above. This repeated many times until charged particle reaches to certain velocity and is directed out of the cyclotron and goes toward the target.

Cyclotrons are designed for specific energy they can supply to the charged particle. Small cyclotron can provide up to 11MeV and are limited to produce fluorine-18 for PET (positron emission tomography). A medium sized cyclotron can produce many different PET radio nuclides like C-11, N-13 and O-15. Larger cyclotrons, typically 30MeV are capable of producing Ga-67, Tl-201and I-123

Radioisotope production with cyclotrons offers many advantages over a nuclear reactor. Firstly, the radioactive waste produced by cyclotrons is far less and much less hazardous than the radioactive waste of nuclear reactors. Secondly, the production is local. Cyclotrons are located in the hospital, by which the delivery of pharmaceuticals to patients is much more secure. Also the risk of transport accidents is practically zero. Thirdly, there are no risks due to nuclear-power accidents, because there is no need for controlled chain reactions.

NUCLEAR REACTOR:

A nuclear reactor is a device that initiate and control a sustained nuclear chain reaction. *Fission* is one of the processes that can produce radio nuclide. In fission heavy nuclei is bombarded with neutrons to split nuclei into two or more lighter nuclei along with kinetic energy, gamma rays and neutrons. These neutrons in turn will hit other nuclei which by splitting will produce more neutrons and soon a chain reaction will start. See Figure 6.2

For example:

$$^{235}_{92}U + ^{1}_{0}n \rightarrow ^{236}_{92}U \rightarrow ^{131}_{53}I + ^{102}_{39}Y + 3^{1}_{0}n \quad \text{where} \quad ^{131}_{53}I \text{ is radionuclide}$$

$$^{235}_{92}U + ^{1}_{0}n \rightarrow ^{236}_{92}U \rightarrow ^{99}_{42}Mo + ^{135}_{50}Sn + 2^{1}_{0}n \quad \text{where} \quad ^{99}_{42}Mo \text{ is radionuclide}$$

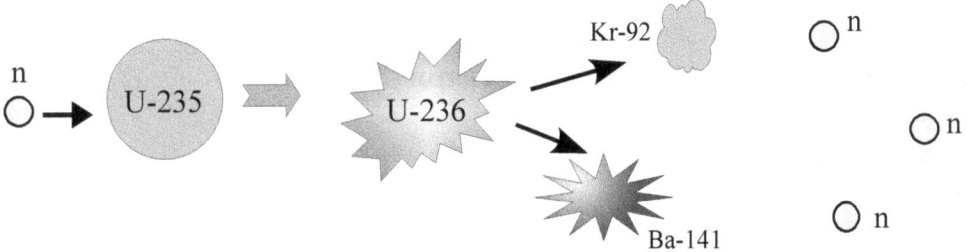

Figure 6.2 An induced nuclear fission event

Neutrons which are by-product of fission are very fast, to control their kinetic energy a process called *thermalisation or moderation* is used. For this purpose heavy water ($^{2}H_2O$) is used. From Figure 6.2 we can see that for every collision, we get three neutrons to produce 3 more collisions and they will in turn produce 9 more neutrons and soon this will produce so high number of collisions which could go out of control and heat produced could melt down the reactor. To prevent this, inside the reactor control rods made from cadmium or boron are inserted to absorb the neutrons.

Neutron activation:

The neutron liberated during the fission reactions in the reactor are used to bombard other stable nuclides which are placed in the nuclear reactor. When target nuclei capture neutrons, they become heavier and enter into excited state. These excited nuclei will eventually decay and often result in the formation of activation product. Such radioactive nuclei can be used in nuclear medicine. This is how cobalt-60 is produced in the nuclear reactor.

$$^{59}_{27}Co + ^{1}_{0}n \rightarrow ^{60}_{27}Co$$

Cobalt-60 has a half-life of 5.27 years and will decay into stable nuclei of Nickel.

The yield N for the radionuclide production over time period t depends on the neutron flux ϕ (cm^{-2} s^{-1}), the cross-section area σ (cm^2) of the neutron capture process, the number of target nuclei n_t and the decay constant λ. This can be expressed by equation 6.4

$$N = \left(\frac{n_t . \phi . \sigma}{\lambda}\right)(1 - e^{-\lambda . t}) \qquad (6.4)$$

When t is greater than 4 half-lives $1 - e^{-\lambda . t}$ approaches to 1 and there is no gain beyond 4

Preparation of Mo - 99: The starting material is 1g of natural MoO$_3$ with 23.78% ^{98}Mo. neutron flux of 2x10^{14} cm^{-2}s^{-1} and kept for the time period of 7 days. This produces ^{99}Mo

$$^{98}Mo + ^{1}n \rightarrow ^{99}Mo + \gamma$$

Total of 37 GBq is produced from 1g of MoO$_3$

GENERATOR:

Desirable properties of radionuclide for generators.

- High radiochemical purity
- Easily and safely transportable
- Daughter nuclide with ideal half life and gamma energy
- High separation efficiency of daughter radionuclide from parent
- High radionuclide purity
- High yields during each elution
- continually available parent radionuclide.

Ideal Generator system

- the output of the generator must be sterile (absence of viable living bacteria) and pyrogen free (pyrogen is any substance that cause fever. The most common type of pyrogens is bacterial endotoxins and exotoxins)

- the chemical properties of daughter must be different than those of the parent to permit easy separation
- To permit separation of daughter from parent, the chemical properties of the daughter *must be* different than those of the parent.
- Generator *should be* eluted with 0.9% saline solution and should involve no violent chemical reactions. Eluents other than 0.9% NaCl may require tedious pH adjustment, associated with a significant radiation dose.
- daughter isotope should ideally have a half life of a few hours and decay via gamma emission
- half life of parent should be short enough so that daughter regrowth after elution is rapid
- granddaughter should be stable by having very long half life. Tc-99 ground state which is the granddaughter of Mo-99, the half-life of 220,000 years guarantees a minimal radiation dose to the patient, regardless of the effective half-life.

Half-life of some radio nuclide's is so short that their manufacturing, quality control and transportation are impractical. A radionuclide generator is a device that allows a weekly supply of a short lived radionuclide to be available on site. A radionuclide generator is comprised of a short lived radioactive daughter with good characteristics for medical imaging of long lived parent nuclei.

This method is typically applied to the separation of $^{99}Tc^m$ with half life of 6 hours from ^{99}Mo with half life of 66 hours. $^{99}Tc^m$ is shipped from nuclear reactors to hospitals. The generator, colloquially known as a *Moly cow*, provides radiation protection during transport and used for extraction of tracer at the medical facility. They must be replaced weekly as the half live of ^{99}Mo is 66 hours.

Technetium-99m is employed in about 80% of all nuclear medicine procedures. This is because it has almost ideal characteristics for a nuclear medicine scan. The chemistry of technetium is so versatile that it can be chemically bound to many different types of biologically active molecule for tracer work. For medical purposes, $^{99}Tc^m$ is used in the form of pertechnetate, TcO_4^- anion.

Figure 6.3 Five 99m Technetium generators

Separation technique for radionuclide

- They must have different physical states
- They must have different chemical properties
- Radionuclide generators have four different methods of separation techniques
  1- Solvent extraction – as they have different solubility
  2- Chromatography – based on different affinities for an ion-exchange resin
  3- Sublimation – based on different volatility
  4- Gel generator.

1- Solvent extraction generator

Parent	^{99}Mo
Parent Radiochemical	$^{99}MoO_4^-$
Daughter	$^{99m}TcO_4^{-1}$
Organic solvent	MEK (methyl ethyl ketone)
Aqueous solvent	KOH (potassium hydroxide)

Solvent extraction refers to the process of selectively removing a solute from a liquid mixture with a solvent. As a separation technique, it is a partitioning process based on the unequal distribution of the solute between two immiscible solvents, usually water and an organic liquid. The solute can be in a solid or liquid form. The extracting solvent can be water, a water-miscible solvent, or a water-immiscible solvent; but it must be insoluble in the solvent of the liquid mixture. The separation is based on selection extraction of TcO_4 into MEK for aqueous alkaline solution of sodium molybdate, Na_2MoO_4. The purification of organic medium is achieved by passing through alumina column. Next is evaporation of organic phase, residue is reconstituted with physiological saline solution and then sterilised to obtain Tc in the form of TcO_4^{-1} suitable for *in vivo* (experiment which is conducted with living organisms in their normal, intact state) use.

Advantages

- High extraction efficiency
- Ability to utilise relatively inexpensive low specific activity
- High radionuclide purity

Disadvantages

- Time consuming separation process
- Introduces operator dependent error which reduces radiochemical purity
- Hazard of handling inflammable solvent
- Difficult and costly to dispose of the mixed waste generated from the large volumes of solvents required.

- The technique also tends to be labour intensive because of the need for multiple extractions using separator funnels.

2 – Chromatography (column) generator

Column chromatography is a method used to purify individual chemical compound from mixtures of compounds. Two methods are generally used to prepare a column: the dry method, and the wet method. For the dry method, the column is first filled with dry stationary phase powder, followed by the addition of mobile phase, which is flushed through the column until it is completely wet, and from this point is never allowed to run dry. For the wet method, a slurry is prepared of the eluent with the stationary phase powder and then carefully poured into the column. Care must be taken to avoid air bubbles. A solution of the organic material is pipetted on top of the stationary phase. This layer is usually topped with a small layer of sand or with cotton or glass wool to protect the shape of the organic layer from the velocity of newly added eluent. Eluent is slowly passed through the column to advance the organic material. Often a spherical eluent reservoir or an eluent-filled and stoppered separating funnel is put on top of the column.

Parent ^{99}Mo as molybdate $^{99}Mo^mTcO_4^{-2}$
Daughter ^{99m}Tc as pertechnetate $^{99m}TcO_4^{-1}$
Absorbent material Al_2O_3 – Alumina (aluminium oxide)
Eluent Saline (0.9% NaCl)
Eluate $^{99m}TcO_4^{-1}$
Decay by β with 1.2 MeV (82%) and γ emission with 740-80 keV

Advantages
- Ease of operation at medical facility
- High elution efficiency
- High purity
- High radioactive concentration

Disadvantages
- High cost due to need for fission produced ^{99}Mo
- Difficult to dispose toxic fission waste generated.

3- Sublimation generators

In sublimation process solid phase of matter is directly transformed into the gaseous phase without passing through an intermediate liquid phase. Sublimation is a technique used by chemists to purify compounds. The sublimation technetium generator uses this process to separate technetium-99 from molybdenum-99. This is possible because compounds such as Tc_2O_7 and pertechnetic acid volatilize at a temperature below the boiling point of molybdenum trioxide. A temperature of 670 °C is sufficient to obtain technetium-99 by sublimation when separating it from a mixture of molybdenum trioxide and plumbous oxide, but higher temperatures are necessary when separating it from molybdenum trioxide alone.

Parent ^{99}Mo
Parent radionuclide $^{99}MoO_3$
Daughter $^{99m}Tc_2O_7$
Boiling point of ^{99m}Tc 310.6 °C
Boiling point ^{99}Mo 1150 °C
Melting point Mo 795 °C

4- Gel generators

A gel technetium generator employs a gel matrix such as zirconium molybdate, in which some of the molybdenum is molybdenum-99. Mo reacts with Zr, in the form of $ZrOCl_2.8H_2O$ in suitable conditions of concentration, pH, temperature and stirring. The gel is further filtered, dry and finally cracked with saline solution. This product can be loaded onto glass columns either dry or wet.

Parent	^{99}Mo
Parent radionuclide	ZrMo gel
Daughter	^{99m}Tc
Absorbent material	Gel + Alumina
Elute	$^{99m}TcO_4^{-1}$

Most molybdenon trioxide is irradiated and then dissolved in basic ammonia. The resulted solution is then added to an aqueous zirconium to obtain zirconium moly precipitate in the form of of gel like matrix which is then separated from the solution by filtration, evaporation, air dried and then sized for use in the generator.

Molybdenum/Technetium Generator

Prior to shipping the generator to a medical facility, ^{99}Mo in the form of MoO_4^{2-} is immobilised on a column of alumina (Al_2O_3) due to its very high affinity for alumina. When the ^{99}Mo decays it forms pertechnetate TcO_4^-, which because of its single charge is less tightly bound to the alumina. Pulling normal saline solution through the column of immobilised ^{99}Mo elutes the soluble ^{99m}Tc, resulting in a saline solution containing the ^{99m}Tc as the pertechnetate, with sodium as the counterbalancing cation. The solution of sodium pertechnetate may then be added in an appropriate concentration to the organ-specific pharmaceutical to be used, or sodium pertechnetate can be used directly without pharmaceutical tagging for specific procedures requiring only the $^{99m}TcO_4^-$ as the primary radiopharmaceutical. Tc generator is shown in Figure 6.4

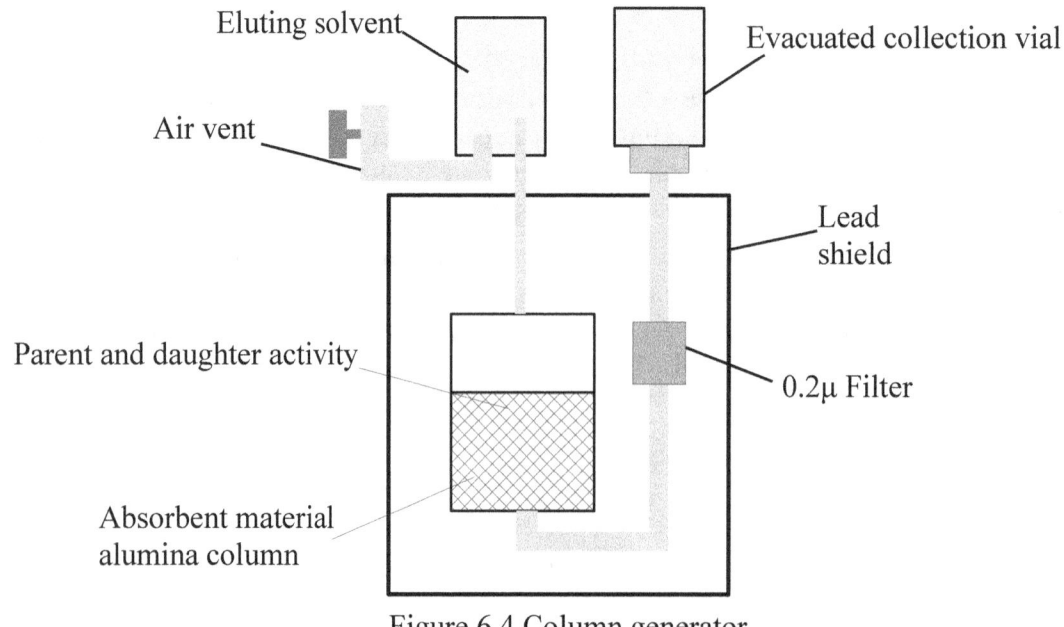

Figure 6.4 Column generator

The Tc-99 is washed out of the alumina column pot by saline solution. This process of extracting radionuclide is called **elution** or *milking* the cow. Typically, a solvent-filled vial is connected to one side of the generator and an evacuated vial is connected to the other side as shown in Figure 6.4. The solvent is then pulled through the generator by opening the air vent which then push the solvent into the alumina column. ^{9m}Tc has different chemical properties to ^{99}Mo and so, as the parent decays to the daughter, the ^{99m}Tc is less tightly bound to the aluminium column. Consequently, the *eluent*, across the aluminium column removes the ^{99m}Tc from the column to produce *eluate* while leaving the ^{99}Mo behind on the column until it decays to more ^{99m}Tc and is subsequently eluted. This goes into the evacuated vial, taking along with it the dissolved radioactive substance to be eluted. The resulting solution is called the **eluate**.

The elute of sodium pertechnetate may then be added in an appropriate concentration to the organ-specific pharmaceutical to be used, or sodium pertechnetate can be used directly without pharmaceutical tagging for specific procedures requiring only the $^{99m}TcO_4^-$ as the primary radiopharmaceutical.

75% of equilibrium activity is reached within two daughter's half life (fig. 6.5) and most of the ^{99m}Tc is produced in the first 3 parent half lives or nearly one week.

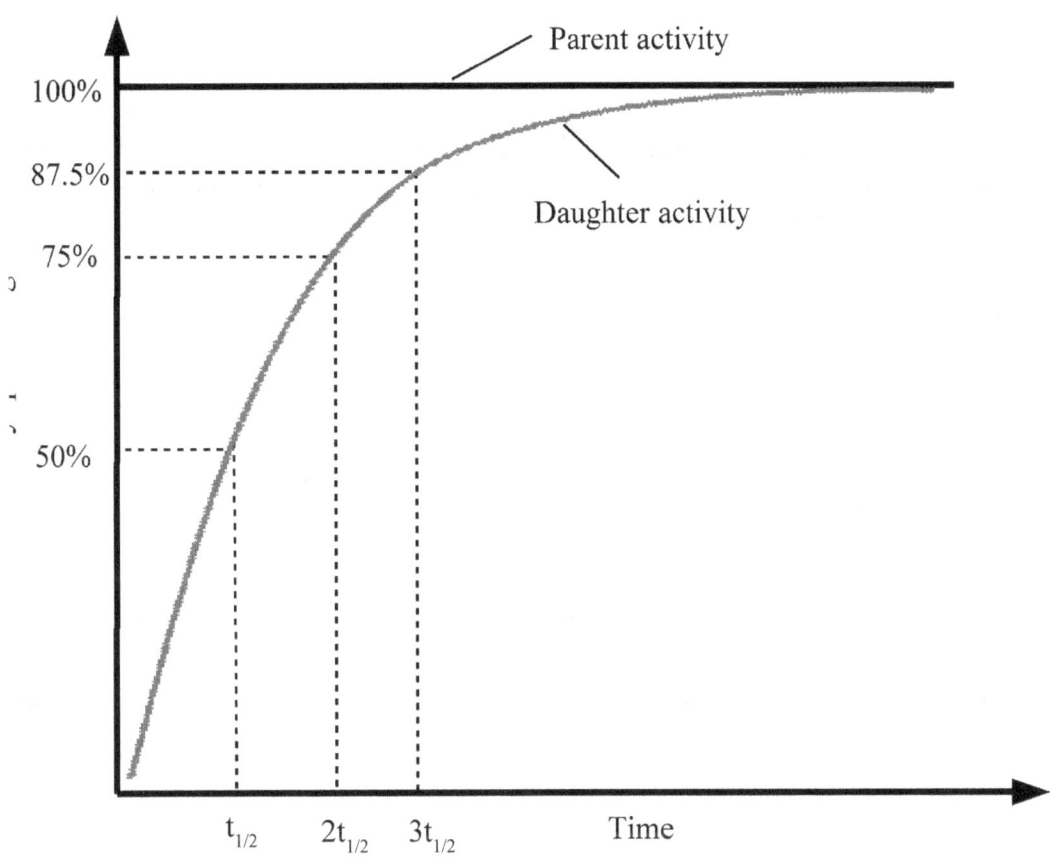

Figure 6.5 Percentage decay

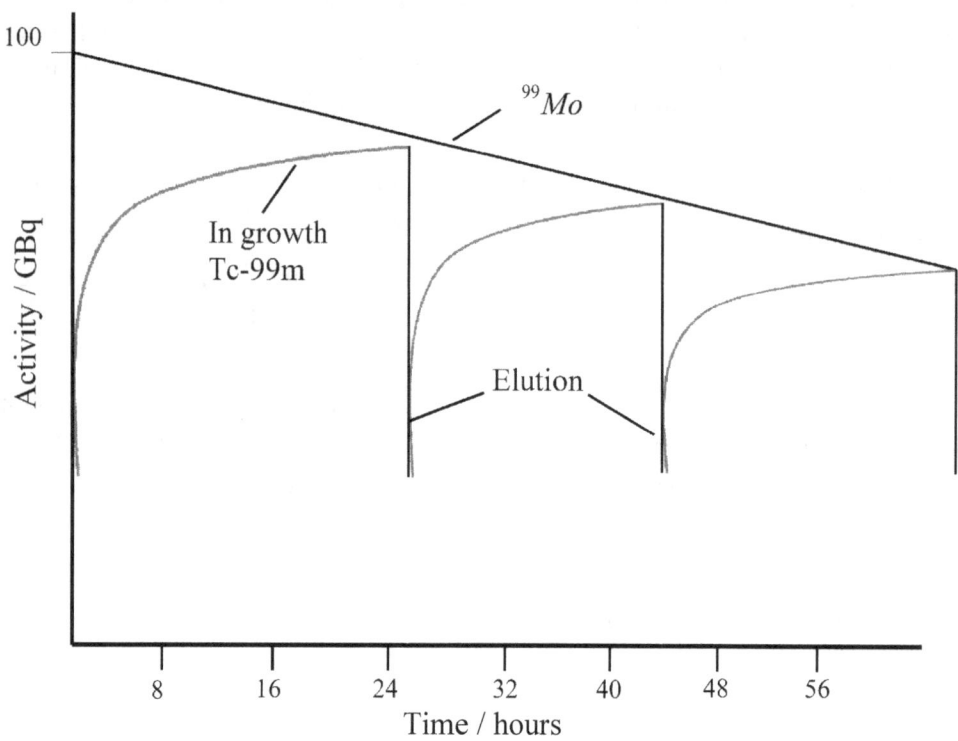

Figure 6.6 Graphical representation of 99mTc elution profile. The y axis is logarithmic.

Equilibrium

Equilibrium is a condition established in radioactive parent-daughter mixture and daughter half life is shorter than that of the parent. If the daughter half-life is greater than parent then the equilibrium will never establish.

The daughter radionuclide N_d, is formed by the decay of a parent N_p, radionuclide. We have studied the decay mechanism in chapter 5.

$$\frac{dN_p}{dt} = -\lambda_p N_p \quad \text{and} \quad N_p = N_p(0) e^{-\lambda_p t} \qquad (6.5)$$

where λ_p and λ_d are decay constant for parent and daughter radionuclide. N_p and N_d represent the number of atoms at time t for parent and daughter radionuclide. $N(0)$ represents number of radionuclide at time t = 0.

The daughter radionuclide is formed at the rate ($\frac{dN_p}{dt}$) at which the parent radionuclide decays. The daughter radionuclide itself decays at the rate of $\frac{dN_d}{dt}$, therefore the net production rate is given by:

$$\frac{dN_d}{dt} = \lambda_p N_p - \lambda_d N_d = \lambda_p N_p(0) e^{-\lambda_p t} - \lambda_d N_d(0) e^{-\lambda_d t} \qquad (6.6)$$

This is a form of linear differential equation whose solution is:

$$N_d = \frac{\lambda_p}{\lambda_d - \lambda_p} N_p(0)(e^{-\lambda_p t} - e^{-\lambda_d t}) + N_d(0) e^{-\lambda_d t} \qquad (6.7)$$

If
$$t_{1/2p} > t_{1/2d} \text{ so that } \lambda_p < \lambda_d$$
this leads to following conclusion:

$e^{-\lambda_d t}$ and $N_d(0)e^{-\lambda_d t}$ is negligible when t becomes sufficiently large. The equation 6.7 simplifies:

$$N_d = \frac{\lambda_p}{\lambda_d - \lambda_p} N_p(0) e^{-\lambda_p t} \quad (6.8)$$

since $N_p = N_p(0)e^{-\lambda_p t}$

so
$$\frac{N_p}{N_d} = \frac{\lambda_d - \lambda_p}{\lambda_p} \quad (6.9)$$

also
$$\frac{A_p}{A_d} = \frac{\lambda_d - \lambda_p}{\lambda_d} = 1 - \frac{\lambda_p}{\lambda_d} \quad (6.10)$$

This ratio of parent to daughter activity is known as equilibrium.

There are two types of equilibriums.

Secular equilibrium

This condition is reached when the physical half-life of the parent is many times (100 times or more) greater than the physical half life of the daughter. For example the parent nuclide of ^{222}Rn with physical half life of 3.8 days is ^{226}Ra with a physical half-life of 1620 years. So after 10 half-lives of daughter since the previous elution the activity of the parent and daughter are the same. At the point at which the activity of the parent and of daughter becomes equal, equilibrium is reached. Parallel lines in the graph in diagram 6.6 represent that equilibrium.

The relationship between half lives in secular equilibrium is

$$t_{daughter}^{apparent} = t_{parent}^{physical} \quad (6.5)$$

This shows us that in the equilibrium mixture, the daughter appears to decay with the half life of the parent, put this simply, they appear equal because daughter cannot decay until its formed. The rate of the formation of daughter becomes equal to the rate of decay of the parent.

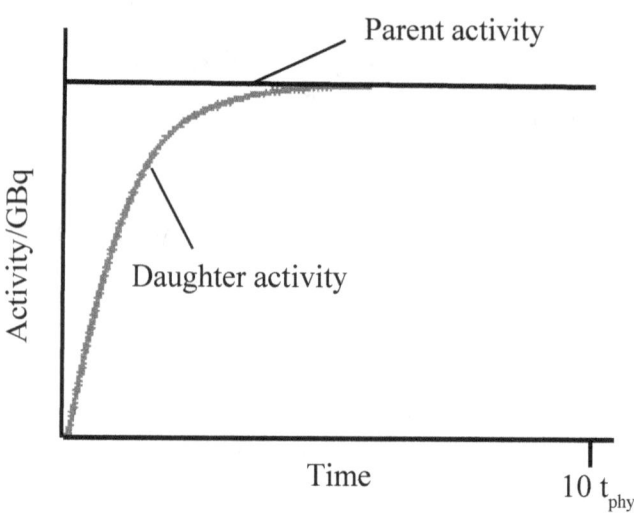

Figure 6.7 Secular equilibrium

Using equation 5.1 we can write the activity equation for parent

$$A_p(t) = A_p(0)e^{-\lambda_p t} \quad (6.6)$$

The equation governing generator system stems from equation 6.6

$$A_d = \frac{\lambda_d}{\lambda_d - \lambda_p} A_p(0)[e^{-\lambda_p t} - e^{-\lambda_d t}] \quad (6.7)$$

where $A_{p(0)}$ is the parent activity at time = 0, t is time since the last elusion. A_d is the daughter product activity. λ_p and λ_d are decay constants of parent and daughter radionuclide respectively. For secular equilibrium $\lambda_p > \lambda_p$ then equation 6.7 becomes:

$$A_d = A_p(0)[e^{-\lambda_p t} - e^{-\lambda_d t}] \quad (6.8)$$

If t is lot less than half life of the parent but greater than seven times the daughter half life, then equation 6.8 becomes even simpler;

$$V_d \simeq A_p^0 \quad (6.9)$$

equation 6.9 shows the equilibrium condition. The growth of the daughter is given by equation 6.10

$$A_d = A_p(0)[1 - e^{-\lambda_d t}] \quad (6.10)$$

both conditions (equation 6.9 and 6.10) are drawn as the graph in Figure 6.7

Secular equilibrium is reached in ~ 6 $t_{1/2}$ of a daughter.

Transient equilibrium

When the parent physical half life is longer (about 10 times) transient equilibrium occurs. This means that the daughter activity increases as the parent decays and reaches to maximum in about four half lives after the previous elution. A classical example is the ^{99}Mo / ^{99m}Tc generator where the half life of the parent is 67 hours compared to 6 hours of daughter's half life. During the 60 hour period which represents 10 half lives of ^{99m}Tc, about 50% of ^{99}Mo has disappeared. This is much more than the case of secular equilibrium.

Between elutions the daughter ^{99m}Tc builds up as parent ^{99}Mo continues to decay. After approximately 23 hours ^{99m}Tc activity reaches a maximum and at this stage the production rate and decay rates are equal. Now parent and daughter are said to be in *transient equilibrium*. After this equilibrium the daughter activity decreases with an apparent half-life equal to the half life of the parent.

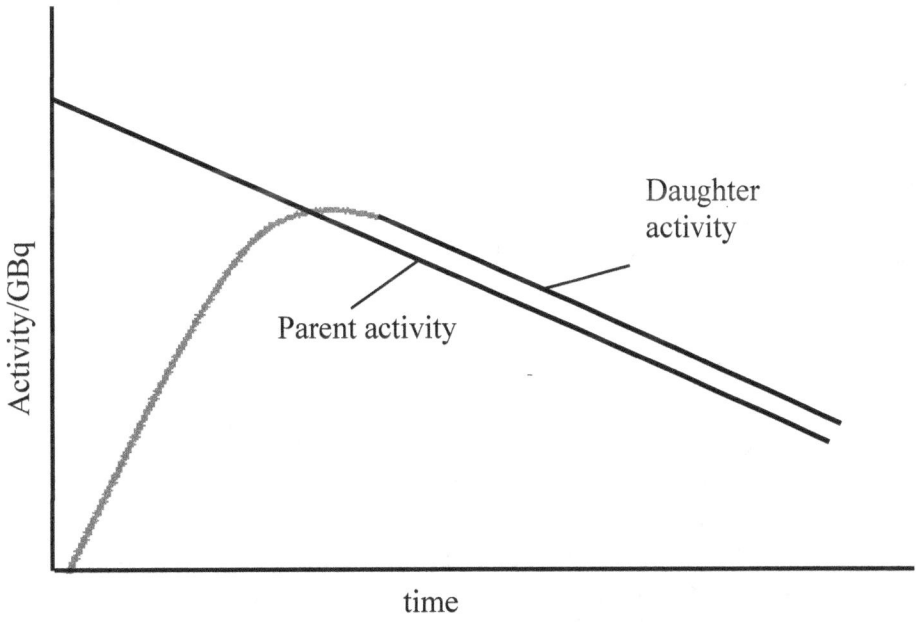

Figure 6.8 Transient equilibrium

Although daughter represents maximum activity, the decay of a parent means that actual equilibrium will not give highest activity for subsequent elutions.

Transient equilibrium is reached in ~ 4 $t_{1/2}$ of a daughter.

The time of maximum activity is given by:

$$t_{max} = \frac{1.44 \times t_p t_d}{t_p - t_d} \times \ln\left(\frac{t_p}{t_d}\right) \quad (6.11)$$

where t_p is the half life of parent and t_d is the half life of a daughter.

For transient equilibrium, by definition $\lambda_d > \lambda_p$, then the equation simplifies

$$A_d = \frac{\lambda_d A_p(0)}{\lambda_d - \lambda_p} \quad (6.12)$$

Physical Half life, t_p

The physical half life is defined as the period of time required to reduce the radioactive nuclei's activity by one half of its original value purely due to radioactive decay. This half life could be as short as nanoseconds or as long as millions of years. The physical half life can be affected by any external influence, like any human tempering, high or low pressure, high or low temperature.

Biological Half life, t_b

The biological half life is defined as the period of time required to reduce the amount of a radioactive drug in the human body by one half of its original value purely due to biological elimination. The biological half life is influenced by external factors like scanning for renal function. If the patient is poorly hydrated will produce less urine and this will make difficult for elimination of a radionuclide that has not localised in the target zone. Each individual organ in the human body has its own biological half life.

Effective Half life, t_e

The effective half life is defined as the period of time required to reduce the level of radioactivity of an internal organ or of the whole body to one half of its original value due to both biological elimination and decay. The effective half life is measured in second to hours. The relationship between effective half-life, biological half-life, and physical half-life is shown in the equation below:

$$\frac{1}{t_e} = \frac{1}{t_p} + \frac{1}{t_b} \quad \text{or} \quad t_e = \frac{t_p \times t_b}{t_p + t_b} \tag{6.13}$$

Ideally, a radiopharmaceutical's effective half life equals approximately 1.5 times the duration of the diagnostic procedures. For example, Tc-99m MDP, which has an effective half-life of 6 hours; since bone imaging is a 4 hour procedure, the ratio of effective half-life to the duration of the test is 1.5:1, considered ideal. Whereas, Tc-99m sulphur colloid has an effective half-life of 6 hours in the liver, but the procedure takes only 1 hour. This 6:1 ratio doesn't mean that a liver scan is a bad procedure to perform, but rather that the compound has a residence time in the liver that is longer than desirable, resulting in an increase in the radiation dose to the target organ which should be avoided. Physician's have now changed Tc-99m sulphur colloid to Tc-99m micro-aggregated albumin decreases the t_{eff} from 6 hours to approximately 3 hours, lowering the ratio of effective half-life to procedure length to 3:1 and decreasing the radiation dose by half, hence minimising the health risk.

There are three special cases to clarify effective half life:

1- $t_p \gg t_b$ this result in $t_e \sim t_b$

Example: Xe-133 radionuclide is used to study pulmonary ventilation study. The physical half life of Xe-133 is 5.3 days and biological half life is 15 seconds. Calculate the t_e'

5.3 day = 5.3 x 24 x 60 x 60 = 457920 seconds

$$t_e = \frac{457920 \times 15}{457920 + 15} \approx 15 \text{seconds}$$

2- $t_b \gg t_p$ this results in $t_e \sim t_p$

Example: Tc - SC is used for liver scan which has t_p = 6 hours and t_b = infinity, calculate the effective half life of Tc-SC.

$$\frac{1}{t_e} = \frac{1}{\infty} + \frac{1}{6} = \frac{1}{6}$$

t_e = 6 hours

3- $t_b = t_p$ this result in $t_e = \frac{1}{2} t_b = \frac{1}{2} t_p$

Example: Tc - MAA is used for pulmonary perfusion imaging which hs $t_p = t_b$ = 6 hours, calcualte the effective half life of Tc – MAA.

$$t_e = \frac{6 \times 6}{6+6} = 3 \, hours$$

Measurement of Radiation

The unit of absolute activity is measured in curie. One curie is equal to 3.7 x 10¹⁰ disintegrations per second. In other words in the decay of C-14 into N-14 via beta emission, one curie of C-14

$$^{14}_{6}C \rightarrow {}^{14}_{7}N + {}^{0}_{-1}\beta + \bar{v}_0$$

activity means that 3.7 x 10¹⁰ C-14 atoms will turn into N-14 within one half life.

1 Ci = 3.7 x 10¹⁰ dps

Another unit of activity is Becquerel, abbreviated Bq. The Becquerel (Bq) is the unit in the International System of Units to replace the curie.

1Bq = 1dps

therefore,   37MBq = 1mCi   or 1MBq = 27μC

Example: A sample measures a 35mCi in a dose calibrator. What will be the count rate in seconds if the detector efficiency is 58%.

count rate = (35 x 10⁻³) x (3.7 x 10⁷ dps) = 1.30 x 10⁶ c/s (dps)

Since the detector efficiency is 58% so the actual count rate is 1.30 x 10⁶ (x 58/100) = 751 x 10³ dps

Example: A count rate of 58 Mcs⁻¹(dps) is obtained from a sample of radionuclide by the detector whose efficiency is 45%. What is the activity of this radionuclide in Curie.

The actual count rate is (58 x 10³) x (100/45) =129 Mcs⁻¹

Count rate = $\frac{129 \times 10^6}{3.7 \times 10^{10}}$ = 3.4 m Ci

Radiation Absorbed Dose Units

Absorb dose is a measure of the energy deposited in a medium by ionising radiation per unit mass. Formally, absorbed dose at a point is defined by the ICRU (International Commission on Radiation Units & Measurements) as:

$$D = \frac{\Delta \varepsilon}{\Delta m} \qquad (6.14)$$

The unit of absorbed dose is Gray (Gy) where one Gray is equal to one Joule per kilogram (Jkg^{-1}).

Equivalent Dose:

Equivalent dose is an average measure of the radiation absorbed by a fixed mass of biological tissue. This account for the different biological damage potential of different ionising radiations. The equivalent dose is measured in units or rem or Sievert (Sv).

Three factors can harm biological tissues:

- the absorbed dose
- types of radiations i.e., alpha, beta, gamma or thermal neutrons
- type of tissue

The equivalent dose is calculated by multiplying the absorbed dose, averaged by mass over an organ or tissue of interest, by a radiation weighting factor appropriate to the type and energy of radiation. To obtain the equivalent dose for a mix of radiation types and energies, a sum is taken over all types of radiation energy dose.

$$H_T = \sum W_R \cdot D_{T,R} \qquad (6.15)$$

H_T is the equivalent dose absorbed by tissue T, W_R is radiation weighting factor as a measure of its biological effect of that tissue and $D_{T,R}$ is the absorbed dose in tissue T by radiation type R. The value of W_R is given in table 6.2

Table 6.2 Weighting factors

Radiation	W_R
Alpha	20
Beta	1
Gamma	1
Thermal neutron	3

Thus for example, an absorbed dose of 1 Gy by alpha particles will lead to an equivalent dose of 20 Sv.

The aim of the radiation weighting factor is to correct the simple deposited energy of the radiation for different biological effect of different types of radiation. An equivalent dose of radiation is estimated to have the same biological effect as an equal amount of absorbed dose of gamma rays. If the equivalent dose is uniform throughout the organism, it will be equal to the effective dose. Otherwise, a weighted average of H_T will have to be taken to average out the radiation dose through the body while correcting for the different sensitivities of different tissues.

Permissible dose

- occupational exposure : 20mSV/year
- lens of eye : 150mSV/year
- skin : 500mSV/year
- hands and feet : 500mSV/year
- fetus (during entire pregnancy) : 2mSV/year

The equivalent dose, H takes three factors into account:

- type of radiations
- energy carried by the radiation
- how much of this energy is absorbed by the tissue

The time of exposure (t) to ionising radiation is also important. An equivalent dose of 300 mSv received in one day is more dangerous than the same equivalent dose received over the course of one month. Therefore,

$$\text{equivalent dose rate} = \dot{H} = \frac{H}{t} \qquad (6.16)$$

with units of Sievert per second.

Effective Dose:

The effective dose in radiation protection and radiology is a measure of the cancer risk to a whole organism due to ionizing radiation delivered non-uniformly to part (s) of its body. It takes into account both the type of radiation and the nature of each organ being irradiated. The effective dose is defined by ICRP (The International Commission on Radiological Protection) Publication 60 as a weighted sum of equivalent doses to all relevant tissues and organ with the purpose "to indicate the combination of different doses to several different tissues in a way that is likely to correlate well with the total of the stochastic effects". This is, therefore, applicable even if the absorbed dose distribution over the human body is not homogeneous. The unit is the joule per kilogram (J kg^{-1}) and is given the special name sievert (Sv). Equivalent dose does not consider the type and amount of tissue exposed, so an equivalent dose applied to only a portion of the body will carry a lower risk than if the same equivalent dose was applied to the whole body. On the other hand an effective dose will carry the same effective risk to the whole organism regardless of where it was applied, and it will carry the same effective risk as the same amount of "equivalent dose" applied uniformly to the whole body.

The *effective dose* of radiation (E) is found by calculating a weighted average of the equivalent dose (H_T) in different body tissues, with the weighting factors (W_T) designed to reflect the different importance of tissue types to the danger to the whole organism.

$$E = \sum W_T \cdot D_{T,R} = \sum W_T \sum W_R \cdot D_{T,R} \qquad (6.17)$$

Radiation Exposure:

Exposure is the amount of radiation travelling through the air. Radiation exposure, is a form of damage to organ tissue caused by excessive exposure to ionizing radiation. The term is generally used to refer to the acute problems caused by a large dose of radiation in a short period, though this also has occurred with long term exposure. Radiation exposure can also increase the probability of developing some other diseases, mainly cancer, tumours, and genetic damage. These are referred to as the stochastic effects of radiation, and are not included in the term radiation sickness.

The unit of exposure are Röentgen ® and coulomb per kilogram (Ckg^{-1}).

Table 6.3 - Units of radiation measurement

	Radioactivity	**Absorbed dose**	**Equivalent Dose**	**Exposure**
Common units	Curie (Ci)	Rad	Rem	Roentgen
SI units	Becquerel (Bq)	Gray (Gy)	Sievert (Sv)	Coulomb/kg

Useful conversions

	1 Rad	0.01 Gy
	1 Rem	0.01 Sv
	1 Roentgen	0.000258 C/kg

Example: A patient of mass 70 kg receives radiotherapy. During the treatment, a tumour of mass 250 g receives 20 J of energy. Calculate the absorbed dose.

$$D = \frac{\Delta \varepsilon}{\Delta m} = \frac{20}{250 \times 10^{-3}} = 80 \text{ Gy}$$

Example: A worker in the nuclear industry receives an absorbed dose of 630μ Gy from thermal slow neutrons. Calculate the total equivalent dose received.

From the above table $W_R = 3$

$$H_{neutrons} = DW_R$$

$$= 630 \times 10^{-6} \times 3$$

$$= 1.89 \text{ mSv}$$

Diagnostic applications of Radiopharmaceutical

Radiopharmaceuticals have revolutionised the medical field by their ability to provide static as well as dynamic images of internal organs in a noninvasive manner as well as by offering efficacious therapy of certain diseases. The use of specific radiotracers for imaging organ function and disease states is a unique capability of nuclear medicine. Radioisotopes are extensively used in medicine for diagnosis, either *in vivo* (within a living organism) or *in vitro* (biological process carried in the laboratory rather than in the living body). Radioisotopes give doctors the ability to look inside the body and observe soft tissues and organs. They are used to derive the detailed description of the morphology and dynamic functioning of the various internal organs of the human body. Radioisotopes techniques are *non-invasive* as they only require administration to the patient, generally intravenously, of a radiopharmaceutical.

Radiopharmaceutical uses a small amount of radioactive material. By tagging radioactive source to some compound that is known to localise in a specific area of the human body. By using gamma camera, doctors can detect the emission given off by the radioactive material and create images of the relative distribution of the radionuclide in the body. The widespread utilization and growing demands for these techniques are directly attributable to the development and availability of a vast range of specific radiopharmaceuticals. Many of the radiopharmaceuticals used for the diagnostic purpose like C14 for pancreatic study and breath test, Cr51used for red cell volume and GFR measurement, Co57 used for gastrointestinal absorption, I123 for thyroid uptake and renal imaging.

Some chemical compounds concentrate naturally in specific organs or tissues in the body. For example, iodine collects in the thyroid and taking the advantage of this proclivity radionuclide can be used as tracer. This tracer is then chemically attached to a compound that will concentrate naturally in a specific organ. This process of attaching a radionuclide to a compound is called *labelling*. The non-radionuclide portion of the diagnostic radiopharmaceutical is often an organic molecule such as a carbohydrate, lipid, nucleic acid, peptide, small protein, or antibody.

Diagnostic nuclear imaging is used principally for bone, brain, kidney, liver, gall bladder, heart and lungs. Radioisotopes with short half-life are preferred for use to minimise the radiation dose. They normally decay into stable elements within a day so that patients could be released from the hospital.

When radiopharmaceuticals are used to help diagnose medical conditions, only small amounts are given to the patient. To evaluate the health of the heart, a radioisotope is injected into a patient's bloodstream. The patient is then asked to exercise on a treadmill. As radioisotope travels towards the heart, this can be monitored on screen. Reduced blood flow through the arteries will give a possible sign of heart disease. Organic radio chemicals labelled with F-18 are injected into the bloodstream to evaluate brain function.

Some radiopharmaceuticals are used in larger amounts to treat certain kinds of cancer and other diseases. In those cases, the radioactive agent is taken up in the cancerous area and destroys the affected tissue. The dosages of radiopharmaceuticals used to diagnose medical problems and will vary from patient to patient and also depend on the type of the test.

For some diagnostic tests, the patient does not require to take radioisotope. Blood or other fluid taken from the patient will undergo a special test procedure called a radio-immunoassay. These tests can detect some disease by identifying and measuring the amounts of enzymes, hormones and vitamins in the body.

Therapeutic applications of Radiopharmaceutical

A great German scientist and physician, Paul Ehrlich (1854–1915) first described "specific chemotherapy" as a search for a chemical substance capable to be taken up by and killing parasites without doing any harm to normal tissue or organism. To reach such an objective, Ehrlich as a student wanted to determine the microscopic and biological distribution of metals in the organism, because he thought metals to be effective therapeutic agents. He also stated that "particles must be attached to something to be effective". These are the fundamental principles of radionuclide therapy and if radio nuclides were known at that time, Ehrlich could have been the father of radionuclide therapy.

In 1936, John Lawrence studied total body irradiation after intravenous administration of P-32 using an animal model of leukaemia mice and various lymphomas in animals. Together with a 29-year old student, who was diagnosed as having myelogenous leukaemia Lawrence performed the first P-32 therapy. After 3 courses with a cumulative dose of 394 MBq (10.64 mCi) P-32 the student, symptomatically and clinically, was normal. In 1940/41 a patient with prostate cancer and painful osteoblastic bone metastases was treated with 8 mCi of Sr-89 with positive effect concerning pain by C. Pecher. About ten years later Friedel reported P-32 therapy to breast cancer bone metastases.[1]

The choice of radionuclide for therapy is well illustrated by bone metastases. The question is whether to treat the invading edge of the metastases interacting with normal bone which generates pain for which a short path length soft radiation agent such as Sm-153 would be appropriate, sparing the marrow; or should one use an agent that also irradiates the marrow where the malignant cells are, to combine a palliative with a potential therapeutic response, but at the cost of damaging the normal marrow cells, such as with P-32 or Sr-89. Should one go for a short-lived radionuclide

---

[1] FISCHER, Therapeutic applications of radiopharmaceuticals, Proceedings of an international seminar: Hyderabad, India, 18–22 January 1999

such as Re-188, a medium lived radionuclide such as Y-90 or Re-186 or a long lived radionuclide such as P-32. The same questions apply to radio-labelled peptides and antibodies. Is the radionuclide carrier bound to the cell surface, or internalised? For the former, a Beta emitter is required, for the latter an emitter of Auger electrons or even alpha particles can be considered. The range of the Beta particle is important. Should one go for cross talk because of heterogeneity of the distribution of cancer cells with a mix of receptor or antigen positive and negative as compared with those that require homogeneity of receptor antigenic expression. The size of the tumour is another consideration, soft Betas for micro-metastases, medium range Betas for small tumours and hard Betas for larger tumours. The site of the therapy is important. For therapy to cavities where there is a chance of escape of the radionuclide into the blood, then a high energy short lived radionuclide may be preferred, so that irradiation is completed before the escape. For direct tumour injection when there is no escape as in a brain tumour, a longer lived radionuclide may be preferred. Where the tumour has good access in the blood such as leukaemia, then an intravenous injection of an alpha or beta emitter may be used, similarly for lymphoma, but for a solid tumour with poor access, a two or three stage approach, pre-targeting the tumour is required to improve the therapeutic ratio.

The use of radiopharmaceuticals for therapeutic applications is increasing. Sodium [131 I] iodide was approved in 1951 for treating thyroid patients. There are currently FDA (Food and Drug Administration, USA) approved radiopharmaceuticals for alleviating pain in patients whose cancer has metastasised to their bones. These include sodium 32 P-phosphate, 89 Sr-chloride, and 153 Sm-EDTMP (where EDTMP stands for ethylenediaminetetramethylphosphate). In February 2002 the first radio-labelled monoclonal Yttrium-90- labelled anti-CD20 monoclonal antibody is used to treat patient with non-Hodgkin's lymphoma.

The treatment of solid tumours is a big challenge for the physicians due to the variety of its histological patterns and its disorganized angiogenesis. Most of these kind of tumours do not respond to the conventional therapy, therefore different alternative therapies are being used. Brachytherapy, that is the placement of radioactive sources in or near the tumour, is a valid alternative for the treatment of this kind of tumours. In this way, a malignant lesion can be irradiated with a planed dose and with negligible or null irradiation to the rest of the organism if the source do not move from the injection point. Different radioisotopes such as 90Y, 198Au, 125I and 32P have been used for the treatment of several diseases.

Radioisotopes may be used internally or externally. If the radioisotopes are used externally or as implants in sealed capsules in a tissue, the dose could be terminated by removal of the sources. If they are given internally as unsealed source, the dose cannot be stopped by removal of the source. The total dose in therapeutic applications may be calculated on the basis of effective half- life of the isotope, concentration of the isotope and the type and energy of radiation emitted.

Significant advances in monoclonal antibody techniques for pretargeting make it likely that radiopharmaceuticals will become an important part of therapy for various cancers. It's now possible that in addition to the use of beta particles, alpha particles are to become mainstream of therapeutic nuclear medicine. With advances in molecular medicine and genetic engineering, new antibodies are being synthesised with greater ease. A major problem encountered in oncology practice is the patient with intractable pain secondary to bone metastases. This is most often seen in cancers of the prostate, breast, and lung. When the pain has become unresponsive to all available analgesics, a multimodal approach involving radiation, hormonal, and even surgical therapy, becomes necessary. The presence of multiple, scattered bone lesions is ideal for radiopharmaceutical therapy since the lesions can be targeted selectively through a single systemic administration of the radiopharmaceutical. The available list of radionuclides has been growing, and now includes phosphorous-32, iodine-131, strontium-89, yttrium-90, rhenium-186/188 and samarium-153.

Gamma Camera

A gamma camera is a device used to image gamma radiations emitted by thye radionuclide in nuclear medicine. One of the earlier techniques that used radionuclide is scintigraphy. When such radionuclide are inserted into the body, after the dispersion, as the isotope decays it emits gamma rays. The gamma camera was designed by American Physicist Hal Anger (1920 – 2005) in 1957. A gamma camera consists of one or more flat crystal of sodium iodide planes called detectors and they are optically coupled to an array of photomultiplier tubes, the assembly is known as a "head", mounted on a gantry. The crystal gives a tiny flash of visible light every time a gamma photon hits it. This flash is picked up by photomultipliers which convert the flash into an electrical signal. The gantry is connected to a computer system that both controls the operation of the camera as well as acquisition and storage of acquired images.

The main parts of the camera are as follows:

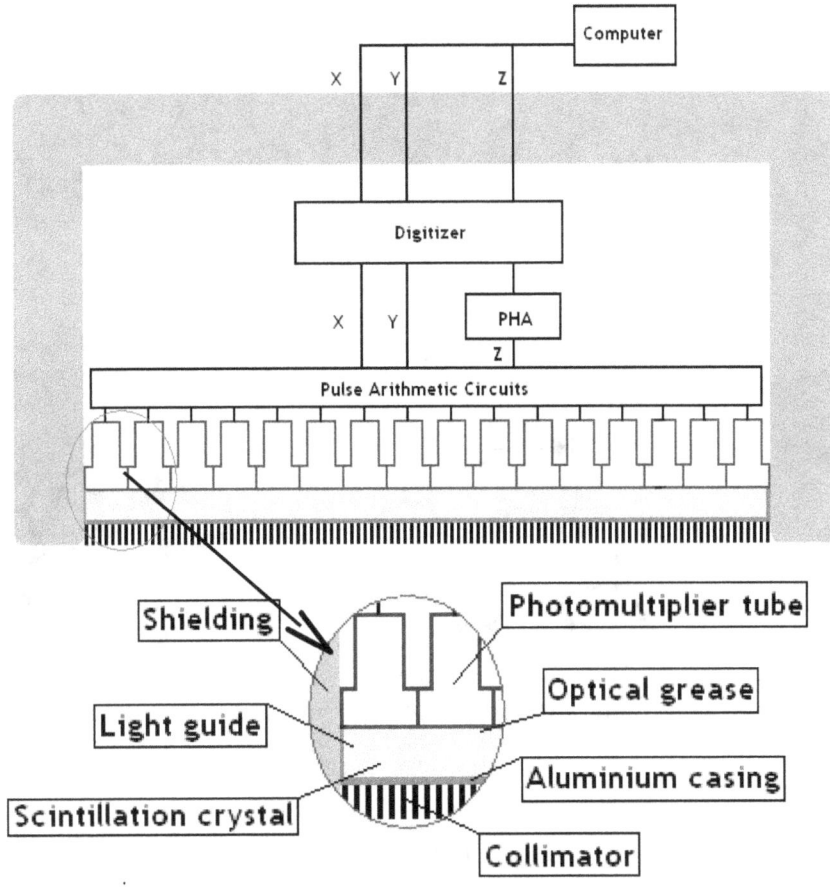

Figure 6.9 Block diagram of gamma camera

Collimator:

This is made of a highly absorbing material such as lead. As gamma rays are emitted in all directions and so they will illuminate the entire crystal and no useful image could be obtained. Collimator consisting of holes in a lead plate can be used to select the direction of the rays falling on the crystal. There are 4 types of collimator in use.

i. Pin hole collimator

ii. parallel hole collimator

iii. Diverging

iv. Converging

Parallel collimator are more commonly used. Collimator are composed of thousands of precisely aligned holes. The collimator conveys only gamma rays travelling directly along the long axis of each hole. Gamma ray emitted in any other direction will be absorbed by *septa*. The collimator attenuates most (>99%) of incident photons and thus greatly limits the sensitivity of the camera system. Large amounts of radiation must be present so as to provide enough exposure to the camera system to detect sufficient scintillation dots to form a picture. The best current camera system designs can differentiate two separate point sources of gamma photons located a minimum of 1.8cm apart, at 5cm away from the camera face. This limits the spatial accuracy of the image: it is a blurry image made up of many dots of detected but not precisely located scintillation. This is a major limitation for heart muscle imaging systems; the thickest normal heart muscle in the left ventricle is about 1.2cm and most of the left ventricle muscle is about 0.8cm, always moving and much of it beyond 5cm from the collimator face. To help compensate, better imaging systems limit scintillation counting to a portion of the heart contraction cycle, called gating, however this further limits system sensitivity.

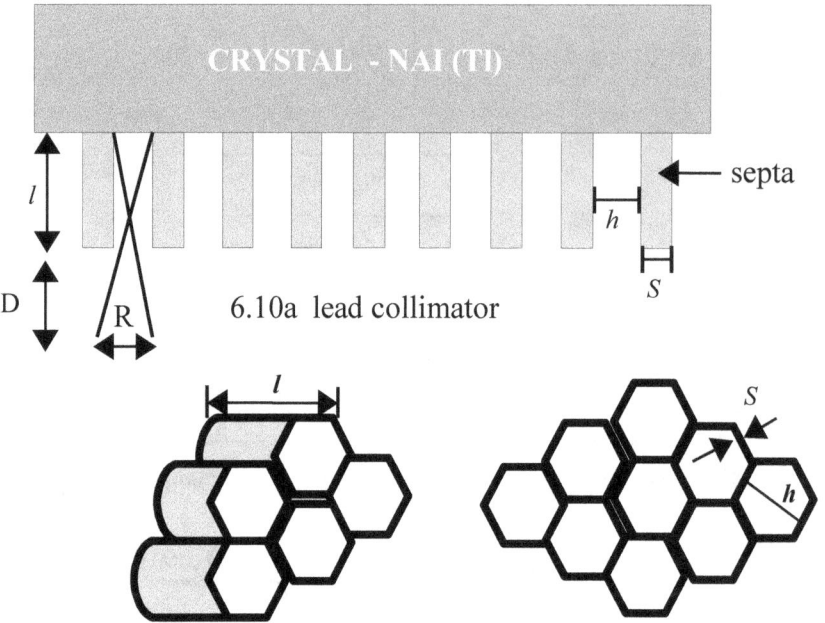

Figure 6.10b Collimator side and front view

The resolution and sensitivity of a collimator depend on a number of factors including:

- hole size (h);
- the thickness of the septa (s), the lead between the holes;
- the length of the holes (l);
- the energy of the gamma rays.

Different collimator are selected for different procedures – e.g. Low energy, high resolution.

A resolution 'R' can be calculated by similar triangle shown in Figure 6.10a.

$$\frac{R}{h} = \frac{D + \frac{l}{2}}{\frac{l}{2}}$$

$$R = h \times \frac{2D + l}{l} = h \times \left(1 + \frac{2D}{L}\right) \qquad (6.18)$$

Most gamma cameras can be connected with different collimator which depends upon the examination. The basic design of collimator is similar but they vary in terms of the diameter of hole (h), the depth of each hole (l) and septum thickness (s). The choice of specific collimator depends on the amount of radiation absorbed by septum and this influence the sensitivity and clarity (spatial resolution) of gamma camera. These two factors are mutually exclusive, means that collimator which will produce image with good spatial resolution will not be very sensitive.

Collimator is a limiting factor in gamma camera. The resolution of the collimator refers to sharpness or details of the gamma ray image projected onto the detector whereas the efficiency of collimator refers to fraction of gamma rays that pass through it. The thickness 's' of septal is designed to prevent gamma rays from penetrating from one hole to another. So high efficiency septal wall thickness must be small. Another factors which affect the performance of collimator are the shape (hexagonal or round), length and diameter. Resolution always is best with the source as close as possible to the collimator.

The other major cause of non uniformity in the image is due to variation of energy (Z) pulse. This is caused by local variations in the crystal which can effect the light generation and transmission to the PM tubes

Scintillation crystal:

A scintillator is a material that exhibits scintillation, the property of luminescence, when excited by ionizing radiation. Luminescent materials absorb the energy of incoming particle and re-emit the absorbed energy in the form of light. All modern gamma cameras use thallium-doped sodium iodide NaI (Tl) as the scintillation crystal. This produces multi-photon flashes of light when an impinging gamma ray interacts with the single sodium iodide crystal of which it is comprised. Sodium iodide is hygroscopic so cannot be left exposed to the air. The front surface is coated with a low atomic number metal that allows the gamma photons to pass through. The rear surface is covered with a glass so that the visible photons can pass through to the photomultiplier tubes. Figure 6.11 shows NaI (Tl) crystal. Scintillation detectors are generally assumed to be linear and this is based on two requirements: (1) that the light output of the scintillator is proportional to the energy of the incident radiation; (2) that the electrical pulse produced by the photomultiplier tube is proportional to the emitted scintillation light. This assumption is usually a good rough approximation, although deviations can occur (especially pronounced for particles heavier than the proton at low energies).

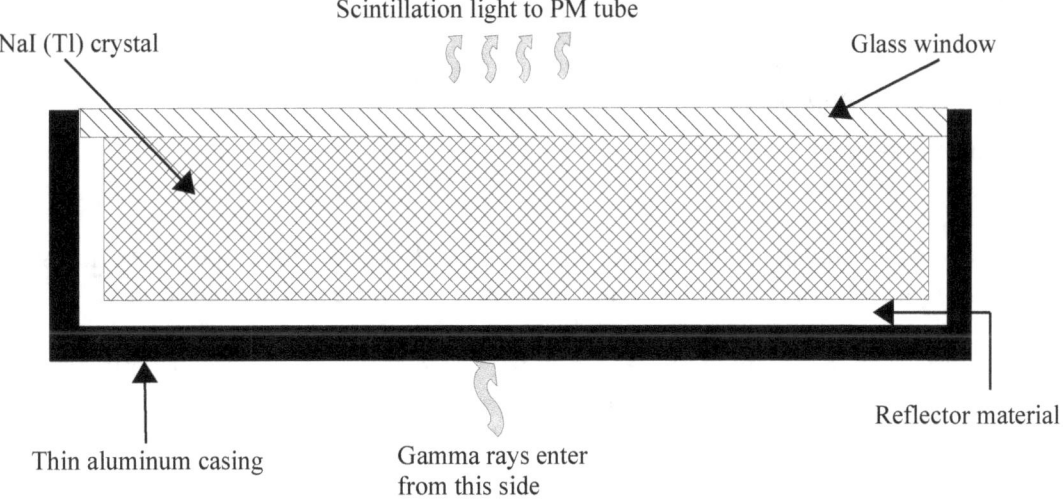

Figure 6.11 Hermetically sealed NaI (Tl) crystal

The passage of ionising radiation through the crystal atoms and excites their electrons, which can subsequently de-excite, emitting a photon.

Photomultiplier Tube:

Photomultiplier tubes that are optically coupled to the crystal to detect these light flashes. The resulting electrical signals output by the photomultiplier tube (PMT) are proportional to the incident gamma ray's energy.

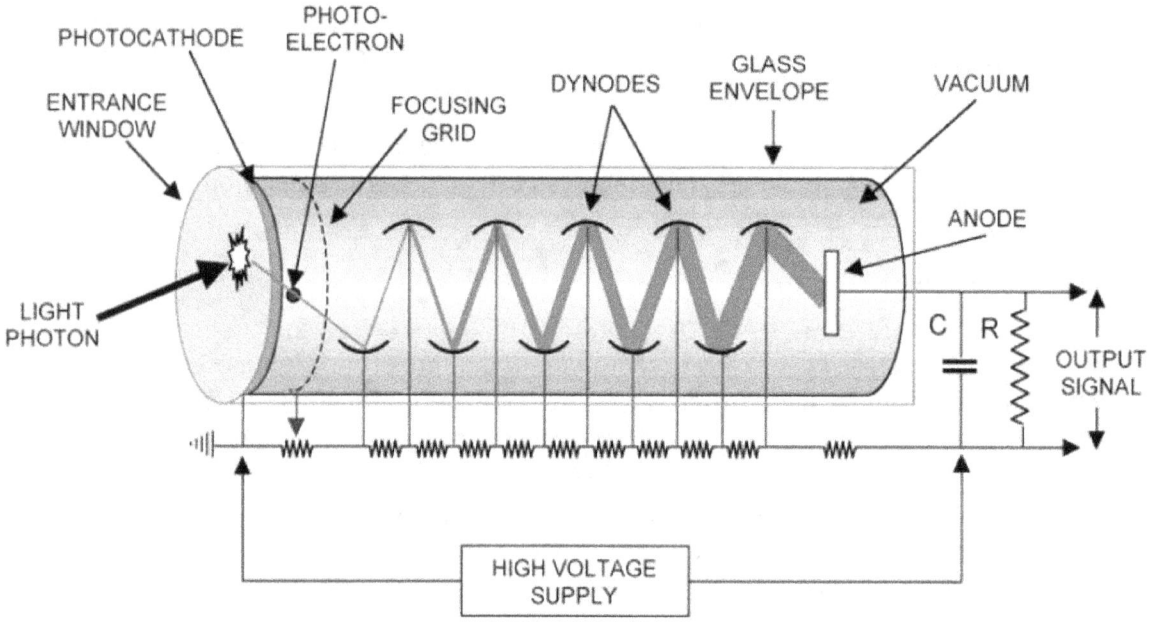

Figure 6.12 Schematic diagram of photomultiplier tube

This is an extremely sensitive photocell used to convert light signals of a few hundred photons from scintillator into a current pulse without significantly increasing noise. A PMT consists of two major elements:

1- a photo-cathode coupled to
2- an electron multiplier

These two components are contained within an evacuated glass envelope. The photo-cathodes can be made of a variety of materials, with different properties. The photo-cathode comprises a photosensitive coating, made of alkali metals, at one end of the evacuated glass tube. This material used in coating is called Bialkali (Sb-K-Cs, Sb-Rb- Cs) caesium-activated antimony-rubidium or antimony-potassium alloy. Light photons liberate low-energy electrons from the photo-cathode. Because the number of photo-electrons produced is about the same as the number of incident light photons, the photo-electrons' total charge will be too small to provide a detectable electrical signal – hence the need for an ***electron multiplier***.

The electron multiplier section consists of an arrangement of **dynodes**. The dynodes are intermediate electrodes located between the photo-cathode and the **anode**. The photo-cathode is kept at high negative voltage and anode at high positive voltage of up to 2000V. Each dynode will have different positive voltage and each one will be more positive voltage than previous one to provide the required environment to produce the electron multiplication effect. Each photo-electron is accelerated towards the first dynode and collide with it. Here its kinetic energy is absorbed and results in the emission of several secondary electrons. This process of electron multiplication continues down the dynode chain, until eventually a large signal is collected at the **anode**. Different gain can be achieved by different applied voltage or different numbers of dynode stages. Negative high voltage supply with positive terminal grounded is preferred, because this arrangement enable the photo-current to be measured at low voltage side of the circuit for amplification by electronic circuit operating at low voltage. Connections to the photomultiplier are made through the photomultiplier base as shown in Figure 6.12, which contains a chain of resistors, connected to a stable power supply, to provide the correct and stable voltages for the cathode, anode and dynodes.

Detection circuitry

Scintillations produced in the crystal are detected by a large number of PM tubes which are arranged in a two-dimensional array. The typical number of PM tubes in gamma camera are between 37 and 91. The PM tubes are divided into horizontal half to obtain X^+ and X^- signals and vertical half to get Y^+ and Y^- signals. These position signals contain information about where the scintillations were produced within the crystal.

Four summing matrix circuits are used to sum up for X^+, X^- and Y^+ and Y^- signals from each PM tube where each of these signals is the product of signal amplitude and position factor.

The position signals also contain information about the intensity of each scintillation. This intensity information can be derived from the position signals by feeding them to a summation circuit (marked $\sum$ in the figure 6.13) which adds up the four position signals to generate a voltage pulse which represents the intensity of a scintillation. This voltage pulse is called the Z-pulse which following pulse height analysis (PHA) is fed as the unblank pulse to the display unit. Z- pulse has amplitude only.

The radiation position is then determined by $X = k(X^+ - X^-)/Z$ and $Y = k(Y^+ - Y^-)/Z$ where k is a scale factor and Z is the total signal amplitude and proportional to the incoming radiation energy.

The positional signal must be normalised by total signal Z because X and Y themselves depend on the both signal and positional factor, different radiation energy give different signal amplitude at the same position

Pulse height analyser (PHA)is used to analyse the Z signal and if accepted, signal will be displayed on the monitor at the position determined by X and Y. When an unblank pulse is generated by the PHA circuit the electron beam of the CRO (cathode ray oscilloscope) is switched on for a brief period of time so as to display a flash of light on the screen. In other words the voltage pulse from the PHA circuit is used to unblank the electron beam of the CRO. The position of the flash of light is dictated by the ±X and ±Y signals generated by the position circuit. These signals are fed to the deflection plates of the CRO so as to cause the unblanked electron beam to strike the screen at a point related to where the scintillation was originally produced in the NaI(Tl) crystal.

Modern designs are a good deal more complex but the basic design has remained much the same as has been described. The most modern approach is to feed the position and energy signals into the memory circuitry of a computer for storage. The memory contents can therefore be displayed on a computer monitor and can also be manipulated (that is processed) in many ways. For example various colours can be used to represent different concentrations of a radiopharmaceutical within an organ.

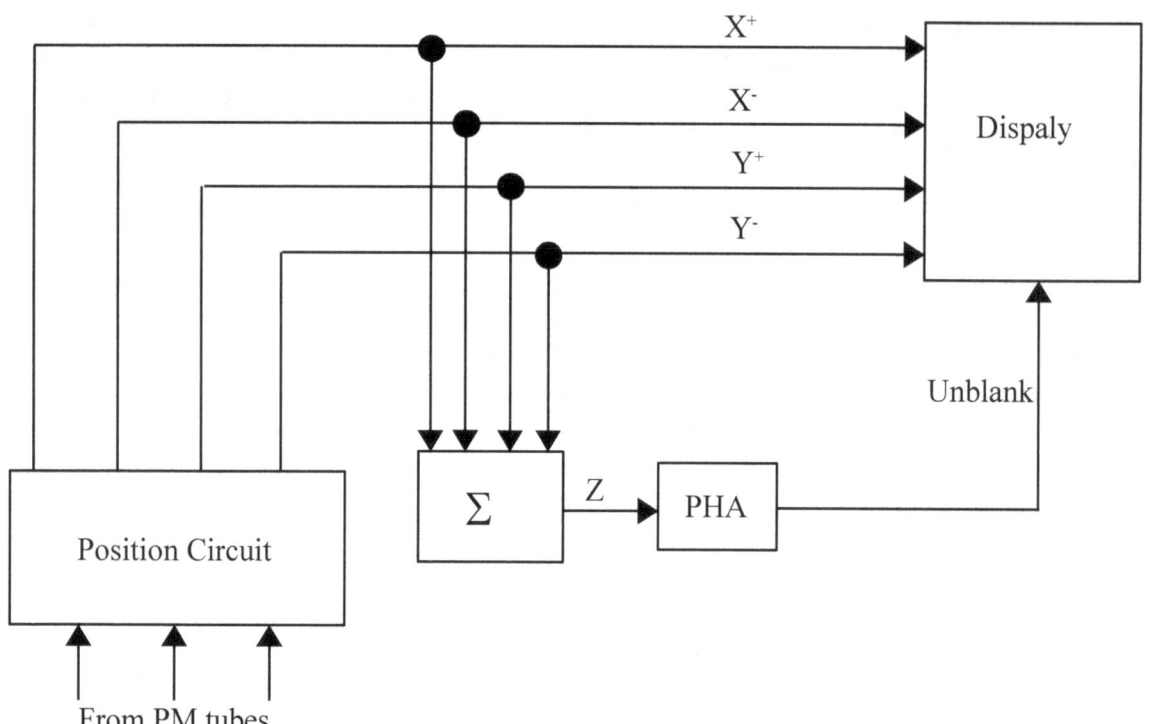

Figure 6.13 Detection and display circuitry for gamma camera

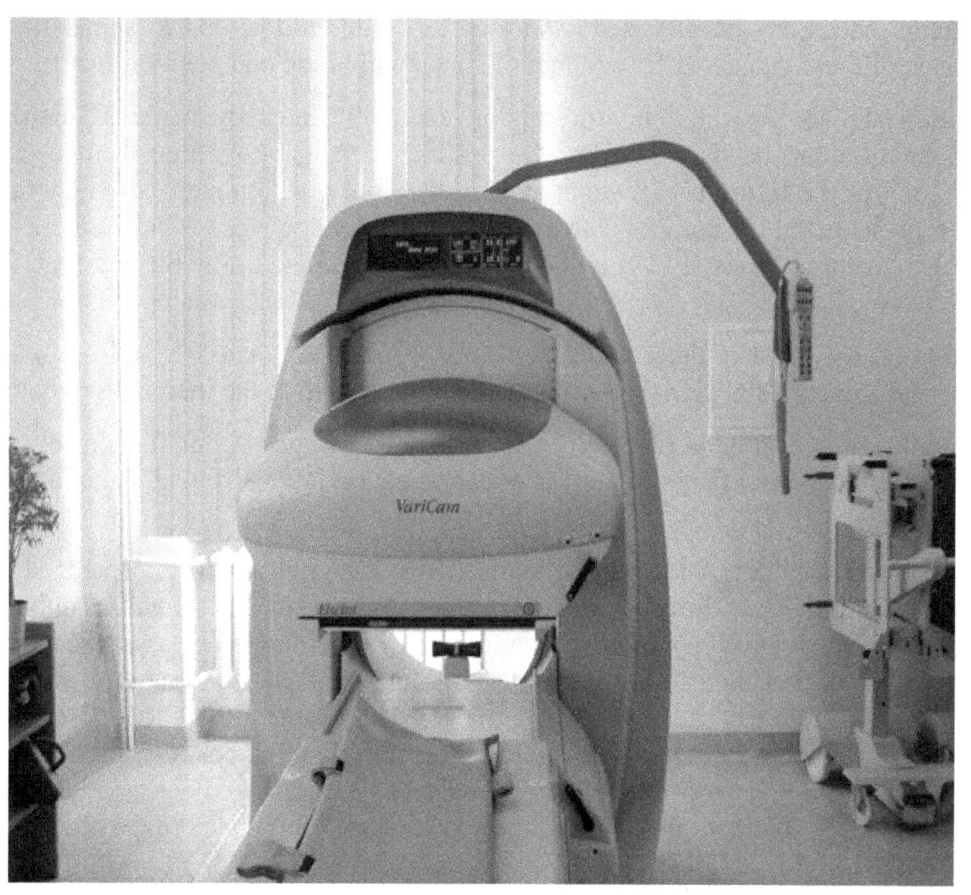

Figure 6.14 A single headed gamma camera

Exercise 6

Q1 – A parent radionuclide decays by gamma emission. Parent and daughter nuclei are:

   a) isotones

   b) isotopes

   c) isobars

   d) isomers

Q2 – Gamma rays are most similar to ……….:

   a) x-ray

   b) high energy electrons

   c) stationary waves

   d) thermal neutrons

Q3 – Which of the following is / are true for transient equilibrium:

   a) at equilibrium, the physical half life of the daughter is ten times as great as the physical half life of the parent

   b) at the equilibrium, the apparent half life of the daughter equals the physical half life of the parent

   c) at the equilibrium, the physical half-life of the daughter equals the physical half life of the parent

   d) at the equilibrium, the apparent half life of the parent equals the physical half life of the daughter

Q4 – Chromatography testing of technetium radiopharmaceuticals is done to check for:

   a) Aluminium Contamination

   b) Molybdenum breakthrough

   c) Radionuclidic purity

   d) Radiochemical purity

Q5 – Which of the following variables does not affect spatial resolution in nuclear medicine imaging systems:

   a) Physical design of the collimator

   b) Material of which the collimator is constructed

   c) Energy of the gamma rays being counted

   d) Byte vs. word acquisition mode

Q6 – The dose rate during preparation of a radiopharmaceutical is 1.66 mrem/hr. With this rate reduced by 2 HVLs, how many hours would it take a worker to accumulate 10 mrems?

   a) 3
   b) 6
   c) 24
   d) 12

Q7 – If a vial containing Tl-201 is found to have an activity of 10.0 mCi at 8 a.m. on Monday, how much activity would remain at 10 a.m. on Friday provided no dose was drawn from this vial (in mCi)? The half life of Thallium-201 is 73 hrs.

   a) 86.5
   b) 4.03
   c) 3.94
   d) 3.10

Q8 – A counter in preparation room is set to collect 15,000 counts or to stop after 5 minutes. The counter stops at 3 minutes. What is the count rate (in cpm)?

   a) 2,000
   b) 5,000
   c) 10,000
   d) 25,000

Q9 – Which of the following accurately defines a millicurie?
   a) 3.7×10^7 counts/sec
   b) 3.7×10^7 counts/min
   c) 3.6×10^7 disintegrations/sec
   d) 3.7×10^7 disintegrations/sec

Q10 – A collimated beam with 4 half-value layers of lead in front of it gives a reading of 10 mR/hr. If 2 half-value layers are removed, what would be the intensity reading?

   a) 5 mR/hr
   b) 40 mR/hr
   c) 4/2 x 10 mR/hr
   d) none of the above

Q11 – A radioactive source has a dose rate of 40 mR/hr at a distance of 10 cm. At what distance from this source would a technician working 40 hours per week for 56 weeks per year receive a total dose of 5,600 mR?

   a) 40 cm
   b) 10 cm
   c) 20 cm

Q12 – Which of the following is/are suitable for thyroid imaging?

    a) Fludeoxyglucose F-18
    b) I^{123} sodium iodide
    c) I^{125} sodium iodide
    d) I^{131} sodium iodide

Q13 – After a molybdenum-99 / technetium-99m generator is eluted with typical efficiency, the ingrowing Tc-99m reaches a maximum

    a) at about 6 hours

    b) at about 6 to 12 hours

    c) at about 12 and 24 hours

    d) at about 24 and 36 hours

Q14 – After an atom has decayed by giving off a negative beta particle and a gamma ray, the remaining atom is:

    a) An atom of a new element having an atomic number one higher than the old and with no or little change in mass number

    b) unchanged except it has now become stable

    c) An atom of a new element having an atomic number one less than the old and with no or little change in mass number

Q15 – A worker in a nuclear power plant receives an absorbed dose of 475μ Gy from slow neutrons and an absorbed dose of 3m Gy from gamma radiation. Calculate the total equivalent dose received

Q16 – A cobalt-60 source emits gamma radiation. A certain distance from this source , the equivalent dose rate is 25μ Svh^{-1}. A thickness of 120mm of lead is needed to reduce the equivalent dose rate to 3μ Svh^{-1}. Calculate the half value thickness of the lead shielding.

Q17– When lead plate is inserted between gamma source and a detector the activity falls to one sixteenth of it's original value to 3k per second. What is the activity of the gamma source?

Q18 - The half value thickness of a material for a gamma source is 12 mm. What thickness of material will reduce the count rate from the gamma source from 60 cps to 15 cps?

Q19 - In a radioactive sample, 30,000 nuclei decay every 5 minutes. What is the activity of the sample?

Q20 - What is meant by absorbed dose? What is the unit of absorbed dose?

Q21 - A patient of mass 70 kg is given radiotherapy. During the treatment, a tumour of mass 250 g receives 7 J of energy from a beta source. What is the absorbed dose?

Q22 - The radiation weighting factor of alpha particles is 10. A patient irradiated with alpha particles receives an absorbed dose of 80 mGy. What is the equivalent dose received by the patient?

Q23 - A tumour of mass 15 g absorbs 5 mJ of energy during a course of treatment. The equivalent dose is 700 mSv. What is the weighting factor of the radiation used?

Q24 – Po-210 has biological half life of 40 days and physical half life of 138 days, calculate the effective half life of Po-210.

Q25 – A radionuclide Iodine-123 has a biological half life of 4 hours and an effective half life of 3.075 hours. Calculate the physical half life of I-123.

Q26 – An isotope of technetium $^{93}_{43}Tc^m$ which is in a metastable state, decays emitting only γ rays. When the isotope is placed 20 cm from a γ ray detector the count rate is 25 counts per second. The background count rate is 120 counts per minute.

Calculate the count rate, in counts per second, when the detector is placed 30 cm from the isotope.

Q27 – Explain why the isotope of technetium $^{93}_{43}Tc^m$, is often chosen as a suitable source of radiation for use in medical diagnosis.

Q28 – Regarding Anger cameras, all of the following are true except:

a) Resolution increases with the number of photomultiplier tubes per unit area.
b) Resolution increases with increasing collimator thickness.
c) Resolution with a parallel hole collimator increases with decreasing hole diameter.
d) Efficiency increases with decreasing collimator length.
e) High-energy collimator have higher efficiency than low-energy collimator.

Q29 – A patient is imaged using a 512 x 512 matrix on a gamma camera with a 300 mm field of view. Calculate the width of each image pixel in mm.

Q30 – How photomultiplier tube converts the flashes into electrical signals.

7 - Body Lever System

A lever is a machine consisting of a beam or rod pivoted at a fixed hinge. A lever amplifies an input force to provide a large output force. The earliest use of lever come dates from the 3rd century BC by Greek physicist Archimedes of Syracuse (c.287BC – c. 212BC). He famously said " *Give me a place to stand and long enough lever, and I shall move the Earth with the lever"*. Our arms, legs and any other part in the body muscles and bones work together as level. A lever allows a given effort (force) to lift a heavier load (resistance) further or faster, than it otherwise could. In the human body, our joints act as a fulcrum, the bones as the lever and the muscle contraction as the effort (force). Load (resistance) is the bone itself, along with overlying tissues and anything else you are trying to move.

All levers follow the same basic principle:

> When effort is further than load from the fulcrum will result in *mechanical advantage*.

This lever system which is also called a power lever, are slower, more stable and used where speed strength is a priority

> When effort is nearer than load from the fulcrum will result in *mechanical disadvantage*.

This lever system which is also called a speed lever, speed and range of movement are gained at the loss of force and this can be distinct beneficial.

A lever comprises of three components.

i. **Fulcrum** or pivot – is the point about which lever rotates.
ii. **Load** – is the point where force is applied by the lever system. Centre of gravity of the lever – also called resistance.
iii. **Effort** – is the point where force is applied by the user of the lever system. Usually muscle insertion – also called force.

The levers are classified into three classes.

First Class Lever System: The force and loads are on opposite side to each other and fulcrum is in the middle. An example is teeter – totters.

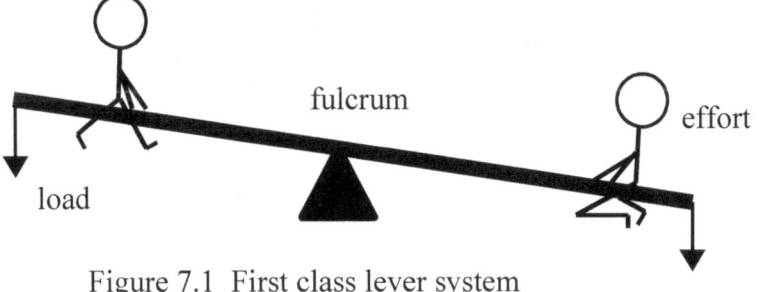

Figure 7.1 First class lever system

A box arrives with its lid nailed down; we take a metal crowbar, use it as a lever, to open the lid. One end of the metal bar is applied to the point of resistance, the lid of the box. At the other end we apply our strength, force. The edge of the box against which the metal bar is worked serves as a fulcrum and lies between the handle where the power is applied and the bevelled edge which moves the resistance or weight.

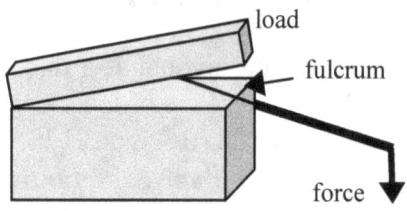

Figure 7.2 Opening box with metal bar as lever (1st order)

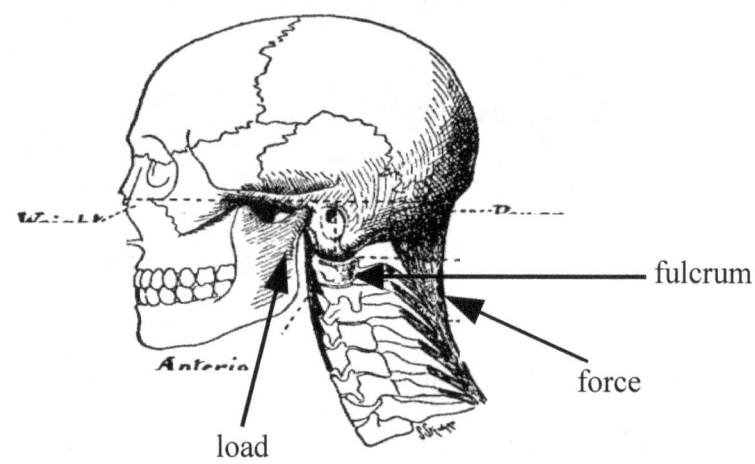

Figure 7.3 Human skull as first class lever

When we move our head up and down, this is an example of first class lever system. Where fulcrum is the atlanto-occipital joint connecting head and spinal column. Load is the mass of the head. Force is the contractions of trapezius muscles at back of neck and shoulders. When someone is working with head down for longer periods then effort (force) is near to fulcrum than the load which is mechanical disadvantage and that is why we complain neck and shoulder pain in such situation.

Second Class Lever System: In the second class lever system the load is between the fulcrum and the force. An example of this system is a wheelbarrow.

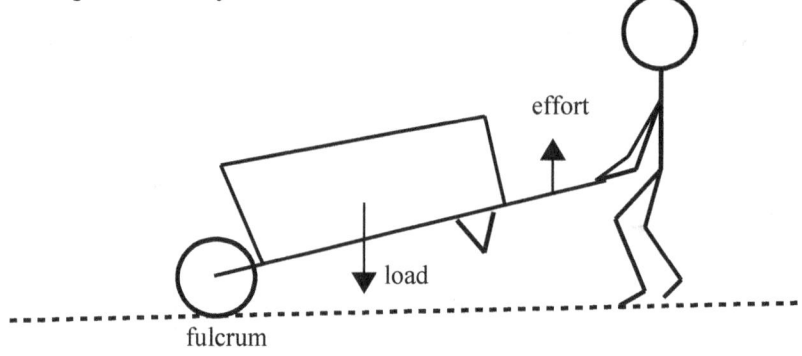

Figure 7.4 Wheelbarrow as second class lever system

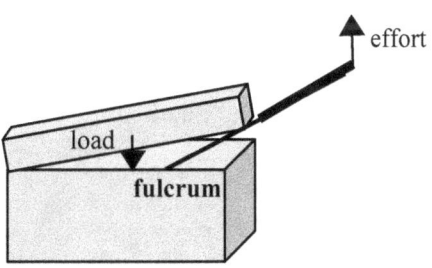

Figure 7.5 Opening box with crowbar as lever (2nd order)

If, however, after inserting the bevelled edge under the lid, we raise the handle instead of depressing it, we change the crowbar into a lever of the second order. The lid is not now forced up on the bevelled edge, but is raised on the side of the crowbar, some distance from the bevelled edge, which thus comes to represent the fulcrum. By using a chisel in this way, we reverse the positions of the weight and the fulcrum and turn it into a lever of the second order. Suppose we push the side of the crowbar—which is 10 cm long—under the lid to the extent of 1 cm, then the advantage we gain in power is as 1 to 10; we thereby increase our strength tenfold. If we push the crowbar under the lid for half its length, then our advantage stands at 10 to 5; our strength is only doubled. If we push it still further for two-thirds of its length, then our gain in strength is only as 10 to 6.6; our power is increased by only one-third. Now this has an important bearing on the problem we are going to investigate, for the weight of our body falls on the foot, so that only about one-third of the lever—that part of it which is formed by the heel—projects behind the point on which the weight of the body rests. The strength of the muscles which act on the heel will be increased only by about one-third.

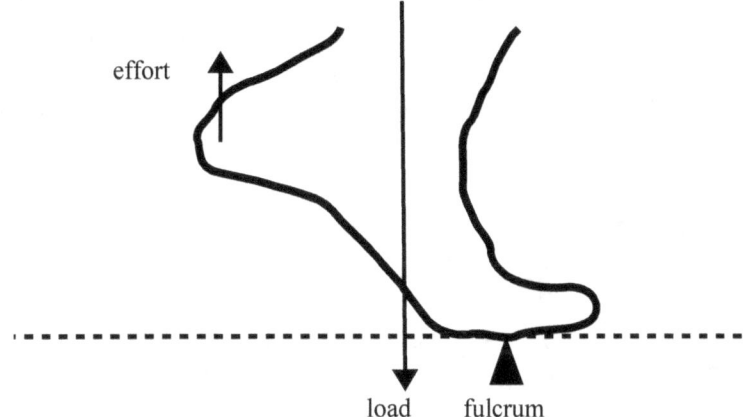

Figure 7.6 Foot when walking act like lever (2nd order)

Let's look closely at the foot lever if we are to understand it. It is arched; the front pillar of the arch stretches from the summit or keystone, where the weight of the body is poised, to the pad of the foot or fulcrum (Fig. 7.7); the posterior pillar, projecting as the heel, extends from the summit to the point at which the muscle power is applied. A foot with a short anterior pillar and a long posterior pillar or heel is one designed for power, not speed. It is one which will serve a hill-climber well or a heavy, corpulent man. The opposite kind, one with a short heel and a long pillar in front, is well adapted for running and sprinting—for speed. Now, we do find among the various races of mankind that some have been given long heels, such as the dark-skinned natives of Africa and of Australia, while other races have relatively short, stumpy heels, of which sort the natives of Europe and of China. With long heels less powerful muscular engines are required, and hence in dark races the calf of the leg is but ill developed, because the muscles which move the heel are small. Whereas Europeans, on the other hand, having short heels, need more powerful muscles to move them, and hence calves are usually well developed.

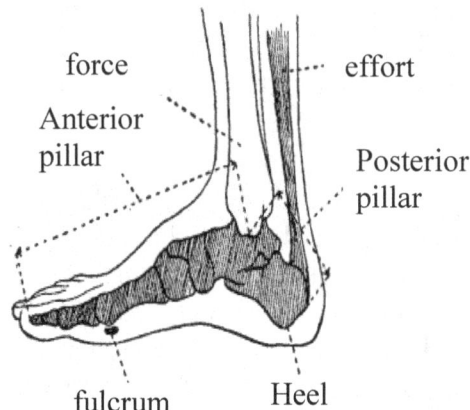

Figure 7.7 The bones forming the arch of the foot, seen from the inner side

Another example of such system in the body is opening of the mouth when teeth are stuck together with gooey toffee.

Third Class Lever System: In the third class lever, the force is between the fulcrum and the load. In this system resistance arm (distance between the load and fulcrum) is greater than force arm (distance between the force and fulcrum), so such system is mechanically disadvantaged. In fact, a large force is actually needed to move a small load. The use of this lever is in the gain in speed of the load. Examples of this lever class include: the coiled spring pulling on a door, fingernail clipper and tweezers.

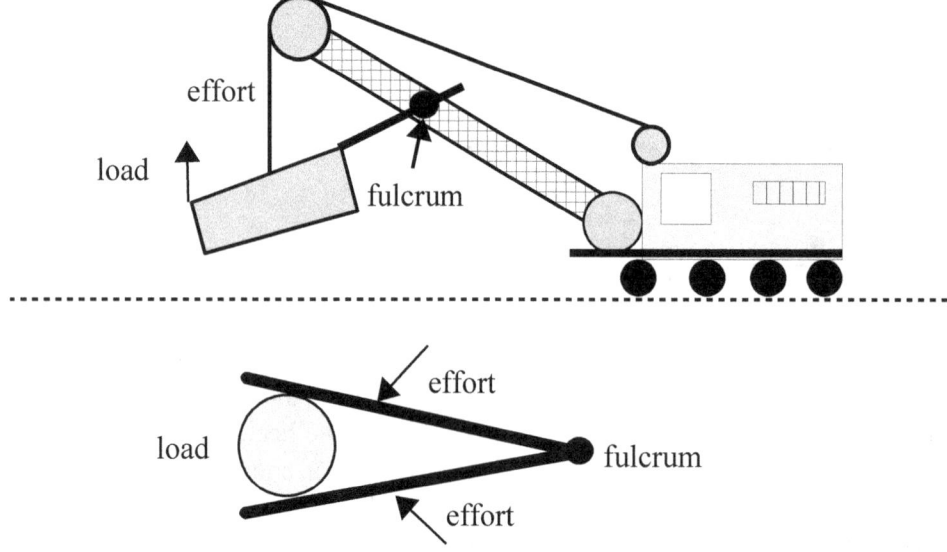

Figure 7.8 Crane and nut craker as examples of 3rd class lever

In levers of the third order, the load is placed at the end of the lever, and the muscle is attached somewhere between the load and the fulcrum as shown below opening box lid with crowbar.

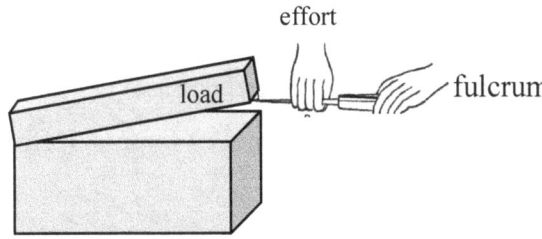

Figure 7.9 Opening box with crowbar using 3rd class lever system

An example of a third class lever in the human body is the elbow joint: when lifting a ball, the elbow joint is the fulcrum across which the biceps muscle performs the work. In levers of the third order, the load is placed at the end of the lever, and the muscle is attached somewhere between the load and the fulcrum.

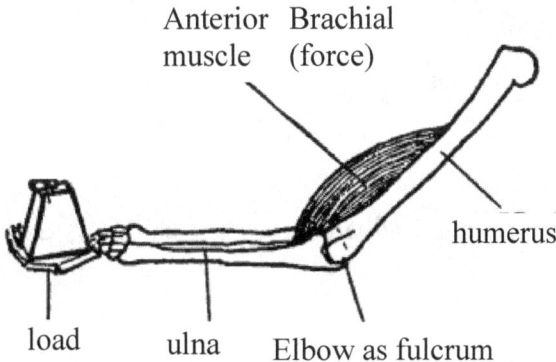

Figure 7.10 The forearm and hand as examples of 3rd class lever

The lever formed by the forearm and hand is shown in Figure 7.10. It is pivoted or jointed at the elbow; the elbow is its fulcrum. At the opposite end of the lever, in the, upturned palm of the hand, we shall place a weight of 1 kg to represent the load to be moved. The force which we are to yoke to the lever is a strong muscular engine, called the *brachialis anticus*, or front brachial muscle. It lies in the upper arm, where it is fixed to the bone of that part—the humerus. It is attached to one of the bones of the forearm—the ulna—just beyond the elbow.

In the second order of levers, we have seen that the muscle worked on one end, while the weight rested on the lever somewhere between the muscular attachment and the fulcrum. In levers of the third order, the load is placed at the end of the lever, and the muscle is attached somewhere between the load and the fulcrum (Fig. 7.10). In the example we are considering, the brachial muscle is attached about 13mm beyond the fulcrum at the elbow, while the total length of the lever, measured from the elbow to the palm, is about 31cm. Now, it is very evident that the muscle or power being attached so close to the elbow, works under a great disadvantage as regards to strength. It could lift a 23.50 kg weight placed on the forearm directly over its attachment as easily as 1 kg weight placed on the palm.

$$\frac{distance\ between\ load-fulcrum}{distance\ between\ force-fulcrum} = \frac{31cm}{1.3cm} = 23.50$$

But, then, there is this advantage: the 1 kg weight placed in the hand moves with 23.50 times the speed of the 23.50 kg weight situated near the elbow. What is lost in strength gains in speed.

Lever Arm Length

Resistance Arm (RA): Distance between fulcrum and point of resistance, R (load)

Force Arm (FA): Distance between fulcrum and point of force, F (effort)

$$F \times FA = R \times RA \qquad (7.1)$$

Example: The distance between the fulcrum and point of resistance is 10 cm with 2 N of load. Calculate the force required if the distance between the fulcrum and force is 2 cm.

$$F \times 2 \text{ cm} = 2 \text{ N} \times 10 \text{ cm}$$

$$F = \frac{2 \times 10 \text{ N } \cancel{cm}}{2 \cancel{cm}} = 10 \text{N}$$

Mechanical advantage: The mechanical advantage of a machine is defined as the ratio of the load to the effort.

$$MA = \frac{load}{effort} = \frac{FA}{RA} \qquad (7.2)$$

- No mechanical advantage if quotient = 1
- If quotient > 1: mechanical advantage in force
- If quotient < 1: mechanical advantage in speed and range of motion

The Muscular System

The muscles in our body are divided into three classes: cardiac, smooth, and skeletal. The muscular system, composed of over 600 muscles, comes in a variety of shapes and forms. The difference between each muscle are recognized by location, function, structure, and the way they are contracted. Skeletal muscles make muscular system in our body and make movement in our bones like leg and arm movement.

Skeletal muscles can be broken down into groups based upon the type of movement they have. The movement of the muscle is based upon the type of joint upon which the muscle works. Skeletal muscles create motion by pulling on tough cords of connective tissue called *tendons*. These tendons in turn pull on the bone which creates motion. Muscles move bones through mechanical leverage. The point where a muscle is connected to a bone is called the point of insertion.

As a muscle contracts, it causes the bone to act like a lever with the joint serving as a fulcrum. Muscle exerts force by converting chemical energy (created during respiration) into tension and contraction. When a muscle contracts, it shortens, pulling a bone like a lever across its hinge just like spring connected to a door will pull the door shut. Skeletal muscles can't expand, but they can contract - a muscle can pull but it cannot push. They generally work in pairs. For example, when we contract our major arm muscle, which is called the biceps, in return the lower arm muscle, called the triceps, extends. So as you contract one muscle the other one extends. These effects can be broken down into groups of their own: Flexion (bending) occurs when contraction causes two bones to bend toward one another, while extension (straightening out) occurs from contraction of muscles, resulting in an increase in angle between two bones. These pairs of muscles are called **antagonistic**.

Often antagonistic muscles are in groups, for example, both the brachialis and the biceps muscles flex the arm at the elbow and antagonize the triceps, but only when the palm is facing upwards. In pairs or groups of antagonistic muscle, one is usually much stronger than the other. The biceps, which flex the arm is larger and more powerful than the triceps which extends it. Flexors bend at the joint, decreasing the interior angle of the joint. The bracius humorous, or biceps, is a flexor of the elbow joint, bringing the fist towards the shoulder. If a flexor appears in either the wrist or ankle joints, it becomes a plantarflexor.

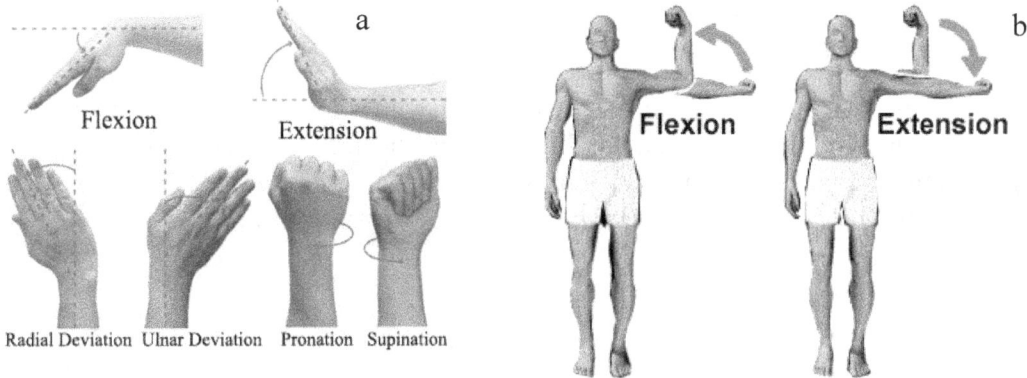

Figure 7.11 Flexion and Extension of muscles bones

The opposites of flexors, extensors unbend at the joint, increasing the interior angle. Extension is a movement of the joints that results in an increased angle between two bones or body surfaces at joints this usually results in straightening of the bones. The tracius humorous, or triceps, is an extensor of the elbow joint, taking the fist farther away from the shoulder. If an extensor is found in the wrist or ankle joints, it becomes a dorsiflexor.

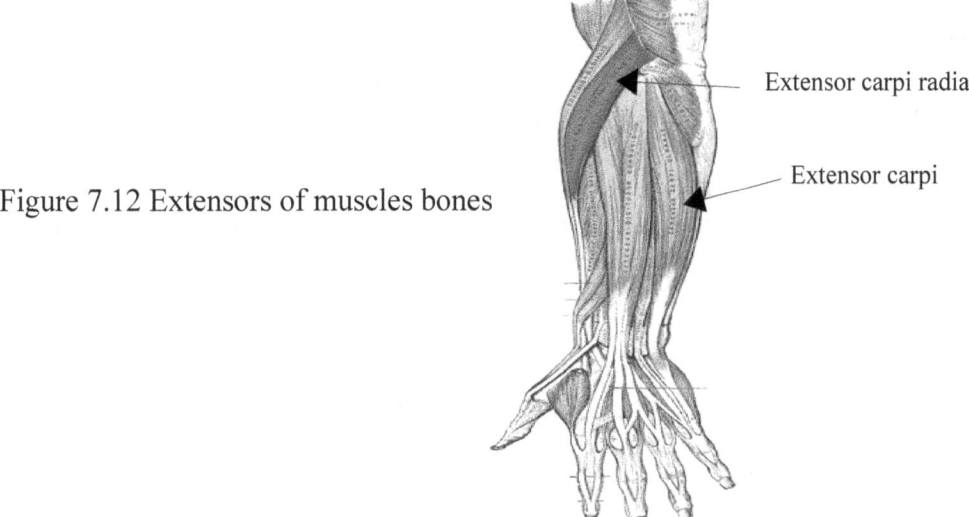

Figure 7.12 Extensors of muscles bones

Abductors take away from the body, like lifting the arm to the side. Abd- means to take away, like abduct and abdicate. Spreading out your fingers uses abductors, because you are taking away your fingers from an imaginary line running down your arm.

Adductors, the opposites of abductors, move toward the body. Add- means to increase or include. By lowing an arm raised to the side, or moving your fingers together while keeping them straight, your muscles are adducting.

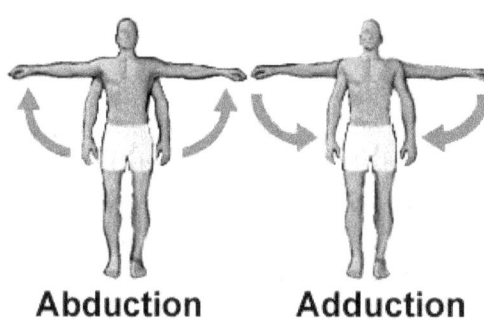

Figure 7.13 Abduction and Adduction of muscles bones

A synergist muscle holds the body in position so that an agonist muscle can operate, thus preventing any unwanted movements that might occur as the prime mover contracts. A fixator muscle by definition is a synergist muscle, but is more specifically referred to as a fixator or stabiliser when it immobilises the bone of the prime mover's origin, thus providing a stable base for the action of the prime mover.

Types of Muscles contractions

During physical activity muscles can contract, stretch called dynamic contraction or remain the same called static contraction.

Static contraction which often is called an isometric muscle contraction, the length of the muscle does not change, whereas the amount of tension increases during the contraction process. For example, during exercise (stretching legs when holding rest of the body on the toes and arms on the ground bending at the elbows) when you exert maximum possible force in fixed position for 10 seconds with 60 seconds of recovery. This also happens when carrying an object in front of you as the weight of the object is pulling your arms down but your muscles are contracting to hold the object at the same level. The amount of force produced by muscles during an isometric contraction depends on the length of the muscle and the point of contraction.

Dynamic muscle contraction is divided into two groups: concentric and eccentric muscle contraction. Concentric muscle contraction is when the muscle is shortened under tension and is a form of isotonic muscle contraction. For example bending the elbow from straight to fully flexed position will result in concentric contraction of Biceps Brachii muscle.

Eccentric contraction involves a muscle lengthening under tension. When a muscle contracts eccentrically it is acting as a brake, thus controlling the movement. For example, during the downward moving part of a jump or squat, the quadriceps muscle group is lengthening under tension and so the work done is called eccentric or negative. Eccentric muscle contraction produces the biggest overload in a muscle, thereby enhancing its development. Sometime both work together, for example, when kicking a football, the Quadriceps muscle contracts concentrically to straighten the knee and the Hamstrings contract eccentrically to decelerate the motion of the lower limb. This kind of activity in sports commonly involved in muscle injuries.

For eccentric contractions, the agonist muscle is the active muscle, which in this case is lengthening. In the case of the landing from a jump or controlled downward movement in a squat, the quadriceps muscle group lengthens under tension, and is therefore the agonist. To be the agonist in this situation, the muscle must be under tension. The antagonist muscle during the example of a downward squatting movement would be the hamstring muscle group, which gets shorter and which relaxes or acts as a fixator for the hip joints.

Isokinetic contractions are similar to isotonic in that the muscle changes length during the contraction, where they differ is that Isokinetic contractions produce movements of a constant speed. It happens very rarely, the best example is breast stroke in swimming, where the water provides a constant, even resistance to the movement of adduction.

The fourth type of muscle "contraction" known as passive stretch. As the name implies, the muscle is being lengthened while in a passive state (i.e. not being stimulated to contract). An example of this would be the pull one's feet in their hamstrings while touching their toes.

Tendons and Ligaments

Muscles wouldn't be very useful alone because they don't directly connect to the bone, so even if they contract, they wouldn't be moving anything. Instead, muscles are connected to tendons, which themselves are connected to the bones. A tendon is a strong band of fibrous connective tissue and is capable of withstanding tension.

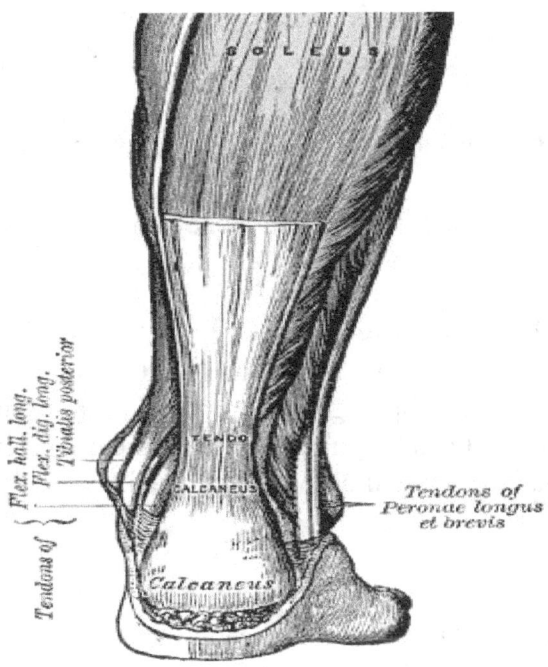

Figure 7.14 The Achilles tendon.

Tendons are composed of parallel arrays of closely packed collagen fibres. They are made of special cells called tenocytes, water and fibrous collagen proteins. A million of these proteins weave together to form a very strong tendon. Tendons then grow into the bones and form a very strong mineralised connection. The mechanical properties of the tendon depend on the collagen fibre diameter and orientation. A body builder will generally have shorter tendons, whereas for athletes to excel in action such as jumping and running must have longer Achilles tendon.

A ligament is the fibrous tissue that connects bones to other bones. They allow most joint to move, control their range of movements and stabilise them. The ligament that helps flex or extend the body parts are called *articulate* ligaments. If the body structure strengthen or support other ligament are called *accessory* ligaments.

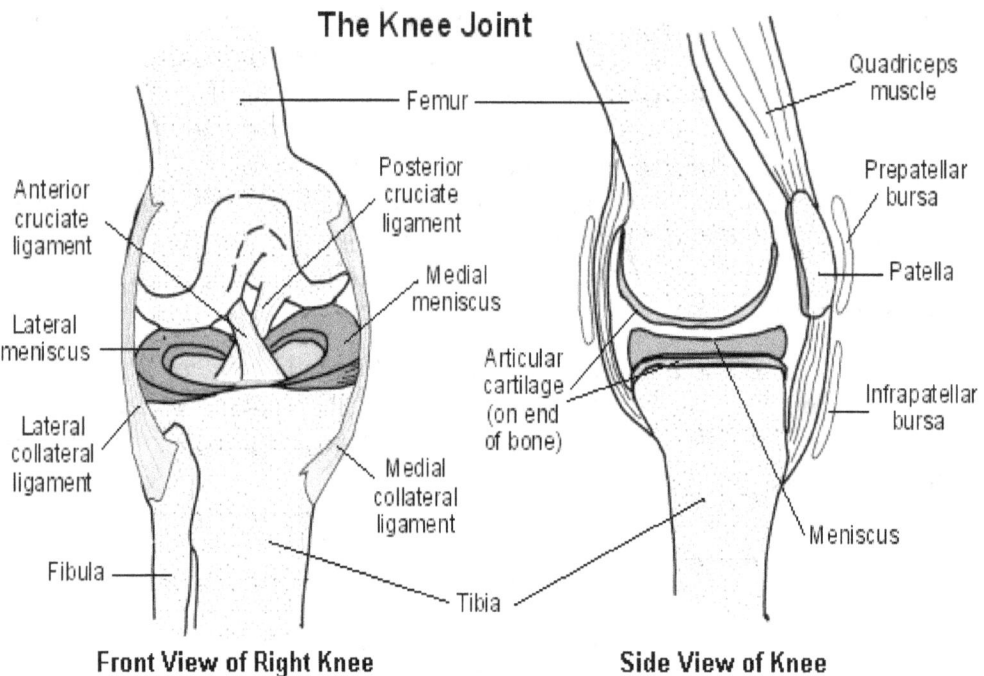

Figure 7.15 Typical bone joint

Without ligaments, instead of bones bending or rotating about each other when muscles contract, they would slide by each other. Ligaments are what hold the bones together. They connect at the ends of muscles and keep them from slipping and sliding, and force them to bend. Regular exercise of stretching can increase the length and flexibility of the muscles and extension of ligaments. If the ligaments lengthen too much, the joint will become weakened.

The Elbow

The elbow joint is considered a "hinge" joint, since it moves like a hinge. This is made up of the *humerus* (upper arm bone), the *radius* (thumb side of the forearm) and the *ulna* (Little finger side of the forearm). Ulna and humerus articulate with one another to form the elbow joint. The elbow has a specific up and down motion with very little rotation involved. The radius and ulna articulate at the elbow in such a way that radius rotate against the ulna, which give a specific rotational movement of the lower arm called *pronation* and *supination*.

There are two ligaments that connect the humerus and ulna at the elbow. The *ulnar-collateral* ligament is found inside the of the elbow, and the *lateral – collateral* is found outside the elbow. The two muscles are responsible for elbow flexion and extension are the triceps and biceps respectively. The biceps is at front of the arm and triceps is found on the back of the upper arm.

With reference to figure 7.10, figure 7.16 represents a simplified lever representation of the arm.

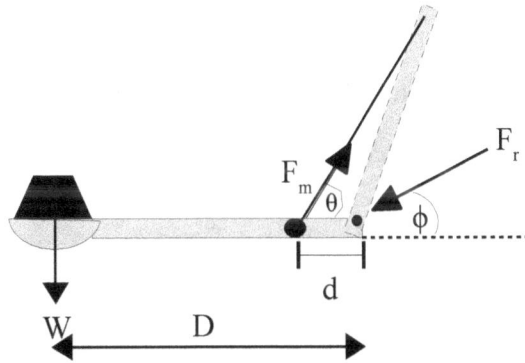

Figure 7.16 Lever representation of fig. 7.10

Figure 7.16 shows an arm lifting a weight, W in the hand with the elbow bent at angle ϕ. The weight this arm holding will rotate the arm counter clockwise if we take arm as a 3rd class lever system. The effort will be made by arm muscle to counter this rotation. This force which is provided by biceps is represented in the diagram as F_m. The reaction force at the elbow which is serving as a fulcrum is shown as F_r.

For equilibrium, the forces in *x* and *y* direction should be zero, otherwise arm will rotate. For this we obtain:

x components of the forces: $F_m \cos \theta = F_r \cos \phi$ (7.3)

y components of the forces: $F_m \sin \theta = F_r \sin \phi$ (7.4)

Also, for equilibrium the torque (turning force) must be zero. We, recall that torque about fixed point A in Figure 7.17 can be calculated using formula 7.5

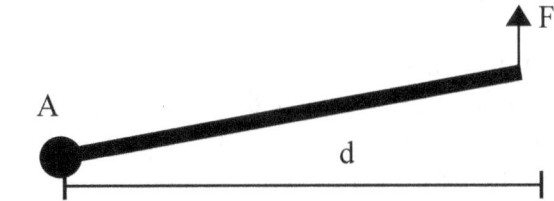

Figure 7.17 Counter clockwise torque at point A

$$T = F \times d \quad (7.5)$$

When we look at figure 7.16, we see two torques at fulcrum: a clockwise torque due to the weight of the arm and a counterclockwise torque due to *y* component of the muscular force F_m. The torque of F_r is zero as the force is acting at that point i.e. the distance of this force from fulcrum is zero. Thus we obtain

$$F_m \sin \theta \times d = W \times D \quad (7.6)$$

Example: Find the force exerted by biceps when arm length, D is 40 cm and the distance between the fulcrum and point of muscle is 30 mm when hand is holding 10 N weight. The angle θ is 75°. Also calculate the reaction force and its angle at the elbow.

Rearranging equation 7.6 $\quad F_m = \dfrac{W \times D}{\sin 75 \times d} = \dfrac{10 \times 0.4}{0.03 \times 0.97} = 138.04\ N$

we can see from the answer that muscle must exert almost 14 time more than the lifting weight.

Using 7.3 and 7.4

$138.04 \times \cos 75° = F_r \cos \phi$ $\quad\quad$ (i)
$138.04 \times \sin 75° = 10 + F_r \sin \phi$ $\quad\quad$ (ii)

$F_r \cos \phi = 35.73$ $\quad\quad$ (iii)
$F_r \sin \phi = 123.34$ $\quad\quad$ (iv)

add (iii) & (iv) and square

$$F_r^2 \cos \phi^2 + F_r^2 \sin \phi^2 = 15212.76$$

$$F_r^2 (\cos \phi^2 + \sin \phi^2) = 15212.76$$

$\cos \phi^2 + \sin \phi^2 = 1$

$F_r^2 = 15212.76$
$F_r = 123.34\ N$

from equation (iii) & (iv), the cotangent of the angle is

$$\cot \phi = \dfrac{35.73}{123.34} = 0.29$$
$$\phi = 73.83°$$

Example: An Olympic weightlifter champion can raise a mass of 40 kg with one hand. His forearm has a length from elbow joint to the palm of the hand as 0.45m, and a weight of 30 N. If the force of biceps muscle acts at a point 0.08m from the elbow (fulcrum), calculate the maximum values of the force in the biceps and reaction in upper arm.

Let's draw the arm lever

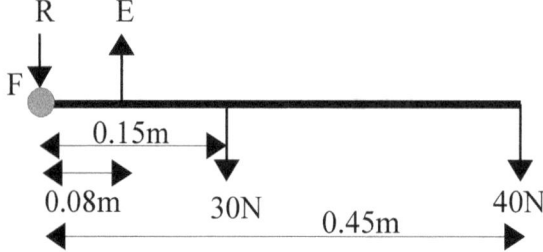

Taking moment about F, fulcrum gives,

E x 0.08 = (30 x 0.15) + (40 x 0.45)
0.08E = 4.5 + 18 = 22.50 N

effort due to biceps $\quad E = \frac{22.50}{0.08} = 281.25N$

Resolving forces vertically

R + 30 + 40 = 281.25
reaction due to upper arm $\quad$ R = 211.25N

Example: 2 painter/decorators weighing 750N and 800N respectively stand on a plank of wood at positions A and B. The plank is resting on 2 trestles (supports) and weighs 400N. The trestles exert upward forces, P and Q, on the plank (known as reactions). Calculate the upward forces P and Q.

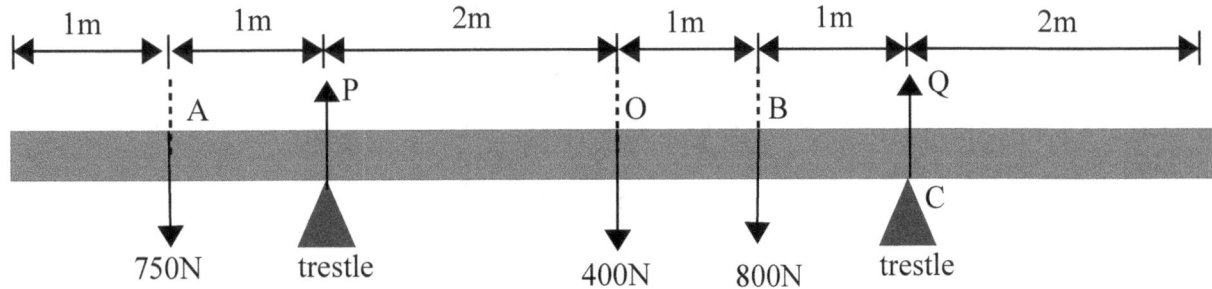

The sum of the forces in one direction equals the sum of the forces in the other direction.

Therefore: P + Q = 750N + 400N + 800N = 1950N

Taking the moment at C gets rid of any moments due to Q.

Now, Clockwise Moment = P x 4m

Anticlockwise Moment = (750N x 5m) + (400N x 2m) + (800N x 1m) = 5350Nm

Since, $\quad$ Clockwise Moment = Anticlockwise Moment

Therefore, $\quad$ P x 4 = 5350Nm

$$P = \frac{5350Nm}{4m} = 1337.5N$$

hence, $\quad$ Q = 1950 – 1337.5 = 612.5N

The Hip

The hip is located lateral (left & right) to the gluteal region. In the adult human body, three of the bones of the pelvis have fused into the hip bone which form part of the hip. The main function of the hip is to support the weight of the body in both standing and walking or running postures. The hip is a synovial joint formed by the articulation of the rounded head of the femur and the cup-like acetabulum of the pelvis which are connected by muscles which is shown in Figure 7.18 as F_m. When a person stand erect this make an angle of 70° with the horizontal plane.

W_L represent the weight of the leg including the foot. W is the weight of the body. We normally take W_L as 17% of the entire body weight, W. W_L acts at the centre of gravity. Force W at the bottom of figure 7.18 represent the reaction of standing on the ground.

From equilibrium conditions we know that horizontal and vertical components of forces add to zero.

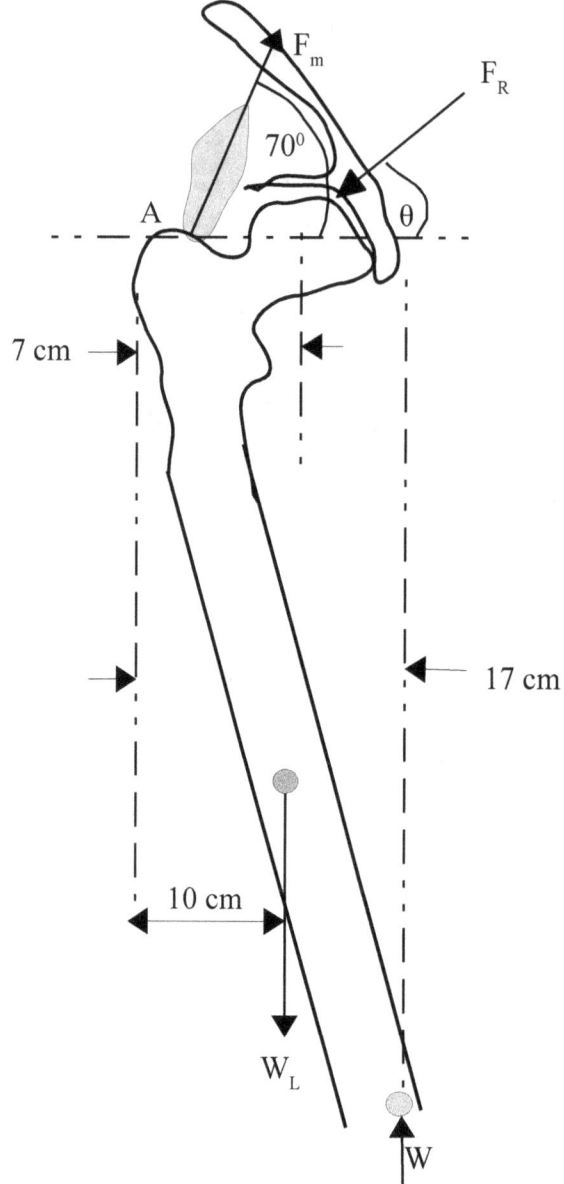

Figure 7.18 Lever system representation of Hip joint and leg

x components of forces: $F_m \cos 70° - F_R \cos \theta = 0$ (i)

y components of forces: $F_m \sin 70° + W - W_L - F_r \sin \theta = 0$ (ii)

If the torque at point A is 0

$$(F_R \sin \theta) \times 0.07 + (W_L \times 0.1) - (W \times 0.17) = 0 \quad (iii)$$

$$W_L = 0.17W$$

substituting in (iii) $(F_R \sin \theta) \times 0.07 + \{(0.17W \times 0.1) - (W \times 0.17)\} = 0$

$$F_R \sin \theta = 2.19W \qquad \text{(iv)}$$

substituting this result in (ii) gives:

$$F_m = \frac{1.37W}{\sin 71°} = 1.45W \qquad \text{(v)}$$

from equation (i) we obtain

$$F_R \cos \theta = 1.45W \cos 70° = 0.50W$$

substituting in (iv) for F_R

$$\frac{0.5W}{\cos \theta} \times \sin \theta = 2.19W$$

$$\tan \theta = \frac{2.19W}{0.5W} = 4.38$$

hence $\theta = 77.14°$

and

$$F_R = 2.24W \qquad \text{(vi)}$$

This result shows that the force on the hip joint is almost two and quarter time the person's own weight.

The Vertebral column

The spine consists of 24 articulating vertebrae and 9 fused vertebrae in that five are fused to form the sacrum and the four coccgeal bones that form the tail-bone. The curvature of the column provides support and balance to the body. The lumber vertebrae graduate in size from 1 through 5. These vertebrae bear much of the body's weight and related bio-mechanical stress. They are also housed and protect the spinal cord in its spinal canal.

Each individual vertebra has unique features depending on the region in which it is found. Every vertebra, has three basic functional parts:
(1) the drum-shaped vertebral body, designed to bear weight and withstand compression;
(2) the posterior (backside) arch, made of the lamina, pedicles and facet joints;
(3) the transverse processes, to which muscles attach.

The top and bottom of the vertebral body are called the *end plates*. The inter-vertebral *disc*, sandwiched between two vertebral bodies, is attached to the end plates.

With each step, a force is exerted by the ground as a reaction to the body weight. If the shock absorbers between the vertebrae did not exist, and if the spine had a straight structure, this force would be transmitted directly to the skull. Consequently, the top of the spine would break into the brain and shatter the skull.

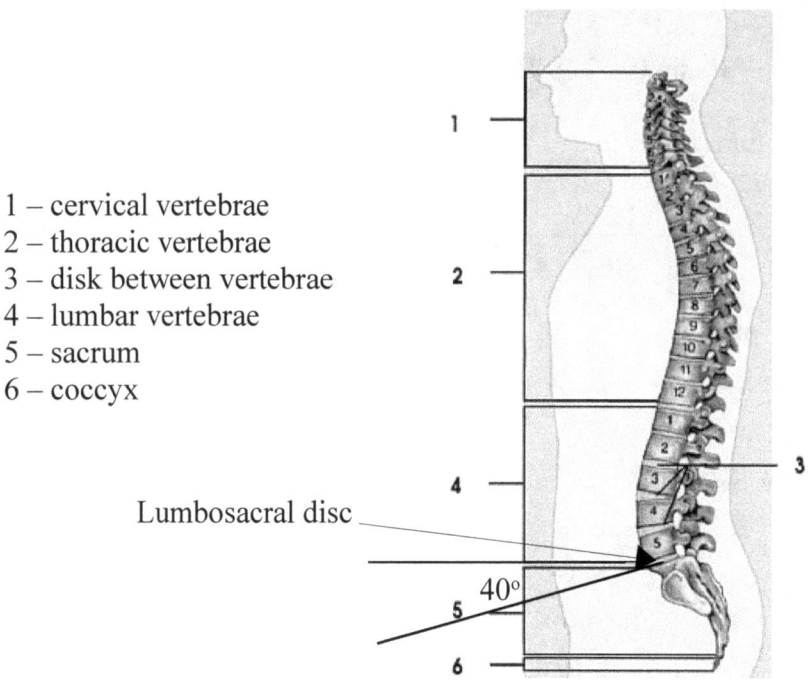

1 – cervical vertebrae
2 – thoracic vertebrae
3 – disk between vertebrae
4 – lumbar vertebrae
5 – sacrum
6 – coccyx

Figure 7.19 The vertebral column

Standing

The disc above the sacrum and below the 5th lumbar vertebrae is called lumbosacral disc which in normal postures lies at an angle of 40° to the horizontal as shown in figure 7.19. This disc support about 60% of body weight. This disc is subject to both compressive and shear (twisting) stress. The compressive stress is perpendicular to the disc and a shear stress parallel to the bone surface, Figure 7.20

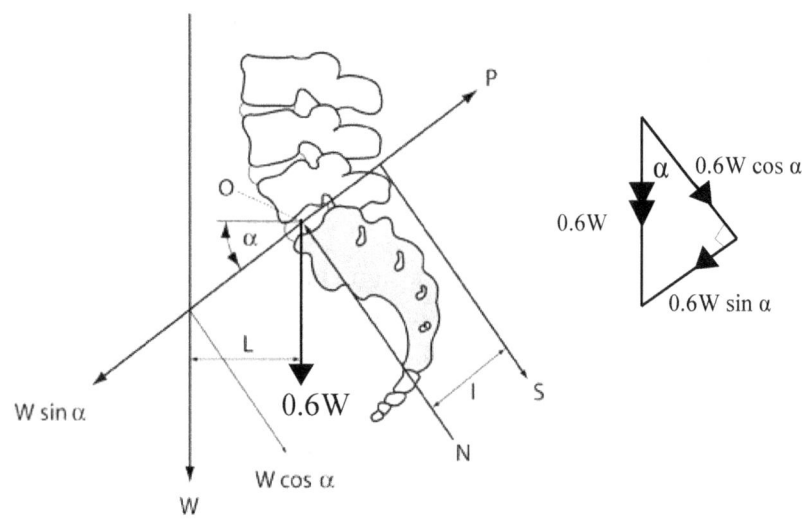

Figure 7.20 Forces on the lumbosacral disc

The disc is at an angle α to the horizontal. There is a downward force of 0.6W, which is balanced by an equal, upward force due to the reaction of the sacrum. Resolving perpendicular and parallel to the surface of the disc gives (taking normal pasture angle 40°)

compressive stress = 0.6W cos α = 0.6W cos 40° = 0.46W

shear stress = 0.6W sin α = 0.6W sin 40° = 0.39W

The disc is designed to withstand compressive stress but not sheer stress. This can be made worse if a person has poor posture so lumbar lordosis angle is greater than 40°. Lordosis is the inward curvature of a portion of the lumbar and cervical vertebral column. These two vertebral columns are normally lordotic. This means that they are set in a curve that has its convexity anteriorly (front) and concavity posteriorly (behind) in human anatomical structure. A common cause of excessive lordosis includes tight low back muscles, excessive visceral fat and during pregnancy.

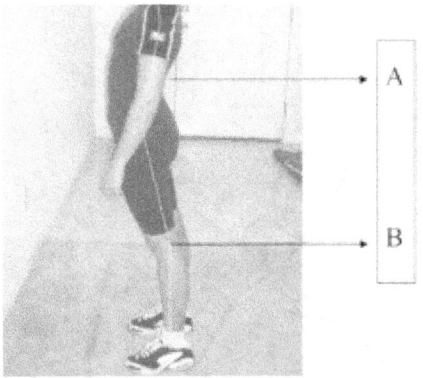

Figure 7.21 Lordosis

A main factor of lordosis is anterior pelvic tilt, which occurs when the pelvic tips forward when resting on top of the femurs. When standing on two legs our body weight is divided between each leg. Where legs connected to the body, the superincumbent weight is about 0.7W. Figure 7.22

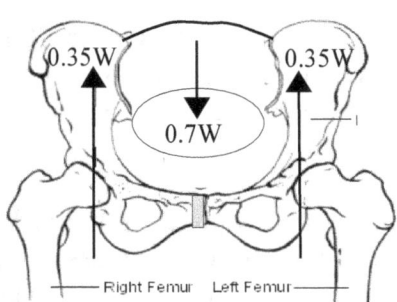

Figure 7.22 Forces at the hip

In normal posture superincumbent weight is equally distributed on each leg – hip joint and that is 35% of body weight. When we stand on one leg the body's centre of gravity will shift so it is more on the side of supporting leg. In this case the increase in the leg – hip joint is 2.5 times the body weight.

Bending and Lifting

Must remember three golden rules for bending to lift an object.

> Bend to lift an object - don't stoop
> Keep your back straight by tucking in your chin
> Lift with the strong leg muscles, not the weaker back muscles

When you bend at your waist and extend your upper body, this will change the centre of mass in the abdomen. The spine has to support both the weight of the upper body and the weight of the load being lifted or lowered. The force on the back can be estimated by calculating the moment and forces created by the weight of the load being lifted and the weight of the upper part of the body.

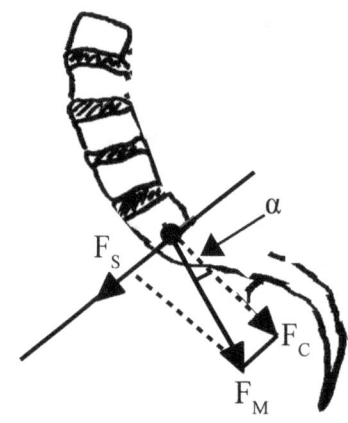

Figure 7.23 Forces at L5-S1 Joint Surface

EFFECT OF FORCES AT L5-S1 JOINT SURFACE

Let say a 75 kg (F_M = 750N) person bends and tries to lift 20 kg (F_G = 200N) mass. We will see what forces will actson his lumbosacral disc which is located at L5 – S1 joint. Gravitational force will act on two masses. The mass superincumbent to the disc which is 60 percent of body weight and the mass he is trying to lift.

The gravitational force will produce a moment at the point of L5 – S1 joint. This depends on the moment arm length. The moment arm length for both masses is shown in figure 7.24 as d_M and d_G. The muscle will produce an extensor moment which is balanced by flexor moment that gravity exerts at joints lateral axis.

therefore,

$$F_M \bullet d_M = F_G \bullet d_G \qquad (7.7)$$

The force on the disc representing lumbar extensor muscle F_M is shown in figure 7.23.

This force has horizontal and vertical components. The component F_C is perpendicular to the disc and represent joint compression and F_S is parallel to the disc and represents joint sheer.

$$F_C = F_M \cos \alpha \qquad (7.8)$$

$$F_S = F_M \sin \alpha \qquad (7.9)$$

also,

$$(F_M)^2 = (F_C)^2 + (F_S)^2 \qquad (7.10)$$

The relative magnitude of sheer and compressive force depends on the angle α. Compressional force can be written as:

$$F_C = F_M \cos\alpha$$

equation 7.7 can be written as

$$F_M = \frac{F_G \times d_G}{d_M}$$

Hence

$$F_C = \frac{(F_G \times d_G) \cos \alpha}{d_M} \qquad (7.11)$$

To reduce the compressional force on the disc:

1- Decrease F_G – this means either lose your own body weight or lift a lighter weight.

2- Decrease d_G – Minimise the moment arms of both the masses on which gravity acts. For this keep the lifting mass closer to your body and keep a posture that ensure your trunks as erect as possible.

3- Increase α – In a squat lift with a lumbar lordosis reduces compressive forces on the lumbosacral disc, but at the cost of the increasing shear force on the disc.

4- Increase d_M – Lean forward to increase moment arm when lifting heavy objects.

Equation 7.7 indicates that a clockwise rotational moment of the torso must be counteracted by a counter-clockwise rotational moment, which is produced by the back muscles, Figure 7.24. Thus, when a person with an upper-body weight of W_b lifts a load with a weight of W_l, the load and upper torso create a combined clockwise rotational moment that can be calculated as

$$M_{\text{load-to-torso}} = \text{Flexion moment} = W_l \cdot d_G + W_b \cdot d_M \qquad (7.8)$$

Flexion moment a = 750×0.025 + 200×0.3 = 18.75 + 60 = 78.75 Nm

Flexion moment b = 750×0.25 + 200×0.4 = 187.5 + 80 = 267.50 Nm

Flexion moment c = 750×0.18 + 200×0.35 = 135 + 70 = 205 Nm

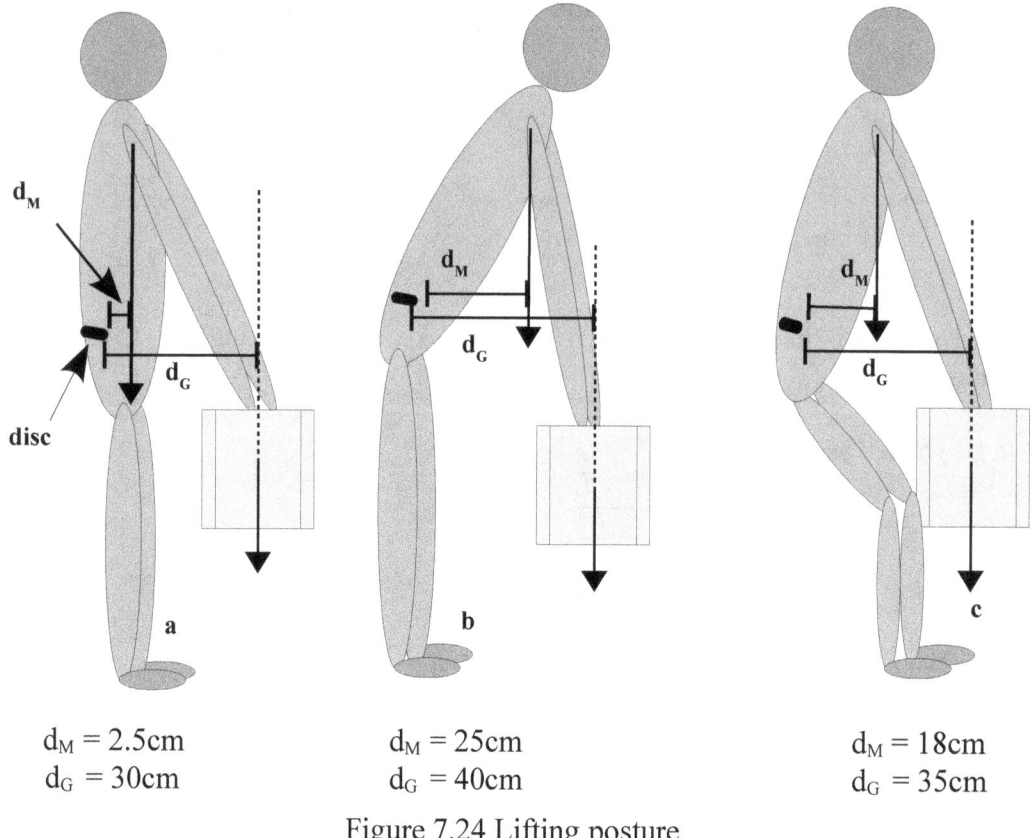

$d_M = 2.5$cm	$d_M = 25$cm	$d_M = 18$cm
$d_G = 30$cm	$d_G = 40$cm	$d_G = 35$cm

Figure 7.24 Lifting posture

Low-back biomechanics of bending and lifting

Low-back pain is often the result of incorrect lifting methods and posture. Repetitive lifting, bending, affect both the degree of severity and frequency of low-back pain. In addition, low-back pain may also be the result of bad lifting habits. To carry out a lifting task, several factors influence the load stress placed on the spine. In this biomechanics model, the weight and the position of the load relative to the centre of the gravity are just two of the factors that are important in determining the load on a spine and lumbosacral disc. Other important factors include the degree of twisting of the torso, the size and shape of the object, and the distance the load is moved.

Consider the spinal cord at an angle of 30° to the horizontal, Figure 7.25. It is hinged at the lumbosacral disc, just below the fifth lumbar vertebra and the sacrum. We can take the axis of torque at this point. There is a reaction force R from the sacrum. The weight, W_1 of the torso (above the hips, excluding arms and head) acts half way down the spinal cord. The weight, W_2 of the arm, head and any object lifted act on the top of the spinal cord. Since torso has 60% of the body weight. It is reasonable to say that $W_1 = 0.4W_b$ and $W_2 = 0.2W_b$, with no load being lifted.

Figure 7.25 shows that M act at an angle that is 30° - 10° = 20° relative to the horizontal. To balance the forces:

balancing horizontally; $\Sigma F_x = R_x - M \cos 20° = 0$ (i)

balancing vertically; $\Sigma F_y = R_y - M \sin 20° - 0.4W_b - 0.2W_b = 0$ (ii)

The torque (rotational force) due to reaction force is zero. The component of M normal to the spinal cord is Msin10°. Torque balance requires:

$$\Sigma \tau = \frac{2L}{3}\sin 10^{o}(M) - \frac{L}{2}\cos 30^{o}(0.4W_b) - L\cos 30^{o}(0.2W_b) = 0 \quad (iii)$$

$$0.116 LM - 0.173LW_b - 0.173LW_b = 0$$

$$0.116M = 0.346W_b$$

$$M = 2.98W_b$$

The body weight of 80 kg will have M = 2.98 x 80 x 9.81 = 2339N

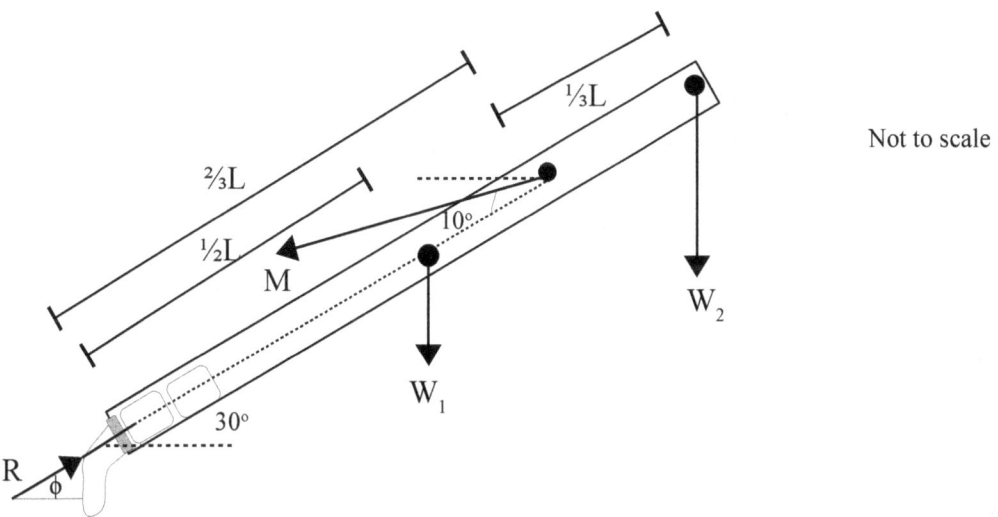

Figure 7.25 Free body diagram of the lumbosacral disc while bending, with the spine modeled as a straight bar at an angle 30°

The reaction force components are:

$$R_x = M \cos 20^{o} = 2198N$$
$$R_y = M \sin 20^{o} + 2(0.4W_b) = 848N$$
$$R = R_x^2 + R_y^2$$
$$R = (2198)^2 + (848)^2 = 2356N$$
$$\phi = \arctan(R_y / R_x) = (848 / 2198) = 19.80^{o}$$

This calculation shows that the spinal cord in bending posture experience around 2356N force which is 3 times larger than body weight.

The above working tells us that the back muscle force would be 2356 N without any load being lifted, which may exceed the capacity of some people. If the same person lifts a load of 450 N, the equation indicates that the muscle force would reach 5,000 N, which is at the upper limit of most people's muscle capability.

Exercise 7

Q1 - A first class lever has the _____ in the middle.
Give an example of a first class lever: _____
Draw a diagram of a first class lever.

Q2 - A second class lever has the _____ in the middle.
Give an example of a second class lever:_____
Draw a diagram of a second class lever.

Q3 - A third class lever has the _____ in the middle.
Give an example of a third class lever: _____
Draw a diagram of a third class lever.

Q4 - Explain how a muscle exerts force.

Q5 - Describe each of the following:
   a. Advantage (mechanical advantage)
   b. Antagonistic
   c. Effort
   d. Extension
   e. Flexion
   f. Force arm
   g. Fulcrum
   h. Insertion
   i. Resistance
   p. Tendon
   q. Weight arm

Q6 - After the students thoroughly examined the force and angular movement with the arm model, they determined that it took more force to raise the forearm than expected. From their observation, it can be concluded that the arm lever system is designed to use …
   A. less force to create more angular movement.
   B. more force to gain more angular movement.
   C. more force to produce less angular movement

Q7 - Joint movement in the body occurs when skeletal muscles _____ the bones.
   A. pull
   B. push
   C. signal

Q8 - What connects bone to bone?

Q9 – The force acting on the forearm illustrated in figure 7.26 are its weight (22N), the weight of the book (12N), If the effort exerted by the biceps muscle is E, and the reaction due to the upper arm is R. Calculate E and R.

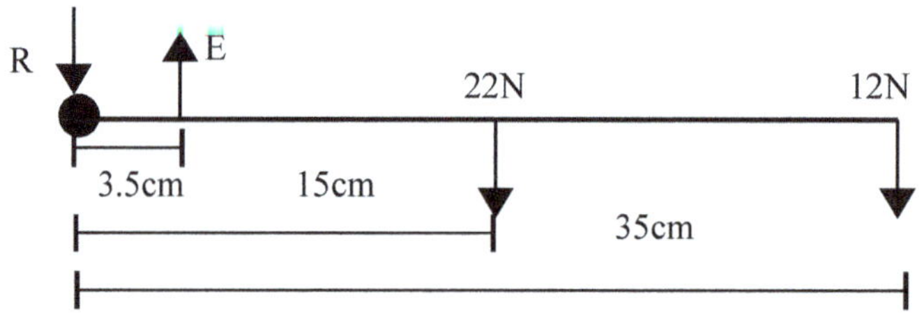

Figure 7.26

134

Q10 – If the load on the hand is doubled, by what factor does the force provided by the biceps muscle increases.

Q11 – Refer to figure 7.16, find the force exerted by biceps when arm length, D is 35 cm and the distance between the fulcrum and point of muscle is 35 mm when hand is holding 35 N weight. The angle θ is 60°. Also calculate the reaction force and its angle at the elbow.

Q12 - Calculate the force exerted by the muscle and the compression force on the fifth lumbar vertebra in Fig. 7.25. Use the following information provided:
a)
   i- body mass is 95 kg
   ii- person is bending with a spinal cord angle to the horizontal is 25°
   iii- the angle of moment to the horizontal is 12°

b) Repeat the calculations in (a) for the case when the person shown in Fig. 7.25 holds a 25kg mass in his hand.

Q13 - For each of the following statements, write the letter of the answer that best completes the statement.

1) _____ When lifting a load you should:
a. get close to the load and tighten the abdominal muscles.
b. stand three feet away and tug hard.
c. both.
2) _____ When turning you should:
a. move the upper body only.
b. move the whole body.
c. neither.
3) _____ When transferring weak persons from a bed to a wheelchair you should:
a. use your knees to brace against their knees to prevent their knees from buckling.
b. lock the wheelchair.
c. both.

Q14 - **True or False.** Write T (true) or F (false) for each of the following statements:

a. _____ When lifting heavy objects, it is best to lift quickly and turn by twisting the upper body.
b. _____ If you know your body, almost everything can be lifted alone.
c. _____ Being overweight doesn't affect lifting or transferring.
d. _____ Abdominal muscles support your spine when you lift.
e. _____ Good posture is also known as body alignment.
f. _____ The natural curves of the spine are the cervical, thoracic, and lumbar curves.
g. _____ The muscles in the hands are used to lift heavy objects.
h. _____ Bending from the waist prevents strains and fatigue.
i. _____ By using your body weight, not your back, when moving someone you will help prevent injury to yourself.

Q15 – Circle around the correct answer.

1- Name the muscles worked together when they cause a movement.
   a) synergists
   b) antagonists

c) fixator
d) insertion

2- What part of the human body act like fulcrum?
a) belly
b) tendon
c) joints
d) arm

3- Where you will find brachialis and brachioradialis muscles?
a) leg
b) thigh
c) trunk
d) arm

4- Which lever class is the typically of most joints of the human body?
a) class I
b) class II
c) class III

Q16 – From figures below (7.27a,b,c), work out the total stress on the lumbosacral disc, for weight of the torso is 750N and disc is at °40 (θ) in normal posture.

i. standing
ii. bending
iii. bending and lifting a 300N load. (hint: as shown in fig 7.27d find the components of stress. S_c is compressive stress parallel to the column and S_s is sheer stress perpendicular to the column)

i- forces on lumbosacral disc when standing. Torso weight and stress act the same point but in opposite direction.

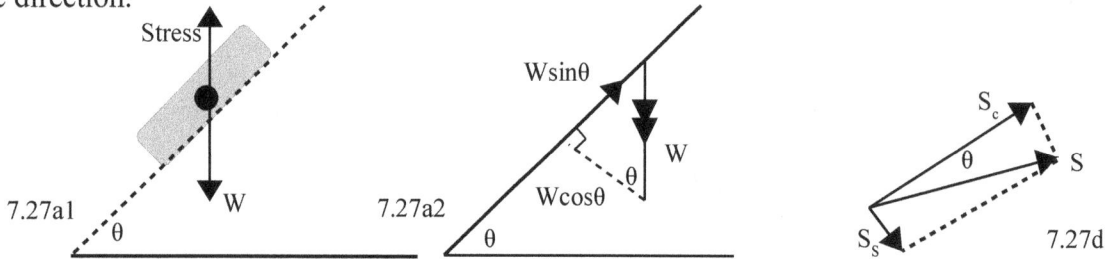

ii – force on lubrosacal disc when bending.   iii – force on lumbosacral disc when lifting

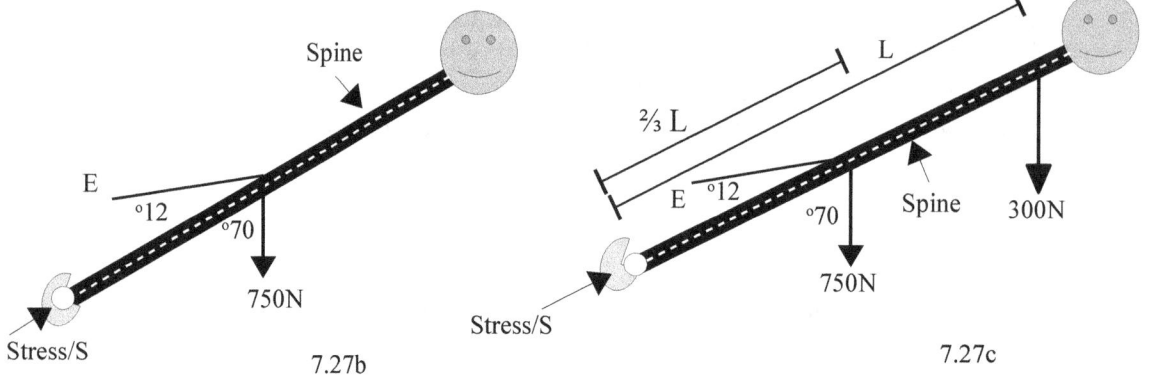

8 – Energy Expenditure

All living systems need energy to function. In animals, this energy is used to circulate blood, obtain oxygen, repair cells, and so on. Even at complete rest, the body needs energy to sustain its life functions. For example, a man weighing 68 kg resting will require around 75 cal per hour. We will discuss here how the basic energy forms that are found in the body and how they are produced and used by the body. How we take energy has biological and behavioural penalties, such as limited growth or loss of lean body mass, excessive accumulation of fat, physical limitations in performing certain tasks.

All living organisms require food. The types of food we need are four sugars, fat, carbohydrates and protein. All four have different nutritional values.

Sugars: All sugars are carbohydrates. Sugars are present in nature; plants make sugars through photosynthesis. There are many different types of sugars. They have 4 kilo calories per gram. The most commonly known sugars include: sucrose, glucose, fructose, and lactose.

- Glucose, fructose and sucrose are found in fruit and vegetables
- Lactose is found in milk
- Sucrose is what you find in your sugar bowl. All of these sugars occur naturally.

Sugars can be divided into two groups: the simple sugars (monosaccharides) and more complex sugars (disaccharides).

Monosaccharides - are the most basic units of biologically important carbohydrate. They are usually colourless, crystalline and water soluble solids. Some of them have a sweet taste. Examples of monosaccharides are fructose (laevulose), glucose (dextrose), sucrose (table sugar) , galactose, xylose and ribose.

Fructose was discovered by French chemist Augustin-Pierre Dubrunfaut (1797 – 1881) in 1847. Natural source of fructose includes fruits, vegetables and honey. Free fructose is absorbed directly by the intestine. When fructose is consumed in the form of sucrose, it is digested (broken down) and then absorbed as free fructose. When sucrose comes into contact with the cell membrane of the small intestine, the enzyme sucrase catalyse the cleavage of sucrose to yield one glucose and one fructose units. Then they are absorbed and enters the hepatic portal vein which direct them toward the liver.

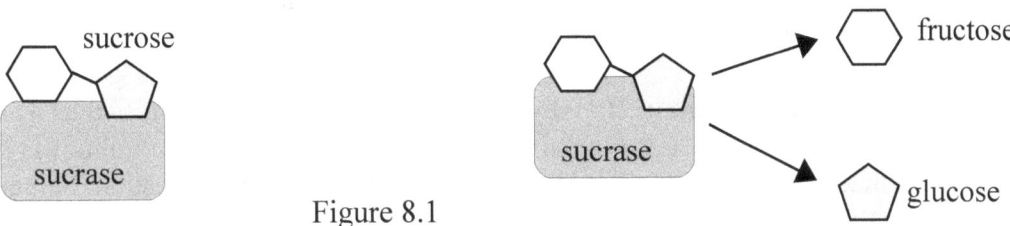

Figure 8.1

In our diet we should consume moderate amounts of fructose. After the process shown in figure 8.1, glucose enter into our bloodstream, the body will release a certain amount of insulin to control it. Fructose is processed in the liver. If too much fructose enters the liver, the liver can't process it fast enough for the body to use it. Instead of this liver will turn surplus fructose into fat and will enter them into the bloodstream as serum triglyceride.

Too much presence of triglycerides in the bloodstream is known as *hypertriglyceridemia*, which is a major cause of cardiovascular disease. The normal level of triglycerides should be about 145 mg/dL. Also fructose circumventing the appetite signals which will result in appetite regulating hormones will not trigger. This means one will still feel hungry after eating and so will result in obesity. High consumption of fructose can also lead to cell being insensitive to the effect of insulin which can lead to type 2 diabetes.

Glucose is one of the primary molecule which serve as an energy source for plants and animals. Glucose is absorbed directly into the bloodstream during digestion. This provides approximately 3.75 kcal of energy per gram. Use of glucose as an energy source in cells is via aerobic or anaerobic respiration. Glucose is oxidized in the body through glycolysis (metabolic pathway that converts glucose) $C_6H_{12}O_6$, into pyruvate, $CH_3COCOO^- + H^+$) and then in the reaction of citric acid cycle (TCAC) will yield energy mostly in the form of ATP. Some of the glucose is converted to lactic acid by astrocytes, which is then utilized as an energy source by brain cells.

Galactose is a type of sugar that is less sweet than glucose. In the human body, glucose is changed into galactose via hexoneogenesis to enable mammary glands to secrete lactose. Galactose is found in dairy products, sugar beets and mucilages and can also be obtained from a protein-free distillate of whey.

Xylose is a natural sugar that is found in some woody materials such as straw, pecan shells, cotton seed hulls, and corn cobs. It is also found in spinach, broccoli, berries and pears. Xylose is safe for use in foods. It is antibacterial and anti-fungal and contains natural healing agents.

Disaccharides - Disaccharides are carbohydrates that are created when two monosaccharides are joined. This happens when monosaccharides undergoes condensation reaction which eliminates water. Three common examples are sucrose, lactose and maltose.

Monosaccharides	Disaccharides
Glucose + Glucose	Maltose + H_2O
Glucose + Galactose	Lactose + H_2O
Glucose + Fructose	Sucrose + H_2O

Maltose is formed during the germination of certain grains, most notably barley where it is the source of the malt used in the manufacture of beer. It is less sweet than glucose, fructose or sucrose. It is formed in the body during the digestion of starch by the enzyme amylase, and is itself broken down during digestion by the enzyme maltase.

Lactose was discovered in milk in 1619 by Fabriccio Bartoletti, and identified as a sugar in 1780 by Carl Wilhelm Scheele. Lactose is the naturally occurring sugar found in milk. It is broken down when consumed into its constituent parts by the enzyme lactase during digestion. Some adults don't have the enzyme that metabolise lactose in their digestive system.

Sucrose is a white, odourless, crystalline powder with a sweet taste. In the humans body, sucrose is broken down into its constituent monosaccharides, glucose and fructose by enzymes sucrase or isomaltase glycoside hydrolases, which are located in the membrane of the micro villi lining duodenum. Sucrose provides a quick energy source and can provoke a rise in blood glucose upon ingestion. High level of sucrose consumption can have health risks, like tooth decay, in which oral bacteria streptococcus *mutans* which lives in plaque convert sucrose into acid that attacks tooth enamel. Another major cause of concern which is linked with high consumption of sucrose is the disease called *diabetes mellitus,* which cause the body to metabolise sugar poorly. This occurs when either:
1. the body attacks the insulin producing cells. Insulin is a hormone which helps metabolising sugar. This is called type 1 diabetes.
2. the body cells exhibit impaired response to insulin which is called type 2 diabetes.

When glucose builds up in bloodstream, it can lead to two health problems:
1. cells become starved in the short terms, as they can't get enough glucose.
2. frequent building up of glucose in the bloodstream will damage many of the body's organs, including the kidneys, eyes, nerves and/or heart.

Another medical condition linked with excess intake of sucrose is *gout*, acute inflammatory arthritis. The high level of insulin will prevent excretion of uric acid from the body, this will result in the concentration of uric acid in joint liquid and begins to precipitate into crystals.

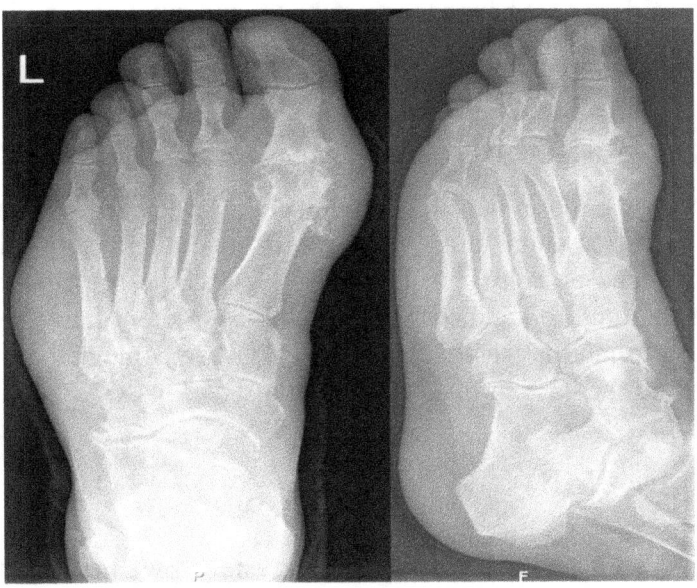

Figure 8.2 X-ray of gout in the left foot. The typical location is the big toe joint. Note the soft tissue swelling at the lateral boarder of the foot.

Fats: Fats are energetically the most concentrated of all sustenance materials. Fats consist of a wide group of compounds that are generally soluble in organic solvents and generally insoluble in water. Fats are triglyceride, triesters of glycerol and of several fatty acids. Fats also serve as energy stores for the body, containing about 37.8 kJ/g. They are broken down in the body to release glycerol and free fatty acids. The glycerol can be converted to glucose by the liver and hence used as a source of energy.

Vitamins A, D, E and K are fat soluble and they can only be digested, absorbed and transported with fats. Fat can also prevent diseases, by acting as a buffer for the host of diseases. When chemical or biological substance reaches to an unsafe level in the bloodstream the body will store it in the fat tissues and will be kept for metabolism at later time or remove it from the body via urination, excretion and sebum excretion.

There are two main types of fats. Saturated fats are solid at room temperature and normally found in the meat, butter, eggs, full fat milk and cheese. These types of fats will raise the cholesterol level which will increase the risk of many chronic diseases, such as stroke, heart diseases and certain cancer. Unsaturated fats are normally liquid at room temperature. They are mostly found in vegetables and oily fish like, salmon, sardines, pilchard and herring. Unsaturated fat contains certain fatty acids which our body cannot manufacture and so we need to get it from food. For example, one quarter of avocado contains 5g of unsaturated fat. Nuts like, cashew, Brazil, pecan and walnut and seed like, sesame, pumpkin and sunflower are also a good source of unsaturated fat.

Protein: Proteins are composed of long chains of amino acids. Amino acids are molecules that are composed of one type of organic compound called an amino group and another one called a carboxylic group. Proteins are essential nutrients for the human body for growth and maintenance. They are one of the building blocks of body tissue, and can also serve as a fuel source. Amino acids are found in meats, milk, fish and eggs, as well as in plant sources such as whole grains, pulses, legumes, soy, fruits, nuts and seeds.

Amino acids can be divided into three categories:
1. essential amino acids cannot be made by the body, and must be supplied by food.
2. non-essential amino acids are made by the body from essential amino acids or in the normal breakdown of proteins.
3. conditional amino acids are usually not essential, except in times of illness.

The amount of protein required in our diet determined in large part by overall energy intake, the body's need for nitrogen and essential amino acids, body weight and composition, physical activity level, individual's energy and carbohydrate intake, as well as the presence of illness or injury. If enough energy is not taken in through diet the body will use protein from the muscle mass to meet its energy needs, leading to muscle wasting over time. Active people and athletes normally require more protein intake due to increase in muscle mass and sweat losses as well as the need for body repair.

Basal Metabolic Rate (BMR)

Basal Metabolic Rate (BMR) is the rate of energy expended by animal or human bodies when at rest. This is the minimum amount of energy needed by the organism to perform essential functions such as breathing, heartbeat and blood circulation, synthesis of molecules at rest. Another way to describe this is energy that is just sufficient for the functioning of the body's tissues and vital organs. In order to keep a constant weight an individual must consume just enough food to provide for basal metabolism plus physical activities. This means that eating too little results in weight loss, while a diet in excess will cause an increase in body fat and therefore weight. The release, and using, of energy in this state is sufficient only for the functioning of the vital organs, the lungs, heart, kidneys, nervous system, liver, sex organs, muscles, intestine and skin.

About 70% of a human's total energy expenditure is due to the basal life processes within the organs of the body (see table). About 20% of one's energy expended comes from physical activity and another 10% from thermogenesis, or digestion of food (*postprandial thermogenesis*). All of these processes require an intake of oxygen along with coenzymes to provide energy for survival (usually from micronutrients like carbohydrates, fats, and proteins) and expel carbon dioxide, due to processing by the Krebs cycle.

BMR decreases with age and this is due to loss of lean body mass. BMR is measured when the person is awake and calm so its sympathetic nervous system isn't stimulated. Therefore, the energy consumed in various activities is usually quoted in Cal m^{-2}hr^{-1}. This rate is known as the *metabolic rate*. For the BMR, most of the energy is consumed in maintaining fluid levels in tissues through osmoregulation, and only about one-tenth is consumed for mechanical work, such as digestion, heartbeat, and breathing. The table-1 below show the BMR for various activities.

Table-1: Metabolic rate for selected activities

Activity	Metabolic Rate Cal m^{-2}hr^{-1}
Sleeping	35
Sitting	50
Standing	60
Walking	140
Running	600
Cycling	250

Note: 1000cal = 1Cal; 1cal = 4.18J; 1Cal/hr = 1.16W

Personal energy requirement = basic energy requirements + extra energy requirements

Basic energy requirements (BER) includes your basal metabolic rate(BMR) and general daily activities.

For every kg of body weight 1.3 Calories is required every hour. (An athlete weighing 60kg would require 1.3 × 24hrs × 60 kg = 1872 Calories/day)

Extra energy requirements (EER)

For each hours training you require an additional 8.5 Calories for each kg of body weight. (For a two hour training session our 60 kg athlete would require 8.5 × 2 hrs × 60 kg = 1020 Calories)

An athlete weighing 60 kg who trains for two hours would require an intake of approximately 2892 Calories (BER + EER = 1872 + 1020)

To obtain the total energy consumption per hour, we multiply metabolic rate with body surface area. To calculate the body surface area following empirical formula can be used:

$$A = 0.202 \times M^{0.425} \times H^{0.725} \qquad (8.1)$$

where, M is the mass of the person in kg and H is the height in meters.

Example: The surface area of a 75kg man who is resting with 1.60m height will have the surface area:

$$A = 0.202 \times (75)^{0.425} \times (1.60)^{0.725} = 1.78 \text{ m}^2$$

His metabolic rate at rest is (60 Cal/m^2-hr) × 1.78 = 107 Cal/hr. This metabolic rate at rest is called *Basal Metabolic Rate* (BMR).

The original work on BMR was done by two American scientists James Arthur Harris (1880 – 1930) and Francis Gano Benedict (1870 – 1957) when they published their findings in 1919. They showed that approximate value could be derived for BMR from body surface area, age and sex along with the amount of oxygen intake and exhale of CO_2.

The hypothalamus regulates the metabolism. The hypothalamus is a part of the brain that contains a number of small nuclei with a variety of functions. The hypothalamus controls body temperature, hunger, thirst, sleep, fatigue and circadian cycles (A cycle which is hard wired into the biology of persons and animals. Our brain wants to sleep when it gets dark and wake when it is light). All of these functions form a survival mechanism that causes us to sustain the body processes that RMR and BMR measures.

Harris-Benedict equation

This equation was the result of research carried out by James Arthur Harris and Francis Gano Benedict which was published in 1919 by the Carnegie Institution of Washington in the monogram "A Biometric Study of Basal Metabolism in Man". This includes keeping your heart beating, inhaling and exhaling air, digesting food, making new blood cells, maintaining body temperature and other metabolic functions in the human body. BMR may vary dramatically from person to person depending on genetic factors. There are people who will eat anything but never gain a gram of fat, they have inherited a naturally high BMR. We have lowest BMR when we are sleeping and not digesting any food.

The original Harris – Benedict equation:

for men, $$P = \left(\frac{13.7516m}{kg} + \frac{5.0033h}{cm} - \frac{6.7550a}{year} + 66.4730\right)\frac{kcal}{day} \quad (8.2)$$

for women, $$P = \left(\frac{9.5634m}{kg} + \frac{1.8496h}{cm} - \frac{4.6756a}{year} + 655.0955\right)\frac{kcal}{day} \quad (8.3)$$

where P is total heat produced while at rest, "m" is the mass of the body, "h" is height and "a" is the age of the person. Using the Harris-Benedict Equation, individuals can take a mathematical approach to weight loss. There are 3500 kilocalories in 1lb (0.45kg) of body fat. Using the Harris-Benedict Principle, if someone has a daily allowance of 2500 kilocalories, but he reduces his intake to 2000, then the calculations show a one pound loss every 7 days.

Example: A 45 years old man weighing 65kg and is 163cm tall. Calculate the BMR for this person and how much power he has per day?

$$P = (13.7516 \times 65) + (5.0033 \times 163) - (6.7550 \times 45) + 66.4730$$

$$= (893.854) + (815.5379) - (303.975) + 66.4730$$

$$= 1471.89 \text{ kcal/day or } 61.33 \text{ kcal/hour}$$

since,
$$1 kcal/h = 1.16W$$
therefore,
$$\text{power} = 61.33 \times 1.16 = 71.14W$$

During the last 100 years our living standard has changed and new types of food have emerged and so in 1984 Allan M. Roza and Harry presented the revised Harris Benedict equation.

Men $\quad P = (\dfrac{13.397m}{kg} + \dfrac{4.799h}{cm} - \dfrac{5.677a}{year} + 88.362) \dfrac{kcal}{day}$ $\quad\quad$ (8.4)

Women $\quad P = (\dfrac{9.247m}{kg} + \dfrac{3.098h}{cm} - \dfrac{4.330a}{year} + 447.593) \dfrac{kcal}{day}$ $\quad\quad$ (8.5)

These formulas are based on body overall mass, which does not take into account the difference in metabolic activity between lean body mass and body fat. Two formulas exist which take into account lean body mass, the Katch-McArdle formula, and Cunningham formula. However, the Cunningham formula is used to predict RMR (Rest Metabolic Rate) instead of BMR.

The Katch-McArdle Formula (BMR):

$$P = 370 + (21.6 \times LBM) \quad\quad (8.6)$$

The Cunningham Formula (RMR):

$$P = 500 + (22 \times LBM) \quad\quad (8.7)$$

LBM is the lean body mass in kilograms, it refers to the sum of the weight of body's bones, muscles and organs... basically the sum of everything other than fat in the body. There are two formulas:

The formula for lean body mass using James's method:

$LBM(\text{men}) = (1.1 \times \text{weight (kg)}) - 128 \times (\text{weight}^2 / 100 \times \text{height(m)})^2$ $\quad\quad$ (8.8)

$LBM(\text{women}) = (1.07 \times \text{weight (kg)}) - 148 \times (\text{weight}^2 / 100 \text{ height(m)})^2$ $\quad\quad$ (8.9)

The formula for lean body mass using Hume's method:

For men over the age of 16 years

$LBM = (0.32810 \times \text{body weight in kg}) + (0.33929 \times \text{height in cm}) - 29.5336$ $\quad\quad$ (8.10)

For women over the age of 30 years

$LBM = (0.29569 \times \text{body weight in kg}) + (0.41813 \times \text{height in cm}) - 43.2933$ $\quad\quad$ (8.11)

These formulas are trusted and highly scientific, based on various types of measurements of human body composition, including dual energy X-ray absorptiometry (DEXA). But every individual is different. For example some can have very dense bone, whereas the other person can have large internal organs.

The next most important thing is to find out the total daily caloric (the heat evolved by food when metabolised) requirement. This is normally found by multiplying your BMR with your level of activities using table-2

This calculation gives us the number of calories we burn in one day at our current level of activity. If you consume the exact amount of calories this should keep your weight on one level. The table-2 shows some of the daily activities.

Table 2

Activity	Factor
Occasional exercise	1.200
Exercise 1 to 3 days per week	1.375
Exercise 3 to 5 days per week	1.550
Exercise 6 to 7 days per week	1.725
Daily exercise plus physical job	1.900

BMR and Weight loss

If someone's BMR is 1760 Cal and his activity factor is 1.550 then his daily calorific requirement is 1760 x 1.550 = 2728 Cal/day. If you know your BMR and number of calories you need to burn, you can make a plan to lose your weight by setting a lower daily calories intake. As we have read that there are 9 calorie in every gram of fat. So to lose 500g of fat we need a deficit of 4500 calories in our diet. You will not actually lose 500g of weight but only 375g, because on average 75% of this is fat and the rest are lean tissue. Another way of losing this mass is to burn these calories by exercise or combination of both. Exercise becomes more important for sustained weight loss.

Normally you should not reduce your calorie intake by more than 20% below your daily calorie requirements at the beginning of your weight loss program and then gradually increase according to your need. Cutting calories too much slows down the metabolic rate, decreases thyroid output and causes loss of lean mass. Once you have achieved your target weight loss, you again need to slowly increase your calorie intake. For example, you have been consuming 2300 calories per day before weight loss program and you gradually reduced your intake to 1300 calories per day and have lost desired weight. Your metabolism is sluggish, an immediate jump to 2300 calories will cause a fat gain because your body has adopted to a lower calorie intake and sudden increase would create a surplus which body will store as fat. The best approach is to increase your daily calories over many weeks.

If you want to remain at the correct weight, you need to monitor your progress. Keep a track of your calorie intake, body weight and body fat percentage. You should then keep a good balance between calorie intake and intensity of your exercise.

Factors that affect basal metabolism?

Age: As we get older our basal metabolic rate becomes increasingly slower on average 2-3% per decade after the age of 21 years. This is mostly due to loss of lean muscle mass as you age and less you use more you will lose. That is why strength training is important throughout life which will help you to protect your muscle tissue as you age. This occurs also due to hormonal and neurological changes that happen as we age.

Growth: Most teenagers eat a lot but put on very light weight. This is because they are growing, and have a higher metabolism. Also infants and children have high energy demands per unit of their body weight as their growth requires more energy.

Hormones: Hormones help to regulate the metabolism. Some of the most common hormonal disorders are concerned with the thyroid. The thyroid hormones triiodothyronine (T3) and thyroxine (T4) are the main regulators of BMR. This gland secretes hormones to regulate many metabolic processes, including energy expenditure. If you have hypo or hyperthryoidism you have been affected by a hormone.

Hypothyroidism - or under-active thyroid and results in a lower metabolism. The metabolism slows because the thyroid gland doesn't release enough hormones. A common cause is the autoimmune condition Hashimoto's disease. It can be more difficult for individuals with this condition to lose weight. People with hypothyroidism should do strength training to boost metabolism. Some of the symptoms of hypothyroidism include unusual weight gain, lethargy, depression and constipation.

Hyperthyroidism - or over-active thyroid. In this case the gland releases more hormones than necessary which then speed up the metabolism. The most common cause of this condition is Graves' disease. Other symptoms of hyperthyroidism include, weight loss, nervousness, increased appetite and diarrhoea.

Some other hormones such as insulin, testosterone and human growth hormone increase the metabolic rate.

The other influential hormone is called Thyroxine Higher levels of Thyroxine lead to a greater BMR. 1 mg of Thyroxine will increase BMR by 1000 calories. Also Thyroxine release can be affected negatively through extreme dieting. These types of diets place the body into starvation mode, causing a down regulation of Thyroxine in an attempt to reduce BMR and thus preserve body weight - yet another cross against crash dieting!

Diet: Eating too few kilojoules encourage the body to slow the metabolism to conserve energy. This becomes a survival mechanism of our body. Most of the dieting programmes will lose up to 25% of your weight in the form of muscle, this will lower your metabolism. After the dieting period is over you will regain weight by increasing intake. So in the diet we should try to burn fat and protect our muscles.

Stress: Stress releases the hormones Adrenaline and Epinephrine, which have the effect of increasing cellular metabolism. But stress is not a good thing to encounter with regularity, so while it can positively affect metabolic rate it comes with other negative health issues.

Body Temperature: BMR increase with the increasing body temperature because of chemical reactions in the body speeds up with increasing body temperature. For every ½ °C increase in internal temperature of the body, the BMR increases by almost 7%. If it is either extremely hot or cold, your body will have to work a little harder to maintain the temperature, and that raises your BMR.

Exercise: Exercise can lead to a chronic increase in metabolism because an increase in lean muscle mass. Aerobic exercise does not have a chronic effect on resting metabolism. Cardiovascular exercise will increase daily energy expenditure. Physical exercise not only influences body weight by burning calories, it also helps raise your BMR by building extra lean tissue. During exercise, the body works harder to supply sufficient energy and nutrients to the muscles due to the increased work load placed on the body.

Caffeine: As coffee is one of the best source of antioxidants and it raises our BMR. People who drink coffee everyday have less chance of strokes, heart attacks, diabetes and Alzheimer's.

Gender: Men normally have a greater muscle mass than women, so they have a higher BMR.

Table 3

	Body energy expenditure / %
Liver	27
Brain	19
Skeletal muscle	18
Kidneys	10
Heart	7
Other organs	19

A person's metabolism varies with their physical condition and activity. Weight training can have a longer impact on metabolism than aerobic training. A decrease in food intake can lower the metabolic rate as the body tries to conserve energy. It is estimated that a very low calorie diet of fewer than 800 calories a day would reduce the metabolic rate by more than 10 percent. The metabolic rate can be affected by some drugs, such as antidepressant, which may produce weight gain. Antithyroid agents, drugs used to treat hyperthyroidism, bring the metabolic rate down to normal and restore euthyroidism. Scientists are focused on developing antiobesity drugs to raise the metabolic rate, such as drugs to stimulate thermogenesis in skeletal muscle. Studies of humans with 100+ year life spans have shown a link to decreased thyroid activity, their resulting lowered metabolic rate is thought to attribute to their increased life expectancy.

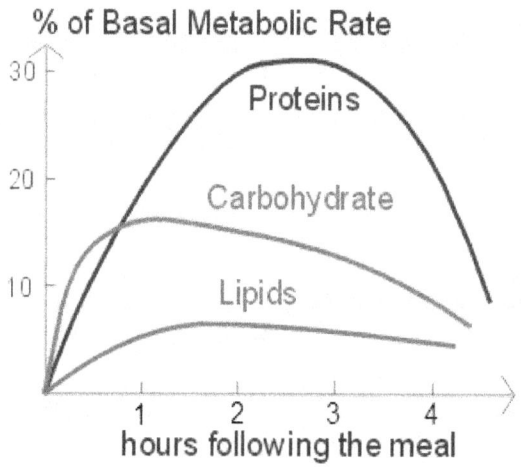

Figure 8.3 *Postprandial thermogenesis* increases in basal metabolic rate occur at different degrees depending on consumed food composition

BMI: The Body Mass Index (BMI) can be easily calculated to assess your body composition.

Your BMI = body weight in pounds x 705 ÷ (height in inches)²   (8.12)

<20 = underweight, 20-25 = Normal, 25-30 = overweight, >30 = Obese

Metabolic equivalent of Task

The Metabolic Equivalent of task, is a physiological measure expressing the energy cost of physical activities and is defined as the ratio of metabolic rate during a specific physical activity to a reference metabolic rate, set by convention to 3.5 ml $O_2 \cdot kg^{-1} \cdot min^{-1}$ or equivalently 1.00 $kcal \cdot kg^{-1} \cdot k^{-1}$ or 4.184 $kJ \cdot kg^{-1} \cdot h^{-1}$ or 58.2 W/m^2. It is commonly used in medicine to express metabolic rates measured during a treadmill test. However, METs is often estimated on the basis of other factors.

METs can be converted to kilocalories consumed per minute: kcal/min. = METs × body weight in kilograms ÷ 60. (8.13)

Oxygen consumption in litres per hour = METs × body weight in kilograms ÷ 0.21 (8.14)

There are two ways to express oxygen uptake (VO_2).

i) *Absolute* (Litter per minute) is used in the form that will yield rate of energy expenditure.

$$1L \text{ of } O_2 = 5 kcal$$

ii) *Relative* ($ml \cdot kg^{-1} \cdot min^{-1}$) is used to compare individuals of different body size and to mostly used to quantify aerobic fitness level.

Table 4

Physical Activity	MET
Light activity	<3
Sleeping	0.9
Reading and Writing	1.8
Walking slowly (0.75 ms^{-1}), Strolling	2.3
Walking fast (1.12 ms^{-1})	2.9
Moderate activity	3 to 6
Very light effort like slow cycling	3.0
Light exercise	3.3
Slow running (1.54 ms^{-1})	3.6
Cycling (4.45 m^{-1})	4.0
Rowing (1.12 ms^{-1})	5.5
Vigorous activity	>7
Jogging	7.0
Push ups, Jumping Jack	8.0
Fast running	8.2
Rope Jumping	10.0

The activities shown in table 4 are experimentally and statistically derived from a sample of persons and are indicative averages. The level of intensity at which a specific person performs a specific physical activity will deviate from the representative experimental conditions used for the calculation of the standard MET values. From a strictly scientific point of view, statistically estimated predictions, such as MET or Body mass index, are inaccurate when used for specific persons and actual energy consumption are highly dependent on physical and environmental factors such as adiposity, physical fitness level, cardiovascular health, or even ambient temperature. Even with these limitations, the MET concept represents a simple, practical, and easily understood procedure for expressing the energy cost of physical activities as a multiple of the resting metabolic rate.

How Does the Human Body Regulate Temperature?

People and other warm-blooded animals must maintain their body temperatures at a nearly constant level. The human body normally maintains a set body temperature. How does this happen? How do we generate additional heat when our body is too cold, and how do we cool off when we are too hot? Humans regulate heat generation and preservation to maintain internal body temperature or core temperature. The normal core temperature at rest varies between 36.5 and 37.5 °Celsius (°C), which is 97.7 to 99.5 °Fahrenheit (°F).

If the human body temperature-regulating mechanisms fail and the temperature rises to 45°C, the protein structures are irreversibly damaged. If body temperature increases significantly above normal, a condition known as *hyperthermia* occurs. If the body is exposed to around 70°C for few hours death is also inevitable. A fall in body temperature below about 28°C results in heart stoppage, is known as *hypothermia*. While core temperature is tightly regulated, skin temperature varies greatly in response to metabolism and the environment. The temperature on the surface of our skin is normally 10°C lower than core body temperature.

The temperature of the body is regulated by neural feedback mechanisms which operate primarily through the hypothalamus, a small portion of the brain that serves as the command center for numerous bodily functions, including the coordination of the autonomic nervous system. The hypothalamus contains not only the control mechanisms, but also the key temperature sensors. It serves as a thermostat, initiating physiological measures to lose or gain heat. Heat is conserved by constriction of the small arteries supplying blood to tiny capillaries near the surface of the skin, this reduces blood flow in the capillaries. To lose heat, small arteries dilate, increasing blood flow in the capillaries as a result heat is lost from them. As sweat is produced the evaporation of the water from the sweat cools the skin. At a skin temperature of 37°C sweating begins and increase as the skin temperature goes up to maintain a constant temperature. If skin temperature falls below 37°C body will respond via number of factors:

- decrease in radiation of heat from skin called Vasoconstriction.
- start shivering to increase heat production in the muscles.
- stop sweating.
- secret norepinephrine, epinephrine, and thyroxine to increase heat production.

But unlike a thermostat, which simply turns the heat on or off until a desired temperature is reached, the hypothalamus must regulate and fine-tune a complex set of temperature-control activities. It maintains this balance by controlling body fluids and salt concentrations, release of chemicals and hormones. In hot conditions Eccrine sweat glands under the skin secrete sweat which travels up the sweat duct, through the sweat pore and onto the surface of the skin. This causes heat loss via evaporative cooling; however, a lot of essential water is lost. The hairs on the skin lie flat, preventing heat from being trapped by the layer of still air between the hairs. This is caused by tiny muscles under the surface of the skin called arrector pili muscles relaxing so that their attached hair follicles are not erect. These flat hairs increase the flow of air next to the skin increasing heat loss by convection. When environmental temperature is above core body temperature, sweating is the only physiological way for humans to lose heat. Arteriolar vasodilation occurs. The smooth muscle walls of the arterioles relax allowing increased blood flow through the artery. This redirects blood into the superficial capillaries in the skin increasing heat loss by convection and conduction.

In cold condition sweat stops being produced. The minute muscles under the surface of the skin called erector pili muscles (attached to an individual hair follicle) contract, lifting the hair follicle upright. This makes the hairs stand on end which acts as an insulating layer, trapping heat. Arteriole carrying blood to superficial capillaries under the surface of the skin constrict, thereby rerouting blood away from the skin and towards the warmer core of the body. This prevents blood from losing heat to the surroundings and also prevents the core temperature dropping further. This process is called vasoconstriction. It is impossible to prevent all heat loss from the blood, only to reduce it. In extremely cold conditions excessive vasoconstriction leads to numbness and pale skin. Muscles can also receive messages from the thermo-regulatory center of the brain to cause shivering. This increases heat production as respiration is an exothermic reaction in muscle cells.

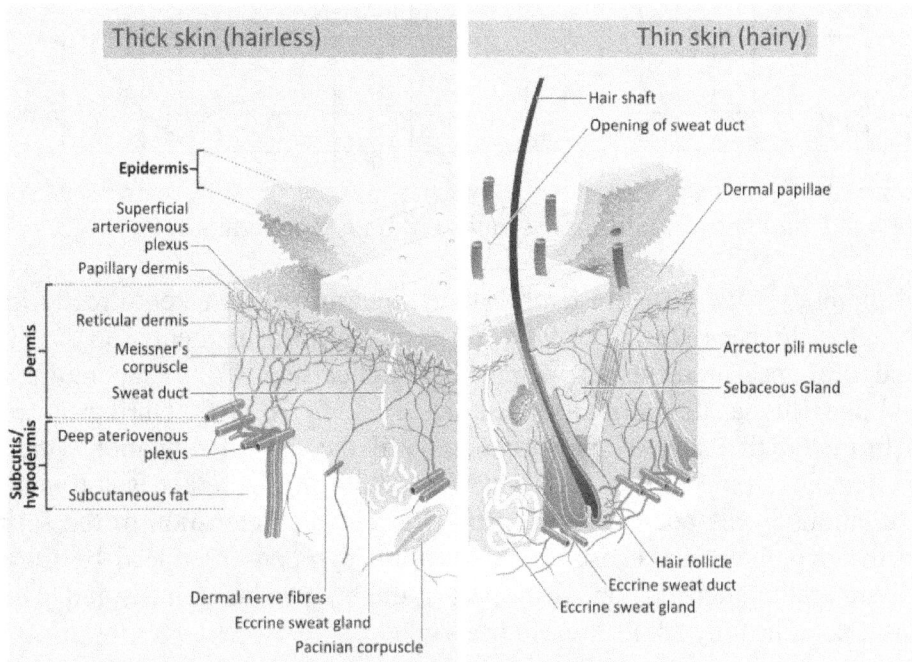

Figure 8.4 Skin layers

The hypothalamus works with other parts of the body's temperature-regulating system, such as the skin, sweat glands and blood vessels. The middle layer of the skin, or dermis, stores most of the body's water. When heat activates the sweat glands, these glands bring that water, along with the body's salt, to the surface of the skin as sweat. Once on the surface, the water evaporates. Water evaporating from the skin cools the body, keeping its temperature in a comfortable range. Thermoregulation is an important aspect of human homoeostasis. Most body heat is generated in the deep organs, especially the liver, brain, and heart, and in the contraction of skeletal muscles. If heat needs to be eliminated, this heat must first be conducted to the skin. For heat to flow from one region to another, there must be a temperature difference (temperature gradient ΔT) between the two regions. Therefore, the temperature of the skin must be lower than the internal body temperature. In a warm environment, the temperature of the human skin is about 35°C. In a cold environment, the temperature of some parts of the skin may drop to 27°C.

Daily variations in human body temperature

Normally our body temperature is highest at the end of our work day, at around about 4pm. Our body temperature starts to drop until we go to bed. It continues to drop until a couple hours before we rise from sleep at around 4am. The variation of the temperature against time of the day is shown in figure 8.5, which represent a particular individual. If you are working hard for longer hours then your temperature will remain high until you finish your shift.

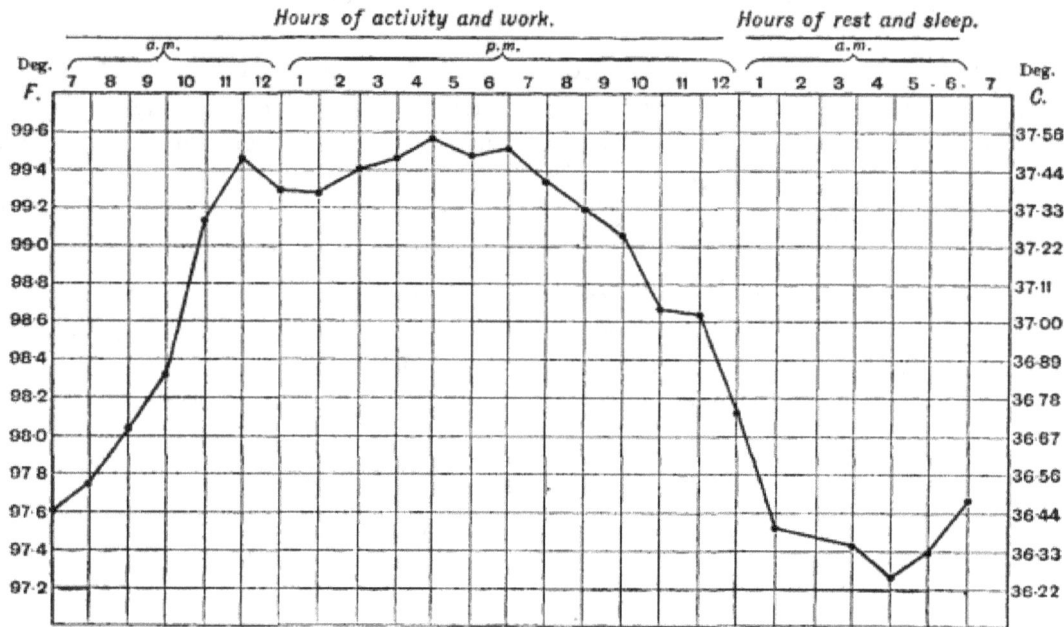

Figure 8.5 Chart showing diurnal variation in body temperature

Different parts of the body have different temperatures, measurements taken directly inside the body, are typically slightly higher (37.0 °C or 98.6 °C) than oral measurements (36.8° ± 0.4 °C or 98.2° ± 0.7 °C), and oral measurements are somewhat higher than skin temperature. In cold countries like Poland and Russia, the normal ranges are 36 °C to 36.9 °C. During the *follicular phase* (which lasts from the first day of menstruation until the day of ovulation), the average temperature in women ranges from 36.45 to 36.7 °C (97.6 to 98.1 °F). Another factor which brings about temperature variations in the body is fever. This is a regulated elevation of the set point of core temperature in the hypothalamus, caused by circulating pyrogens produced by the immune system. The temperature reading depends on which part of the body is being measured. The typical daytime temperatures among healthy adults are as follows:

- Temperature under the arm (axillary) is about 36.5 °C (97.7 °F)
- Temperature in the mouth (oral) is about 37.0 °C (98.6 °F)
- Temperature in the anus (rectum/rectal), vagina, or in the ear (otic) is about 37.5 °C (99.5 °F) (Source: Elert, Glenn (2005). *"Temperature of a Healthy Human (Body Temperature)"* The Physics Factbook)

Cooling of Human Body

First we will look at human body when it needs to cool when the surrounding ambient temperature is lower than the skin temperature. Even when we are sitting we lose heat at the rate of around 90 watts (77.39 Cal/hr). Normally, the temperature of the human skin is about 35°C. Sweating begins almost at a skin temperature of 37°C. The external heat transfer mechanisms are *radiation, conduction and convection*. This becomes a problem when the ambient temperature is higher than the body temperature. All three heat transfer mechanism will work against this heat loss by transferring heat in the body. The nature has given a solution to this problem that the body cools by the *evaporation of perspiration*.

At a temperature of above 45°C the evaporation process must supersede the transfer of heat into the body and give off enough energy to maintain a 90 watt outward energy flow rate.

Conduction

We have studied that most of the body heat is produced deep inside the body. To get rid of any excess heat it must be conducted to the skin. Heat will flow through solid object if there is a temperature difference between two faces or more scientifically *thermal conductance* is the quantity of heat that passes in unit time through a plate of *particular area and thickness* when its opposite faces differ in temperature by one kelvin.

The rate of heat loss through the air surrounding the skin (this is equivalent to the thickness of the block in the main conductivity equation, here its approximately the distance between skin and the ambient temperature) is given by:

$$H_{cd} = \frac{kA(T_1 - T_2)}{d} \tag{8.15}$$

This equation can be used for heat conducted by tissues and convection by air just above the skin.

where A is the surface area of the human body taken on average as 2 m² = 2 x 10⁴ cm² or cross-section area of the tissue, d is the thickness of the tissue or air, The constant k is the *coefficient of thermal conductivity*. K is usually given in units of $\frac{cal}{sec.cm.C^o} = \frac{Cal.cm}{m^2.hr.C^o}$

Example: A tissue (without blood) having cross sectional area of 2 m² and thickness of 2.5 cm has a value for coefficient of thermal conductivity k is 18 Cal.cm/m².hr.°C, with a temperature difference between the inner body and the skin of 3°C. Calculate the amount of heat lost. Also covert into watts.

$$H_{cd} = \frac{18 \times 2 \times 2}{3} = 24 \, Cal/hr$$

now,

$$1 Cal/hr = 1.163 W$$

$$H_{cd} = 24 \times 1.163 = 27.91 W.$$

If the body wants to conduct more heat, let say 200 Cal/hr, the temperature difference between the inner body and skin should be 17°C, and this high temperature could be fatal for the human body. Our human body possesses another efficient method of heat conduction from inside the body. The body uses the blood circulatory system to conduct heat as car engine uses water as a coolant. The circulatory system carries the heated blood near to the skin. The heat is then transferred to the outside surface by conduction. When body wants to keep the heat inside, blood carrying capillaries near the surface become constricted and the blood flow to the surface is greatly reduced.

Convection

When the skin is exposed to the air or some other fluid like water (as we may take showers in hot summer), heat is removed from it by convection currents. Convection denotes energy transfer between a surface and a fluid moving over the surface. The dominate contribution is due to the bulk motion of fluid particles.

The rate of heat removal is proportional to the exposed surface area and to the temperature difference between the skin and the surrounding air. Newton's law of cooling for heat transfer between a surface of arbitrary shape, area A and temperature T and a fluid is given by;

$$Q_C = h A (T_s - T_f) \qquad (8.16)$$

where Q_C is the heat lost, A is the area of the skin exposed to the air or fluid, T_s is the skin temperature and T_f is the temperature of the air or fluid, h is the coefficient of convection, which has a value that depends primarily on the prevailing wind velocity. The value of 'h' against air velocity is shown in figure 8.6. Initially the coefficient increases sharply with air velocity and then settles at higher speed.

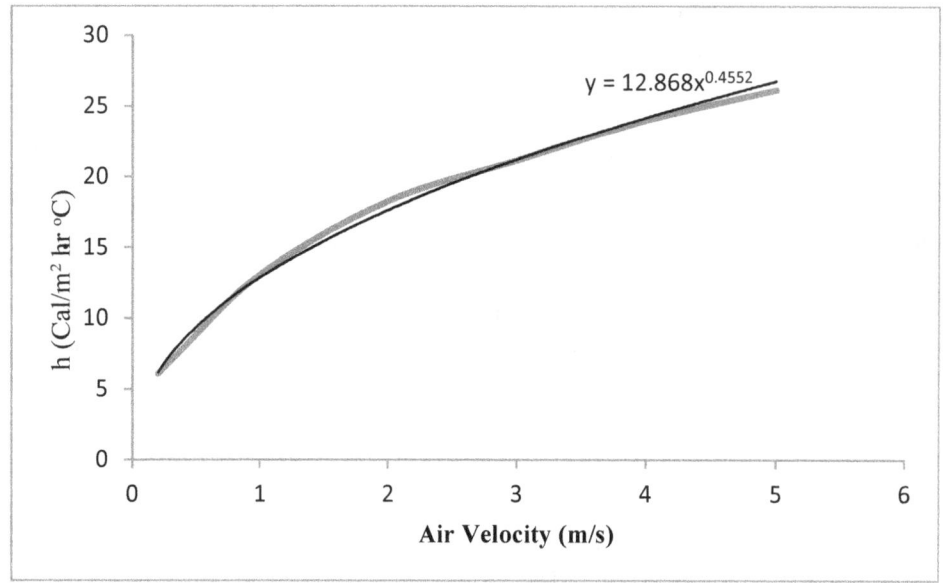

Figure 8.6 Convection coefficient as a function of air velocity

There are three modes of convection. If the motion of flow is generated by external force, such as a fan or pump, it is referred to as forced convection. If it is driven by gravity forces due to the temperature gradient, it is called natural (or free) convection. When the external means or not strong and gravitational forces are strong, the resulting convection heat transfer is called mix convection.

Example: Consider a naked person whose total surface area is 2m². Standing straight, with legs together and arms close to the body. If the air temperature is 25°C and the average skin temperature is 35°C, calculate the amount of heat removed by the air at 2.5 ms⁻¹

With the above body posture only 80% of total body will be exposed to the air, that is;

$$\frac{80}{100} \times 2 = 1.6 m^2$$

Next, we need to calculate the value of 'h' at the airspeed of 2.5 ms⁻¹. From the graph the equation is

$$h = 12.868 \, x^{0.4452} \qquad (8.17)$$

hence, $\qquad h = 12.868(2.5)^{0.4552} = 19.53$ Cal/ m² hr °C

$$Q = 19.53 \times 1.6 (35 - 25) = 312.48 \text{ Cal/hr}$$

Radiation

Radiant Cooling is based on the physical principle that bodies with varying temperatures exchange thermal radiation until an equilibrium is achieved. These are electromagnetic radiation emitted from the surface of a human body as a result of our body temperature. Since radiation can travel through a vacuum at the speed of light, it is not dependent on air movement to transfer large heat loads. When the core body temperature is too high, an area of the brain is signalled which causes certain internal changes. More blood is sent to the skin from within the body, and heat can then be released from the body as *radiation*. Heat is absorbed by radiation as well as given off, in the form of infrared rays. This is why human bodies can be "seen" by infrared cameras. When the environment is cooler than the body, the body is radiating heat. But if the environment is warmer than the body, the body will be absorbing heat.

When we go into the frozen food department of a supermarket, we feel cold. Why do we feel cool in the freezer section? It's because the temperature of the frozen compartment in relation to our skin is lower so our body releases its heat via radiation to the cooler surface. This is the reason we feel cold. Our body at rest transfer 50% of its via radiation.

The flux rate is a function of the temperature differential between 2 surfaces and the emissivity of the body materials (ability to absorb and radiate energy). If a hot object is radiating energy to its cooler surroundings the net radiation heat loss rate can be expressed as

$$Q_R = \epsilon \sigma A (T_h^4 - T_c^4) \tag{8.18}$$

where,

ϵ = *emissivity of the object*, 0.97 for human skin

σ = Stefan-Boltzmann constant 5.67×10^{-8} Wm^{-2} K^{-4} (6 Cal m^{-2}. hr. °C)

A = human body surface area – typically 2m^2

T_h = Skin temperature, T_c = temperature of nearby radiating surface.

Example: Calculate the amount of heat lost via radiation from a naked person in the convection problem.

$Q_R = 0.97 \times 2 \times 5.67 \times 10^{-8} (308^4 - 298^4)$

= 122.43J

or

$Q_R = 0.97 \times 2 \times 6 (35 - 25)$

= 116.40 Cal/hr

If the radiating surface is warmer than the skin surface, the skin is heated by radiation. In the extreme case, when the skin is illuminated by the sun or some other very hot object like a fire, the skin is heated intensely.

Evaporation of Perspiration

In a warm climate, when the surrounding temperature is high above the skin temperature, the radiation, conduction and convection all transfer heat into the body rather out. When water boils in a pan on the stove, we keep on providing energy, which will not raise the temperature of the water but will turn into steam. In general when water evaporates it requires heat energy. The amount of heat energy required is called the latent heat of vaporisation. Let say if the water is not in the pan and the stove is not that supplies the energy, the energy must come from some place else. When that water is not boiling on a stove evaporates, it gets the required heat energy from the nearby environment. Hence evaporation is a cooling process that cools the nearby environment.

Perspiration is primarily a means of thermoregulation which is achieved by the secretion of the eccrine glands. The perspiration or sweat rate of an adult human being can be up to 2-4 litres per hours (10-15 g min^{-1}. m^{-2}). At normal skin temperatures, the latent heat of vaporization for water is 0.580 Cal/g. Therefore, about 580 Cal of heat is removed for each liter of sweat that evaporates from the skin. The skin will start sweating almost precisely at 37°C and perspiration will increase with increasing skin temperature. About 600 g/day of insensate loss of moisture from the skin. To calculate the rate of heat lost per day is

Power = (600 g/day)(580 cal/g)(4.186 J/cal)(1day/24 hr)(1hr/3600s) = 17 W.

Evaporation of sweat from the skin surface has a cooling effect due to the latent heat of evaporation of water. Sweating causes a decrease in core temperature through evaporative cooling at the skin surface. As water evaporates from the skin, releasing energy absorbed from the body, the skin and superficial vessels decrease in temperature. Cooled venous blood then returns to the body's core and counteracts rising core temperatures. If the atmospheric humidity is low, this water evaporates easily. The heat energy needed to evaporate the water comes from our bodies. So this evaporation cools our bodies, which have too much heat. For the same reason taking shower on ourselves when it is hot feels good. Being wet during cold weather, however can excessively chill us because of this same evaporation effect. However, when it is very humid, our sweat does not evaporate as easily. With the body's primary cooling processes not working efficiently, we feel hotter. That is why a hot humid day is more uncomfortable than a hot dry day.

The evaporation cooling rate is given by

$$\frac{q}{t} = \frac{mL_v}{t} \qquad (8.19)$$

where,

m = mass of water in g at 100 °C
L_v = 539 cal/g at 100 °C
t = time in seconds

There are two situations in which the nerves will stimulate the sweat glands, causing perspiration: during physical heat and during emotional stress. In general, emotionally induced sweating is restricted to palms, soles, armpits, and sometimes the forehead, while physical heat-induced sweating occurs throughout the body.

Exercise 8

Q1 - In this activity, you will calculate your approximate metabolic rate in kilo calories per day. Please fill in the following information and then answer the questions below:

Height: _____ cm   Weight: _____ kg   Age _____ years

Now you're going to calculate your *Basal Metabolic Rate*, which refers to the energy burned when your body is at rest. Please fill in the appropriate numbers and calculate your BMR in the space below. The unit for BMR is *Calories per day*.

BMR for men = 66 + [13.7 x weight in kg] + [5 x height in cm] – [6.8 x age in year] = ...Cal/day

BMR for women = 655+[9.6 x weight in kg]+[1.8 x height in cm]-[4.7 x age in year] = ...Cal/day

Q2 - To calculate your *Active Metabolic Rate*, you need to select an appropriate conversion factor from the table below. This conversion factor is based on your activity level:

Activity Level	Description	Conversion factor
Sedentary	Sitting, standing, short walk	1.200
Light	Light exercise, slow running 1-3 days/week	1.375
Moderate	Moderate exercise, fast running 3-5 days/week	1.550
Strenuous	Heavy exercise, running for 1 hour 7 day/week	1.725

Active Metabolic Rate = BMR x conversion factor = ... Cal/day

Q3 - Most doctors and nutritionists suggest that we get 30% of our calories from fat, 50-60% of our calories from carbohydrate, and 10-20% of our calories from protein. You may like different diet, choose what works best for you. Please enter the percentages that you'd like to work with:

_____% of my calories should come from fat
_____% of my calories should come from carbohydrate
_____% of my calories should come from protein

Now, multiply these percentages by your *Active Metabolic Rate* to calculate the *number of calories* you'll need from fat, carbohydrate, and protein each day.

Calories from Fat =
Calories from Carbohydrates =
Calories from Protein =

Calculate the number of GRAMS of each nutrient you need each day. Use the conversion factors below:
• Protein and Carbohydrate = 4 Calories per gram   Fat = 9 Calories per gram

Grams of Fat per day = …..Grams of Carbohydrate per day = …….Grams of Protein per day = …..

Q4 - Eating 1 chocolate chip cookie (65 kcal) each day for 1 year how many pounds of weight you will put on? Take 1 pound of fat = 3500 kcal

Q5 - If you maintain your kcal intake and run an extra 5 miles per day, 5 days per week. How much weight would be your net weight loss. (take expenditure of energy of 90 kcal/mile). Compare your answer to Q4.

Q6 - Calculate your BMI using equation 8.12. Following conversion could be useful.

$$1 \text{ inch} = 2.54 \text{ cm}, \quad 1 \text{ kg} = 2.2 \text{ pounds}$$

Q7 - You have calculated your active metabolic rate (also known as estimated energy requirement EER) in Q2. To calculate your carbohydrate intake allowance you need to use the following equation.

$$A = EER \times 0.55 = \ldots\ldots\ldots\ldots\ldots \text{kcal/day of carbohydrate}$$

carbohydrate required (g per day) = A ÷ 4 kcal/day

Q8 - Use the table below to identify your daily activity and using the formula given below to calculate the amount of protein you need to take daily.

Activity level	Protein factor (g)
Low to Moderate	0.5
Endurance training	0.6-0.7
Strength training	0.8

Protein requirement (g per day) = Body weight in pounds x protein factor

Q9 - You have calculated your active metabolic rate (also known as estimated energy requirement EER) in Q2. To calculate your fat intake allowance you need to use the following equation.

$$F = EER \times 0.30 = \ldots\ldots\ldots\ldots\ldots \text{kcal/day of fat}$$

fat required (g per day) = F ÷ 9 kcal/day

Q10 - A fluid flows over a plane surface 1 m by 1 m with a bulk temperature of 50 °C. The temperature of the surface is 20 °C. The convective heat coefficient is 2000 W/m² °C. Calculate the rate of energy transfer.

Q11- Calculate the heat loss per square meter of skin surface at −30 °C in moderate wind (about 0.5m/sec, $h = 10$ Cal/m²-hr- °C).

Assume that the skin temperature is 30 °C

Q12 - Select the correct answer from the text below for each question

a) The process of keeping internal conditions constant is called...............
        static   regulation   homoeostasis   sweating   osmosis

b) _____ is the process of keeping the body at a constant temperature.
        Global regulation   Osmoregulation   glucoregulation   thermoregulation

c) Which organ is important in regulating body temperature? ....
        kidney   brain   bladder   skin

d) To help us cool down blood vessels near the surface of the skin can swell or dilate. This is called....
        vasoconstriction   veinodilation   vasodilation   varicoseveins

e) If the body is too hot, glands under the skin secrete sweat onto the surface of the skin in order to increase heat loss by ....
        condensation   evaporation   vaporisation   liquidation

Q13 - Select the correct option from the questions below

a) When a person's body temperature is high, more sweat is produced. This cools the body as it...

i- evaporates
ii- condenses
iii- freezes

b) Which word best describes sweat glands and muscles?

i- effectors receptors
ii- processors
iii- receptors

c) The part of the brain which monitors blood temperature is called the...

i- cortex.
ii- hypothalamus.
iii- cerebrum.

d) Which statement best describes what happens as the body temperature rises?
i- increased blood flow to skin reduces energy loss
ii- reduced blood flow to skin reduces energy loss
iii- increased blood flow to skin increases energy loss

e) Which of the following words describes the body temperature falling below 35 °C?
i- hypoglycaemia hypothermia
ii- hypothermia
iii- hypotension

Q14 – What are those two things which human body does to keep cool.
  i. Exercise, Vasodilation
  ii. Vasodilation, Sweating
  iii. Sweating, Vasoconstriction

Q15 – What is the role of insulin in human body

  i. turn glucose into glycogen

  ii. maintain body temperature

  iii. clean blood

Q16 – When listed on nutritional information tables, fats are generally divided into six categories: total fats, saturated fatty acid, polyunsaturated fatty acid, mono-unsaturated fatty acid, dietary _____, and trans fatty acid.

  i. atherosclerosis

  ii. cholesterol

  iii. hypertension

Q17 – Glycogen can be rapidly converted to _____ when energy is required for sustained, powerful contractions.

  i. fructose
  ii. sucrose
  iii. glucose

Q18 – Muscles also keep a storage form of glucose in the form of _____.
  i. glycogen

  ii. sucrose

  iii. carbohydrate

Q19 – How does skin help to regulate the body temperature?

Q20 – The secretion of which substance helps to control the the body's temperature.

Q21 – Size matters for marathon runners. Big athletes produce more heat and find it harder to keep cool. Shape matters too - a tall, thin runner has fewer problems keeping cool than a short, tubby runner of the same body mass. A 65 kg athlete running a marathon in 2 hours 10 minutes in reasonably dry conditions can avoid overheating at air temperatures up to 37 °C, but in humid conditions the same level of performance is possible only at temperatures below about 17 °C.

(a) Explain how athletes produce heat when they run.

(b) Why does a 'tall, thin runner have fewer problems keeping cool than a short, tubby runner of the same body mass'?

(c) Explain why runners are more likely to overheat in humid conditions.

(d) Describe how the body responds to a rise in core body temperature.

9 – The Human Eye and Vision System

Human Vision

Our brain is a very complex organ in the human body. Brain has approximately 10 billion cells and is divided into 4 lobes. Visual processing centre of the brain is in *occipital lobe* which contains most of the anatomical region of the visual cortex. The main visual cortex is Brodmann area 17, commonly called V1(visual one). V1 is also called the striate cortex as it can be identified by a large strip of myelin. Of the 4 lobes, the parietal, temporal and occipital lobes are involved in vision, whereas frontal lobe is for higher thoughts.

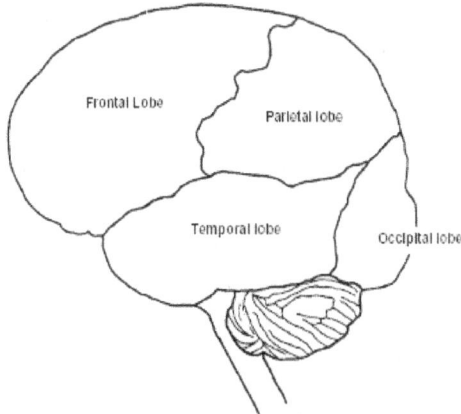

Figure 9.1 Lobes of brain

There are two main streams or pathways for vision as each V1 transmits information to two primary pathways, called the dorsal stream and the ventral stream

1. The VENTRAL pathway begins with V1, goes through visual area V2, then through visual area V4, and to the inferior temporal cortex (Precedes from occipital to temporal lobe). This is associated with form recognition and object representation. It is also associated with storage of long term memory.
2. The DORSAL pathway begins with V1, goes through Visual area V2, then to the dorsomedial area and Visual area MT (also known as V5) and to the posterior parietal cortex. This is associated with motion, representation of object locations, and control of the eyes and arms, especially when visual information is used to guide saccades(the simultaneous movement of both eyes in the same direction)

There is a visual cortex in each hemisphere of the brain. The left hemisphere visual cortex receives signals from the right visual field and the right visual cortex from the left visual field. The ventral stream is critical for visual perception whereas the dorsal stream mediates the visual control of skilled actions. As visual information passes forward through the visual hierarchy, the complexity of the neural representations increases. Whereas a V1 neuron may respond selectively to a line segment of a particular orientation in a particular retinotopic location, neurons in the lateral occipital complex respond selectively to complete object (e.g., a figure drawing), and neurons in visual association cortex may respond selectively to human faces, or to a particular object. Along with this increasing complexity of neural representation may come a level of specialization of processing into two distinct pathways: the dorsal stream and the ventral stream.

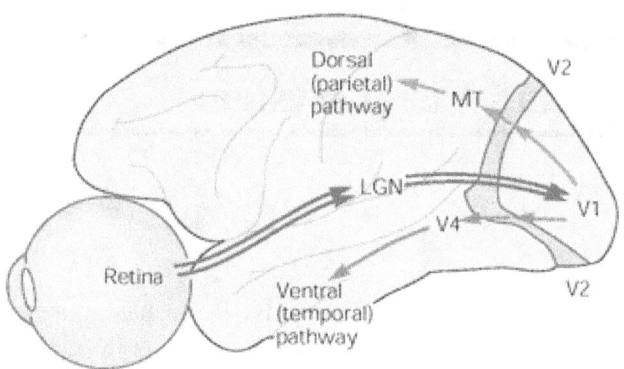

Figure 9.2 The pathway of Dorsal and Ventral originate from visual cortex

Nerve cells are called neurones. A neurone has a cell body with extensions leading off it. There are roughly 10^{10} neurons and there are roughly 104 connections per neuron in the brain. Each neuron has inputs, some processing and then information transmitted through the axon. Neurones have specialized projections called dendrites and axons. Numerous *dendrons* and *dendrites* provide a large surface area for connecting with other neurones, and carry nerve impulses towards the cell body. The one dendron is connected to axon which carries the nerve impulses away from the cell body. Most neurones have many companion cells called Schwann cells, which wrap their cell membrane around the axon many times in a spiral to form a thick insulating lipid layer called the myelin sheath. Information from one neuron flows to another neuron across a *synapse*. The synapse contains a small gap separating neurons. A nerve is a discrete bundle of several thousand neurone axons. Figure 9.3

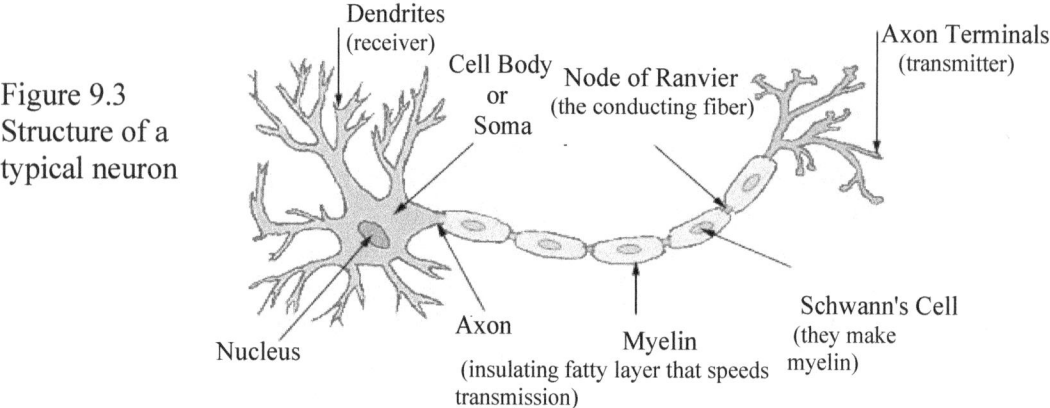

Figure 9.3 Structure of a typical neuron

Some other parts of the brain important for visual system:
  i. AXON - the primary transmission line of the long part of the neuron that conducts the signal. They vary in length from millimetre to meter. They take information away from the cell body.
  ii. DENDRITES - Dendrites are the branched projections of a neuron that act to conduct the electrochemical stimulation received from other neural cells to the cell body, or soma, of the neuron from which the dendrites project. Dendrites bring information to the cell body.
  iii. SYNAPSE - the method of communication between neurons. When an impulse reaches a synapse causes it to release neurotransmitter, which diffuse across the gap and triggers an electrical impulse in the next neuron.
  iv. NODES OF RANVIER - these nodes are areas where the action potential is amplified using a high density of sodium Na^+ ions which are passed along the axon.
  v. ACTION POTENTIAL - electrical signal transmission, sent on the axon
  vi. GRADED POTENTIAL - change in potential of neuron caused by absorption of neurotransmitters.

Most axons carry signals in the form of action potentials, which are discrete electrochemical impulses that travel rapidly along an axon, starting at the cell body and terminating at points where the axon makes synaptic contact with target cells. The defining characteristic of an action potentials is that it is "all-or-nothing" every action potential that an axon generates has essentially the same size and shape. This characteristic allows action potentials to be transmitted from one end of a long axon to the other without any reduction in size. There are, however, some types of neurons with short axons that carry graded electrochemical signals, of variable amplitude.

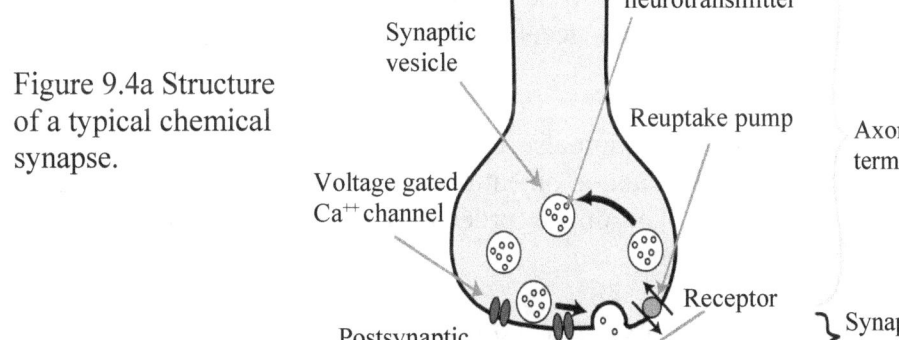

Figure 9.4a Structure of a typical chemical synapse.

When an action potential reaches a presynaptic terminal, it activates the synaptic transmission process. This rapidly open calcium ion channels in the membrane of the axon, allowing calcium ions to flow inward across the membrane. The resulting increase in intracellular calcium concentration causes vesicles (tiny containers enclosed by a lipid membrane) filled with a neurotransmitter chemical to fuse with the axon's membrane and empty their contents into the extracellular space. The neurotransmitter chemical then diffuses across to receptors located on the membrane of the target cell. The neurotransmitter binds to these receptors and activates them. Depending on the type of receptors that are activated, the effect on the target cell can be to excite the target cell, inhibit it, or alter its metabolism in some way. This entire sequence of events often takes place in less than a thousandth of a second. Afterward, inside the presynaptic terminal, a new set of vesicles are moved into position next to the membrane, ready to be released when the next action potential arrives.

Axons allow for the conduction of information from one part of the body to another. Ion channels produce and move an action potential through the cell. There are two types of channels, voltage-gated ion channel and ion channel pumps. Neurons send impulses through an action potential. This takes place from the dendrite and all the way to the axon ends. The first step is sending a signal to an axon called "depolarization". A neuron will fire an action potential when depolarization occurs at -55 mV (milivolts). An action potential is a change of voltage within the axon. In other words, the negative state off the inner axon turns positive when the impulses comes by. This happens by the use of a sodium and potassium pump when ion channel pumps use energy to actively transport ions from one side to another.

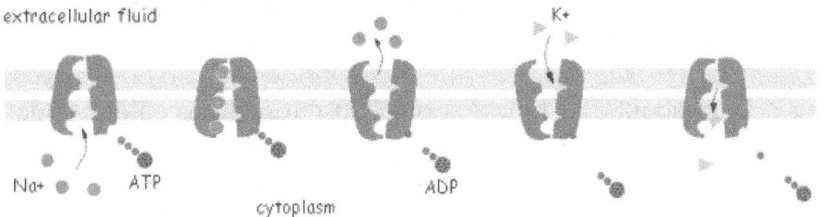

Figure 9.4b Sodium Potassium pump

Sodium Na⁺ ions surrounds the axon with a positive charge, while the potassium K⁺ ions are within the axon see figure 9.4. As an impulse enters at the axon hillock, the sodium, potassium pump puts positive sodium into the axon while it puts negative potassium out of the axon. The exchange of ions between the membrane of neurons produces an action potential. Repolarization, when sodium ions enters a negative neuron (the inside of the neuron is always negative) it becomes positive due to the positive charge in the sodium and the neuron gets depolarized. Due to this positive impact of the sodium the potassium channels rush into the cell and this process closes the sodium channels. At this point,the action potential gets forwarded back to -70 mV, which simply means that the action potential has reached to the process of "repolarization". When potassium channels stay open for too long it reaches to the level of hyper-polarization, since the action potential goes past -70 mV.

As more sodium enters the potential of the impulse changes from -70 mV to +40 mV, (a difference of 110 mV) This change is called an action potential. This spike in voltage moves down the axon. The duration of the spike tends to be on the order of 1 millisecond. The 'spike rate' indicates something about activity of the neuron.

As the impulse leaves the axon, it is reverted to a normal state which is called the resting potential. At the end of the axon, the impulse or stimulus enters the synapses and is called a post synaptic potential. From here, the impulse is transferred into neurotransmitters, some of which are chemicals called epinephrine and dopamine. These neurotransmitters flow into the fluid filled gap called the synaptic cleft and enter the dendrites. And, again, the process is repeated.

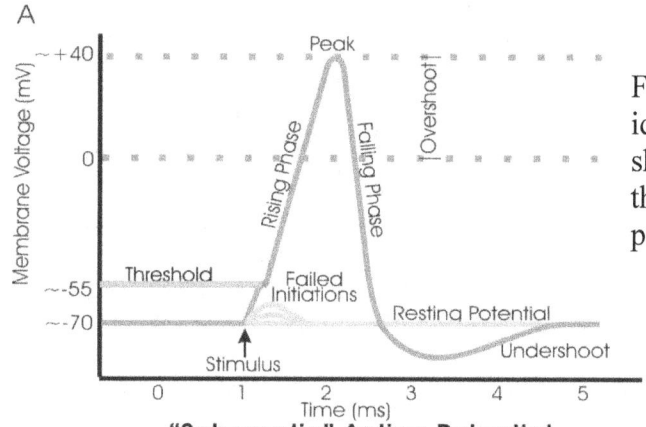

Figure 9.5 A view of an idealised action potential shows its various phases as the action potential passes a point on a cell membrane.

HUMAN VISUAL SYSTEM

Two eyes
Optic nerve: 1.5 million bers per eye (each ber is the axon from a neuron)
125 million rods
6 million cones

The eye can deal with a large amount of variation in illumination: approximately 14 orders of magnitude. At the low end of the scale, a rod can respond to a single photon. Typically it takes at least 6 photons for the brain to distinguish the signal from the noise.

Visible light ranges from about 400 nm to about 700 nm and the colour is determined by the wavelength. Most light sources, however, output light that is a mixture of several wavelengths, normally with one wavelength dominating. The colours of the spectrum that are most visible to the human eye are violet (390–430), blue-violet (460–480), cyan, green (490–530), yellow (550–580), orange (590–640), and red (650–800).

Human visual system enables organisms to process visual details. Our visual system is a complex network which deals with reception of light; the building of a binocular perception from a pair of two dimensional projections; assessing distances to and between objects etc.

Eye Anatomy

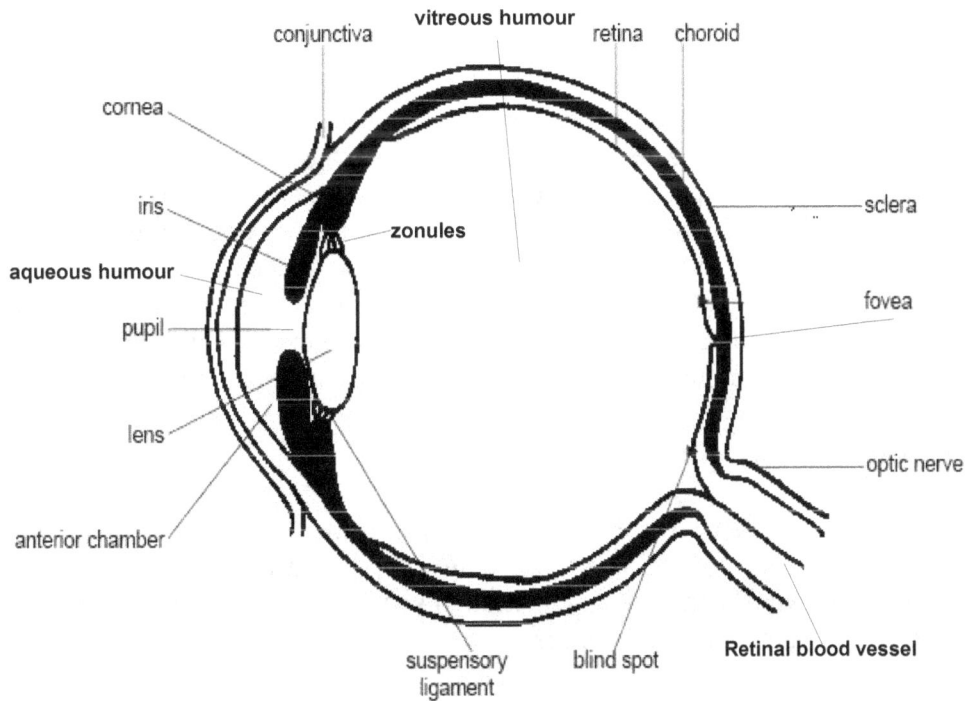

Figure 9.6 Diagram of eye anatomy

1- The cornea is the transparent front surface of the outer eye that covers iris, pupil and anterior chamber. Because there are no blood vessels in the cornea, it is normally clear and has a shiny surface and is a very sensitive part of the body-there are more nerve endings in the cornea than anywhere else in the human body. The cornea has a diameter of about 11.5 mm and a thickness of 0.5–0.6 mm in the cent and 0.6–0.8 mm at the periphery. The functions of the cornea are, to acts as a barrier preventing germs, dirt and other harmful material from entering the inner eye. The cornea also acts as the eye's outermost lens. Cornea provides up to 75 percent of the eye's total focusing power, because light first strikes the cornea which then refracts the incoming light onto the lens. The refractive power of the cornea is 43 dioptres, whereas the refractive index of the cornea is 1.376.

The human cornea has five layers.
i. Corneal epithelium- about 5-6 cell layers thick fast growing and quickly regenerates when the cornea is injured and is kept moist with tears.

ii. Bowman's membrane lies just beneath the epithelium is a condensed layer of collagen mainly type I collagen fibrils. Because this layer is very tough and difficult to penetrate, protects the corneal stroma (plates of collagen fibrils in the cornea) . This layer is eight to fourteen micrometer thick.

iii. Stoma (also known as *substantia propria*) is the thickest, transparent middle layer. It is composed 200 of tiny collagen fibrils that run parallel to each other with sparsely distributed interconnected keratocytes cells which are responsible for general repair and maintenance. Each layer is 1.5 to 2.5 micrometer. This special formation of the collagen fibrils gives the cornea its clarity.

iv. Descement's membrane (also known as *posterior limiting membrane*) lies between the stroma and the endothelium is a thin cellular layer. This layer is composed mainly of collagen type IV fibrils, less rigid than collagen type I fibrils, and is around 5-20μm thick, and this thickness is age dependent. This layer pumps water from the cornea, keeping it clear. If damaged or diseased, these cells will not regenerate.

v. Corneal endothelium consists of mitochondria-rich cells. These cells are responsible for regulating fluid and solute transport between the aqueous and corneal stromal compartments. Tiny vessels at the outermost edge of the cornea provide nourishment, along with the aqueous and tear film. The cells of the endothelium do not regenerate. Instead, they stretch to compensate for dead cells which reduces the overall cell density of the endothelium and has an impact on fluid regulation. If the endothelium can no longer maintain a proper fluid balance, stromal swelling due to excess fluids and subsequent loss of transparency will occur and this may cause corneal edema thus impairing the image formed.

Diseases and disorders affecting the cornea

Keratitis- This is an inflammation and ulceration of the cornea due to infection after bacteria or fungi enter the cornea after an eye injury normally from wearing contact lenses. The normal symptoms are pain, corneal discharge and reduction in visual clarity. It's not very serious and can be treated with antibiotic or anti-fungal eye drops.

Keratoconus- is a progressive disease and arises when the middle of the cornea thins and gradually bulges outward, forming a rounded cone shape. This abnormal curvature changes the cornea's refractive power, producing moderate to severe distortion (astigmatism) and usually near sightedness. Keratoconus may also cause swelling and a sight-impairing scarring of the tissue and eventually loss of vision.

Keratoconus stems from one of several possible causes:

- inherited.
- Eye injury (e.g. excessive rubbing your eyes or long-term use of hard contacts)
- Eye diseases, such as retinitis pigmentosa, retinopathy of prematurity, vernal kerato conjunctivitis
- Systemic diseases, like Down syndrome, osteogenesis imperfecta, Addison's disease, Leber's congenital amaurosis.

Special corneal testing called topography provides the doctor with detail about the cornea's shape and is used to detect and monitor the progression of the disease. A pachymeter may also be used to measure the thickness of the cornea.

This condition can be corrected by wearing special glasses or soft contact lenses. If astigmatism gets worse then you need to wear rigid gas permeable contact lenses and problem will normally be stabilised after a few years. In some cases the cornea will eventually become too scarred or will not tolerate a contact lens. If either of these problems occur, a corneal transplant may also be needed.

Ocular Herpes- Also called Herpes of the eye, is a recurrent viral infection of the eye that is caused by the herpes simplex virus I (HSV-I) but it can also result from the sexually transmitted herpes simplex II virus (HSV II) that causes genital herpes.

Herpes produce a painful sore on the surface of the cornea and, in time, the inflammation can spread deeper into the cornea and eye. Prompt treatment with anti-viral drugs helps to stop the herpes virus from multiplying and destroying epithelial cells. However, the infection may spread deeper into the cornea and develop into a more severe infection called stromal keratitis, which causes the body's immune system to attack and destroy stromal cells. Recurrent episodes of stromal keratitis can cause scarring of the cornea, which can lead to loss of vision and possibly blindness.

Herpes Zoster (Shingles)- This infection is produced by the varicella-zoster virus, the same virus that causes chickenpox. After having chickenpox, After the initial exposure, herpes zoster lies dormant in certain nerve fibres. It may become active as a result of many factors such as: ageing, stress, suppression of the immune system, and certain medications.

Because of the layout of the nerves that herpes zoster resides in, it only affects one side of the body. It begins as a rash that lead to blisters or lesions and sores on the skin. When the nerve branch that supplies the eye is involved, the forehead, nose, and eyelids may also be affected. Sores on the nose are a key signal of possible eye involvement. This acute painful phase usually lasts several weeks.

When effected, the doctor will perform a thorough examination with a slit lamp microscope and an ophthalmoscope. Visual acuity and intraocular pressure are also monitored. Herpes zoster is treated with anti-viral, pain and anti-inflammatory medications.

Map-Dot-Fingerprint Dystrophy- This dystrophy occurs when the epithelium's basement membrane develops abnormally, the epithelial cells cannot properly adhere to it. This, in turn, causes recurrent epithelial erosions, in which the epithelium's outermost layer rises slightly, exposing a small gap between the outermost layer and the rest of the cornea. As this membrane that separates the epithelium and stroma grows irregularly (thicker in some places, thinner in others), the findings in the cornea appear, resembling maps, dots and small fingerprints.

Map-dot-fingerprint dystrophy usually affects adults between the ages of 40 and 70, or children as a result of heredity. It is usually painless and causes no vision loss, and sometimes clears up without treatment. The cornea's normal curvature may be altered causing astigmatism and nearsightedness. Treatment may include an eye patch, eye drops, and ointments.

Fuchs' Dystrophy- Fuchs' dystrophy is the gradual deterioration of endothelial cells for no apparent reason. The endothelium functions as a pump mechanism, constantly removing fluids from the cornea to maintain its clarity. Patients gradually lose these endothelial cells as the dystrophy progresses. Once lost, the endothelial cells do not grow back, but instead spread out to fill empty spaces. The pump system becomes less efficient, causing corneal clouding, swelling and eventually, reduced vision, haze and small blisters on the corneal surface may also appear.

As a slowly progressing disease, signs of Fuchs' dystrophy begin to appear in people 30 to 40 years of age, but it doesn't normally affect vision until about 20 years later. It is more common in women than men. An early sign of Fuchs' dystrophy includes awakening with blurred vision that gradually clears up during the day. As the disease worsens, swelling becomes more consistent and vision blurs.

Fuchs' cannot be cured; however, with certain medications, blurred vision resulting from the corneal swelling can be controlled. Salt solutions such as sodium chloride drops are often prescribed to draw fluid from the cornea and reduce swelling. Another simple technique that reduces moisture in the cornea is to hold a hair dryer at arm's length, blowing air into the face with the eyes closed. This technique draws moisture from the cornea, temporarily decreases swelling, and improves the vision.

Pink Eye (*Conjunctivitis*)- This term describes a group of diseases that cause swelling, itching, burning, and redness of the conjunctiva, the protective membrane that lines the eyelids and covers exposed areas of the sclera. The three most common types of conjunctivitis are viral, allergic, and bacterial. Each requires different treatments. With the exception of the allergic type, conjunctivitis is typically contagious.

Viral conjunctivitis: Watery discharge, Irritation, Red eye and Infection usually begin with one eye but may spread easily to the fellow eye

Allergic conjunctivitis: Usually affects both eyes, Itching, Tearing and Swollen eyelids

Bacterial conjunctivitis: Stringy discharge that may cause the lids to stick together, especially after sleeping, Swelling of the conjunctiva, Redness, Irritation and/or a gritty feeling and usually affects only one eye but may spread to the fellow eye.

At its onset, conjunctivitis is usually painless and does not adversely affect vision. The infection will clear in most cases without requiring medical care. But for some forms of conjunctivitis, treatment will be needed. If treatment is delayed, the infection may worsen and cause corneal inflammation and a loss of vision. For the allergic type, cool compresses and artificial tears sometimes relieve discomfort in mild cases. In more severe cases, non-steroidal anti-inflammatory medications and antihistamines may be prescribed. Some patients with persistent allergic conjunctivitis may also require topical steroid drops. Bacterial conjunctivitis is usually treated with antibiotic eye drops. Viral conjunctivitis usually resolves within 3 weeks

2- The Conjunctiva is the thin transparent membrane lining the eyelid. It begins at the outer edge of the cornea, covering the visible part of the sclera, and lining the inside of the eyelids. It is reflected in the upper and lower fornices onto the anterior surface of the eyeball. It fuses with the cornea along the conjunctival limbus forming a circular 'opening' for the cornea. The conjunctival limbus are situated about 1 mm anterior to the edge of the corneal limbus (where the transparent cornea stops and the opaque sclera starts).

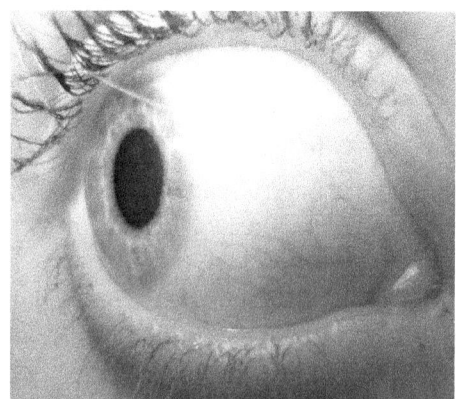

Figure 9.7 Image of a human eye clearly showing the blood vessels of the conjunctiva

The conjunctiva is composed of three sections.
- the palpebral - which lines the under-surface of the eyelids and is moderately thick
- the bulbar - which covers the front, external eyeball, is very thin and movable, easily slide back and forth over the front of the eyeball it covers.
- the fornix - which forms the junction between the eyelid and eyeball.

Because the bulbar is clear, it is easy to see underlying blood vessels. Within the bulbar, conjunctivae are "goblet cells" that secrete an important component of the pre-corneal tear layer, which protects and nourishes the cornea. It is a highly vascular membrane. It is nourished by tiny blood vessels that are nearly invisible to the naked eye. The conjunctiva also secretes oils and mucous that moisten and lubricate the eye.

Diseases and disorders affecting the conjunctiva

Pinguecula is a benign, yellowish growth that forms on the conjunctiva and is often referred to as fatty degeneration of the conjunctival tissues, because nearly transparent collagen fibres of the conjunctive degenerate and are replaced by thicker, yellowish, more durable fibres containing calcium crystals. Typically, pinguecula are located on either side of the cornea but is more common on the nasal side. Pinguecula are very common in adults who spend a lot of time in sunny and/or windy environments. Most recent studies indicate that pinguecula may be stimulated by ultraviolet light from the sun.

There is no effect on vision from a pinguecula, but may cause irritation if it becomes elevated. In rare cases. Pingueculitis describes a condition in which these pinguecula become inflamed and swollen. This inflammation is represented by increased redness and irritation of the eye.

Pinguecula rarely require treatment, if extremely inflamed with irritation, the doctor may prescribe artificial tears for lubrication and mild anti-inflammatory medication to reduce swelling.

Pterygium although produced by the same things which causes a pinguecula but does not emerge from it. Pterygium is a progressive and will affect the cornea if not treated on time. It commonly grows from the nasal side of the sclera. It is triangular in shape, with the base of the triangle located in the conjunctiva an the apex of the triangle encroaching onto the cornea, Figure 9.8

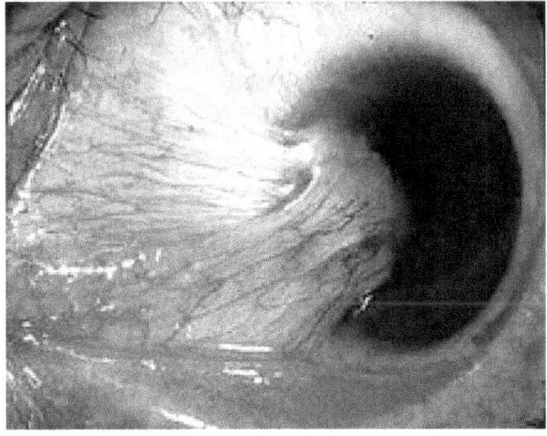

Figure 9.8 Pterygium growing onto the cornea

The exact cause is unknown, but it is associated with excessive exposure to sunlight, wind and sand because it occurs in the areas near the equator. In addition, pterygia are twice as likely to occur in men than women. Symptoms of pterygium include persistent redness, inflammation, foreign body sensation, tearing, which can cause bleeding, dry and itchy eyes. In advanced cases the pterygium can affect vision. There is no treatment which can reduce or prevent pterygium and only require surgery unless it grows to such an extent that it covers the pupil.

3- The iris is a circular muscle near the front of the eye. This colour is genetically determined. There are two pigments which determine the eye colour. Melanin (brown) deposition is controlled by a gene on chromosome 15. Lipochrome (yellowish-brown) deposition is controlled by a gene on chromosome 19. It is made up of three layers of connective tissue and muscle fibres: endothelium, stroma and the epithelium.

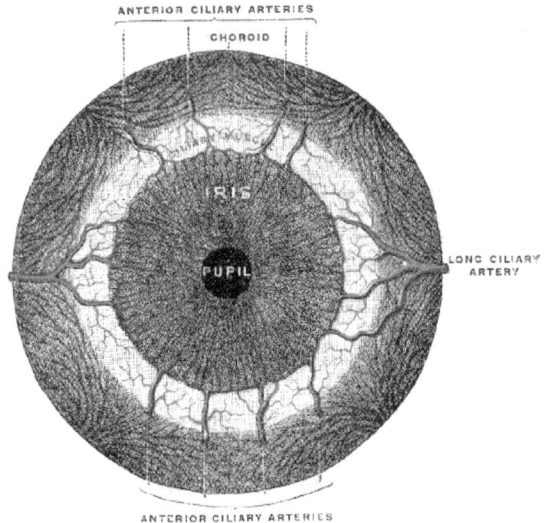

Figure 9.9 Front view of the iris

It controls light levels inside the eye by controlling the diameter and size of the pupil and thus the amount of light reaching the retina. The iris is embedded with tiny muscles that dilate (widen) and constrict (narrow) the pupil size. The sphincter muscle lies around the very edge of the pupil. In bright light, the sphincter contracts, causing the pupil to constrict. The dilator muscle runs radially through the iris, like spokes on a wheel. This muscle dilates the eye in dim lighting.

Diseases and disorders affecting the iris

Iritis is inflammation predominantly located in the iris of the eye. Iritis may affect one or both eyes. There are two main types of iritis: acute and chronic. Acute iritis heal independently within a few weeks without any treatment. Chronic iritis can exist for months or years before recovery occurs. Chronic iritis is also accompanied by a risk of serious visual impairment.

Iritis is related to a disease or infection in another part of the body. Diseases such as arthritis, tuberculosis, or syphilis can contribute to the development of iritis. Infection of some parts of the body (tonsils, sinus, kidney, gallbladder and teeth) can also cause inflammation of the iris.

The symptom of iritis are red eyes, pain ranging from soreness to intense discomfort, blurred vision and small specks or dots moving in the field of vision.

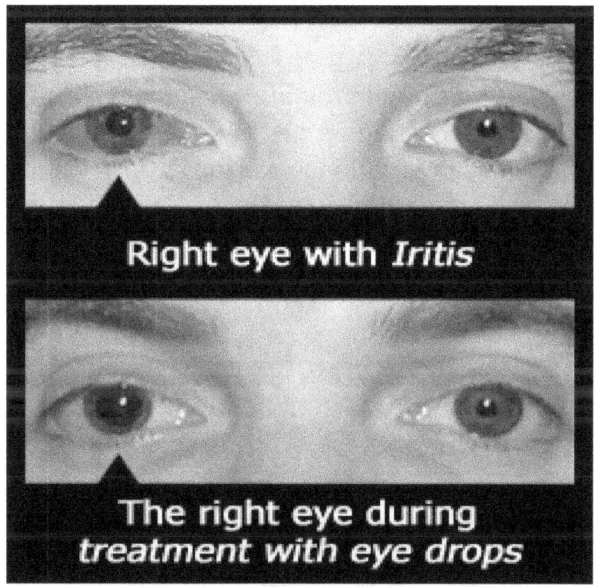

Figure 9.10 Eye with iritis symptom

To establish iritis doctor will look for white blood cells (leukocytes) as they are shed into the anterior chamber of the eye with inflamed iris. They will be floating in the convection currents of the aqueous humour. These cells can be counted and form the basis for rating the degree of inflammation. This is measured on a scale of 1-4, four being a very high cell count and severe iritis.

Iritis can readily be treated with corticosteroids and anti-inflammatory drops. Dilating eye drops and ointments also make the eye more comfortable by relaxing the muscle that constricts the pupil. The application of hot packs may also provide relief from the symptoms of iritis. In severe cases, oral medications and injections may be necessary to treat the condition.

4- The Pupil is a hole located in the centre of the iris that allows light to enter the retina. It appears black because all light rays entering the pupils are absorbed. Normal pupils are round, and when reacts to more intense light becomes smaller, so let less light entering the eye. In the dark, same pupil will dilate, to allow more light to enter the eye. People with light coloured iris and short sightedness have larger pupil than people with dark coloured iris and far sightedness.

5- The Lens is a transparent, ellipsoid and the biconvex crystalline structure of the eye, located just behind the iris. Its purpose is to refract light to be focused onto the retina. The nucleus, the innermost part of the lens, is surrounded by softer material called the cortex. The lens is encased in a capsule-like bag and suspended within the eye by tiny "guy wires" called zonules. It is composed of fibres that come from epithelial cells. The cytoplasm of these cells makes the transparent substance of the lens. The anterior surface is less curved than the posterior.

Diseases and disorders affecting the Lens

Cataract is a clouding of the lens in the eye that affects vision, varying in degree from slight to complete opacity and obstructing the passage of light. As described above that the lens is contained in a capsule. As old cells die they become trapped within the capsule. Over time, the cells accumulate causing the lens to cloud, making images look blurred or fuzzy.

Cataracts typically progress slowly to cause vision loss. Age-related cataracts can affect your vision in two ways:

a) The lens is mostly made of water and protein. When the protein clumps up, it clouds the lens and reduces the light that reaches the retina. In its early stage the cloudiness affects only a small part of the lens. Gradually it will increase in size and vision will get bluer.
b) When the cataract has progressed the colour of the lens will gradually be yellowing which will add a brownish tint to vision. The opacification of the lens will reduce the perception of blue colour.

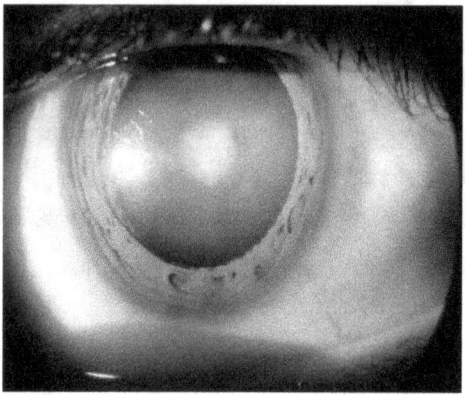

Figure 9.11 Eye with cataract

If cataract is untreated can cause phacomorphic glaucoma, in which the optic nerve is damaged in a characteristic pattern. This can permanently damage vision in the affected eye(s) and lead to blindness.

The risk of cataract increases as one gets older. Other risk factors for cataract include:

- Long-term exposure to ultraviolet light and/or ionising radiation
- Certain diseases such as diabetes, trauma and hypertension.
- Congenital cataract (just before or after birth) is genetic.
- Iodine deficiency

The early cataract may be improved with new eyeglasses, anti-glare sunglasses, or magnifying lenses. If these measures do not help, surgery is the only effective treatment. Surgery involves removing the cloudy lens and replacing it with an artificial lens. There are two methods of surgery.

i. Phacoemulsification, A small incision is made on the side of the cornea, a tiny probe is inserted into the eye. This device emits ultrasound waves that soften and break up the lens so that it can be removed by suction. Most cataract surgery is done by phacoemulsification, also called "small incision cataract surgery."
ii. Extracapsular surgery. a longer incision on the side of the cornea is made to remove the cloudy core of the lens in one piece. The rest of the lens is removed by suction.

The lens is replaced by an artificial lens called an intraocular lens (IOL). Cataract surgery slightly increases risk of retinal detachment.

6- **Aqueous Humour** is a clear, gelatinous fluid similar to plasma that fills and helps form the anterior and posterior chambers of the eye. It contains low protein concentrations. It is produced by ciliary epithelium and the flows into the anterior chamber, out through a spongy tissue at the front of the eye and into a drainage canal. It nourishes the lens and the cornea, removes excretory products from metabolism, transports neurotransmitters and also give the eye its shape.

7- **Vitreous Humour** comprises a large portion (about 2/3 of the eye's volume) of the eyeball. The vitreous is the transparent, colourless, gelatinous mass that fills the space between the lens of the eye and the retina lining the back of the eye. Vitreous humour has the following composition:

I. water – 99%
II. collagen fibrils
III. hyaluronic acid and ascorbic acid
IV. peripheral cells – hyalocytes
V. inorganic salt
VI. sugar

The vitreous humour is produced by cells in the non-pigmented portion of the ciliary body driven from embryonic mesenchyme cells which then degenerate after birth. The vitreous has a viscosity two to four times that of water, giving it a gelatinous consistency. It also has a refractive index of 1.336.

Unlike the aqueous humour which is continuously replenished, the gel in the vitreous chamber is stagnant. Therefore, if blood, cells or other byproducts of inflammation get into the vitreous, they will remain there unless removed surgically.

Diseases and disorders affecting the Vitreous Humour

Posterior vitreous detachment (PVD)

As we age, the vitreous humour change from a gel to a liquid, so the vitreous mass gradually shrinks and collapse and separating and falling away from the retina. This is called posterior vitreous detachment. A person suffering from PVD will see flashing lights and/or floaters in his field of vision. The flashing occurs due to vitreous tugs on the sensory layer of the retina, as the vitreous humour is detaching and retinal nerve can be stimulated. The floater are cells or debris released when vitreous detaches. This normally occurs around the age of 40 years and more common in people with nearsightedness. This normally last for a few weeks and if the frequency of this occurrence increases or occurs with a sudden onset of a large number of floaters, then this could be the sign of retinal detachment which is much more serious and sight-threatening condition.

Retinal detachment is a disorder of the eye in which the thin lining (retina's sensory and pigment layers) at the back of your eye called the retina begins to pull away from the blood vessels that supply it with oxygen and nutrients. The initial detachment may be localized, but without rapid treatment the entire retina may detach, leading to loss of vision or total blindness. Retinal detachment is considered an ocular emergency that requires immediate medical attention and surgery. It is a problem that occurs more frequently in the middle-aged and elderly.

The retina is a thin layer of light sensitive tissue in the back wall of the eye. The optical system of the eye focuses light on the retina much like light is focused on the film or sensor in a camera. The retina translates that focused image into neural impulses and sends them to the brain via the optic nerve. A trauma or injury, PVD to the eye may cause a small tear in the retina. The tear allows vitreous fluid to seep through it under the retina, and peel it away like a bubble in wallpaper.

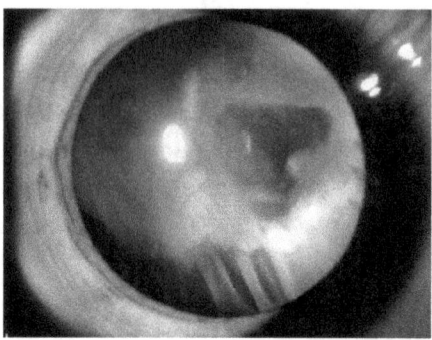

Figure 9.12 Slit lamp photograph showing detachment in Von Hippel-Lindau disease.

There are three types of retinal detachments. The most common type occurs (as described above) when there is a break in the sensory layer of the retina, and fluid seeps underneath, causing the layers of the retina to separate. This is called Rhegmatogenous retinal detachment. Retinal breaks are divided into three types - holes, tears and dialyses. Holes form due to retinal atrophy, tears are due to vitreoretinal traction and dialyses which are very peripheral and circumferential may be either tractional or atrophic.

The second most common type occurs when strands of vitreous or scar tissue (fibrovascular tissue, caused by an inflammation, injury or neovascularization [proliferation of blood vessels in tissues not normally containing them]) create traction on the retina, pulling it loose. This is called tractional retinal detachment. Patients with diabetes are more likely to experience this type.

The third type called exudative, serous or secondary retinal detachment happens when fluid collects underneath the layers of the retina, causing it to separate from the back wall of the eye. This type usually occurs in conjunction with another disease affecting the eye that causes swelling or bleeding. Sometime a tumour can grow due to cancer called choroidal melanoma.

SIGNS AND SYMPTOMS

- Curtain obstructing vision
- Light flashes
- Wavy vision
- Floaters like spider web
- Sudden decrease of sight

There are a number of ways to treat retinal detachment. The appropriate treatment depends on the type, severity and location of the detachment. Pneumatic retinopexy is one type of procedure to reattach the retina. With a local anaesthesia, the surgeon injects a small gas bubble into the vitreous cavity. The bubble presses against the retina, flattening it against the back wall of the eye. Since the gas rises, this treatment is most effective for detachments located in the upper portion of the eye. The gas bubble slowly absorbs over the next 1-2 weeks.

A surgery technique called scleral bucket is an established treatment in which the eye surgeon sews one or more silicone bands to the sclera (the white outer coat of the eyeball). The bands push the wall of the eye inward against the retinal hole, closing the break or reducing the fluid flow through it and reducing the effect of vitreous traction thereby allowing the retina to re-attach.

Another useful treatment is called vitrectomy. It involves the removal of the vitreous gel and is usually combined with filling the eye with either a gas bubble (SF_6 or C_3F_8 gas) or silicon oil. An advantage of using gas in this operation is that there is no myopic shift after the operation and gas is absorbed within a few weeks. A disadvantage is that a vitrectomy always leads to more rapid progression of a cataract in the operated eye.

8- Choroid is the pigmented, highly vascular layer of the eye, containing connective tissue, and lying between the retina and the sclera. It is composed of layers of blood vessels that nourish the back of the eye. The choroid provides oxygen and nourishment to the outer layers of the retina.

The structure of the choroid is generally divided into four layers:

- Haller's Layer - outermost layer of the choroid consisting of larger diameter blood vessels
- Sattler's layer - layer of medium diameter blood vessels
- Choriocapillaris - layer of capillaries
- Bruch's membrane - innermost layer of the choroid

The main function of the choroid is to nourish the outer layers of the retina but it is also thought to regulate retinal heat, to assist in the control of intraocular pressure and the pigment absorbs excess light so avoiding reflection.

Diseases and disorders affecting the Choroid

Choroidal detachment and haemorrhage is a serious ocular condition, which may be associated with permanent loss of sight. The choroid detaches from the underlying sclera with an associated accumulation of serum-like fluid or blood. Both limited and massive choroidal haemorrhages may occur as complications of most forms of ocular surgery, as well as from trauma. Choroidal haemorrhage may occur when a fragile vessel is exposed to sudden compression and decompression events. Systemic conditions which may serve as risk factors for expulsive choroidal haemorrhage include advanced age, arteriosclerosis, hypertension, diabetes mellitus, blood dyscrasias, and obesity.

As soon as the diagnosis is made, topical steroids, cycloplegics and mydriatics are given. Intra-ocular pressure-lowering drugs are also given. Surgery may follow depending on the nature of the detachment. Operative choroidal haemorrhage tends to occur during surgery and is managed there and then in operation theatre.

Choroidal Nevus: A nevus is typically a flat, benign, pigmented area that may appear inside the eye or on its surface. A benign choroidal nevus rarely causes symptoms. Rarely a choroidal nevus can leak fluid or are associated with the growth of abnormal blood vessels. These related changes can cause a localized retinal detachment/degeneration, flashing lights and loss of vision. Choroidal nevus is typically a pigmented tumour of the blood vessel layer (choroid) beneath the retina. A choroidal nevus is typically grey but can be brown, yellow or variably pigmented.

A benign choroidal nevus requires no treatment and there is no way to safely remove them. Since a choroidal nevus can turn into a choroidal melanoma, it is reasonable to have it periodically observed eye care specialist.

9 – Sclera is a tough white covering called the sclera protects the eye containing collagen and elastic fibre. Part of the white sclera can be seen in the front of the eye. A clear, delicate membrane called the conjunctiva covers the sclera. It is made up of three divisions: the episclera, loose connective tissue, immediately beneath the conjunctiva; sclera proper, the dense white tissue that gives the area its colour; and the lumina fusca, the innermost zone made up of elastic fibres.

10 – The Retina is a multi-layered, light sensitive tissue that lines the back of the inner eye. It acts like the film in a camera, images come through the eye's lens and are focused on the retina. The retina then converts these images to electric signals and sends them via the optic nerve to the brain. The retina is normally red due to its rich blood supply.

The retina is a layered structure with several layers of neurons interconnected by synapses. It contains millions of photoreceptor cells that capture light rays and convert them into electrical impulses. These impulses travel along the optic nerve to the brain where they are turned into images. There are two types of photo-receptors in the retina: rods and cones.

Rods function mainly in dim light and provide black-and-white vision. There are approximately 125 million rods. They are spread throughout the peripheral retina and function best in dim lighting. The rods are responsible for peripheral and night vision. Cones support daytime vision and the perception of colours and the retina contains approximately 6 million cones. The cones are contained in the macula, the portion of the retina responsible for central vision. They are most densely packed within the fovea, the very central portion of the macula. A third, much rarer type of photoreceptor, the photosensitive ganglion cell, is important for reflexive responses to bright daylight.

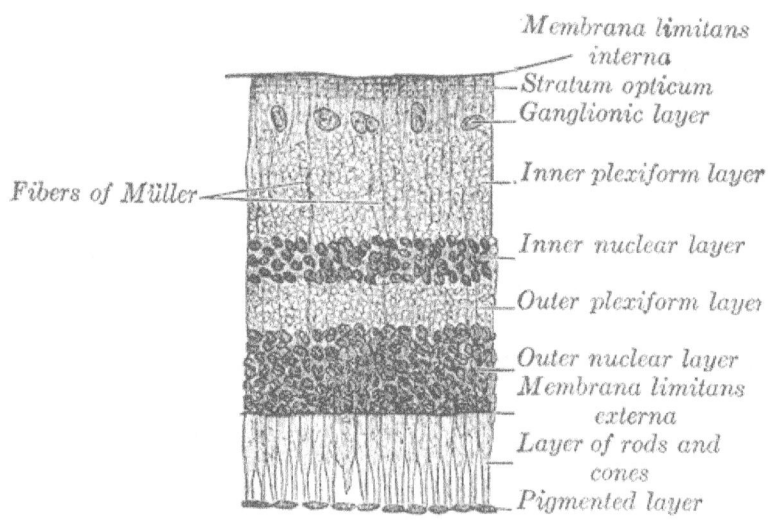

Figure 9.13 cross section of retina

The retina has ten distinct layers.

i. *Inner limiting membrane* - is the boundary between the retina and the vitreous body, formed by astrocytes and the end feet of Müller cells. It is separated from the vitreous humour by a basal lamina.
ii. *Nerve fibre layer* - axons of the ganglion cell nuclei

iii. *Ganglion cell layer* - contains nuclei of ganglion cells, the axons of which become the optic nerve fibres for messages and some displaced amacrine cells.
iv. *Inner plexiform layer* - contains the synapse between the bipolar cell axons and the dendrites of the ganglion and amacrine cells.
v. *Inner nuclear layer* - contains the nuclei and surrounding cell bodies of the bipolar cells.
vi. *Outer plexiform layer* - is a layer of neuronal synapses in the retina of the eye. It consists of a dense network of synapses between dendrites of horizontal cells from the inner nuclear layer, and photoreceptor cells inner segments from the outer nuclear layer.
vii. *Outer nuclear layer* - cell bodies of rods and cones.
viii. *External limiting membrane* - layer that separates the inner segment portion-of the photoreceptors from their cell nucleus
ix. *Photoreceptor layer* – Rod and cones
x. *Retinal pigment epithelium* - single layer of cuboidal cells

If we take the shape of the eye as sphere with diameter of about 22 mm then retina is approximately 72% of the sphere.

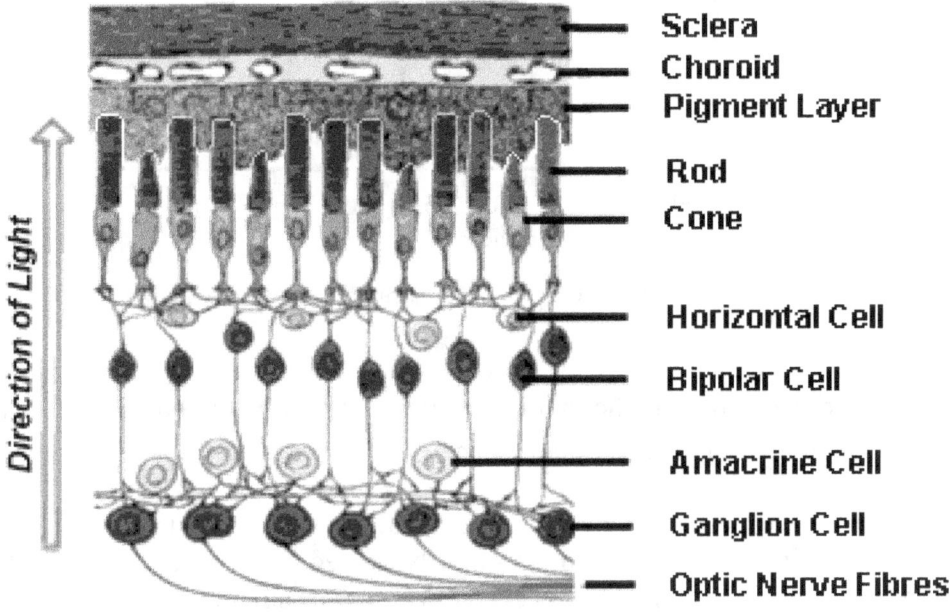

Figure 9.14 Rods, cones and nerve layers in the retina.

The front (anterior) of the eye is on the left. Light passes through several transparent nerve layers to reach the rods and cones. A chemical change in the rods and cones send a signal back to the nerves. The signal goes first to the bipolar and horizontal cells, then to the amacrine cells and ganglion cells, then to the optic nerve fibres. The signals are processed in these layers. First, the signals start as raw outputs of points in the rod and cone cells. Then the nerve layers identify simple shapes, such as bright points surrounded by dark points, edges, and movement.

An image is produced by the excitation of the cones and rods in the retina. The excitation is processed by the neuronal system and various parts of the brain working in parallel to form a representation of the external environment in the brain. The cones respond to bright light and mediate high-resolution colour vision during daylight and is called photopic vision.

The eye uses three types of cones to sense light in three respective bands of colour. The pigments of the cones have maximum absorption values at wavelengths of about 420 nm (blue), 534 nm (Bluish-Green) and 564 nm (Yellowish-Green). Their sensitivity ranges overlap to provide vision throughout the visible spectrum.

The eye uses scotopic vision under low-light conditions, and mesopic vision in intermediate conditions. In the eye cone cells are nonfunctional in low light – scotopic vision is produced exclusively through rod cells which are most sensitive to wavelengths of light around 498 nm (green-blue) and are insensitive to wavelengths longer than about 640 nm (red). Scotopic vision occurs at luminance levels of 10^{-2} to 10^{-6} cd/m².

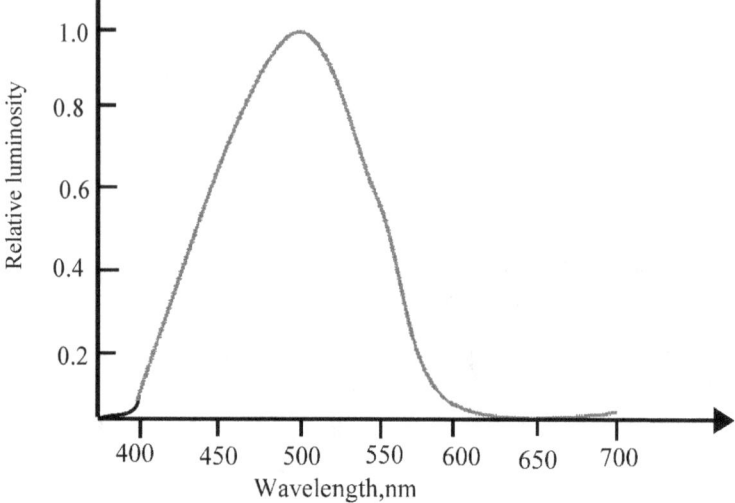

Figure 9.15 A graph of the scotopic luminosity curve

Mesopic vision is a combination of photopic and scotopic vision in low but not quite dark lighting situations. Mesopic light levels range from luminance of approximately 0.001 to 3 cd m^{-2}.

A rod cell is sensitive enough to respond to a single photon of light. Rod cells are almost entirely responsible for night vision. Cone cells, on the other hand, require tens to hundreds of photons to become activated. Additionally, multiple rod cells converge on a single inter-neuron, collecting and amplifying the signals and this convergence reduces visual acuity (resolution), as pooled information from multiple cells are less distinct.

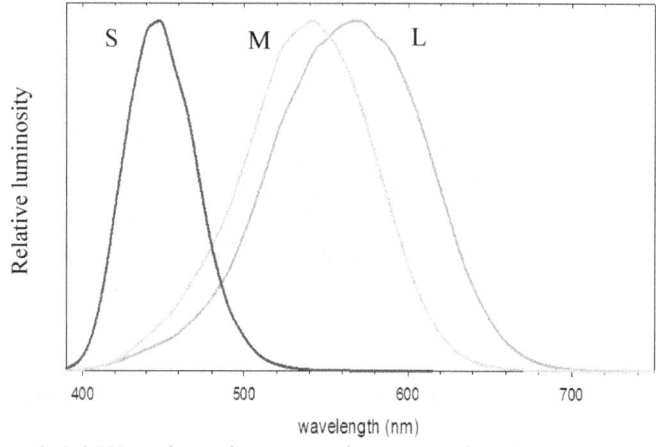

Figure 9.16 Wavelength responsiveness of rods compared to that of three types of cones. Short, medium and long wavelength.

Retinal scan is a biometric technique that uses the unique patterns on a person's retina to identify them. It is not to be confused with other ocular-based technology, iris recognition. Retinal scans use a low-intensity infrared light and a delicate sensor to scan unique pattern of blood vessels of the retina, which is the nerve tissue that lines the back of the eye. There is little chance that retinal patterns can be replicated or forged. Retinal scans are therefore considered to be among the least violable of biometric security measures. (Fingerprints, by comparison, are relatively easy to forge.)

Because of the complex structure of the capillaries that supply the retina with blood, each person's retina is unique. This network in the retina is so complex that even identical do not share a similar pattern. Although retinal patterns may be altered in diseases, the retina typically remains unchanged from birth until death. It's this unique and unchanging nature, the retina appears to be the most accurate and reliable biometric system. The error rate for fingerprint identification is sometimes as high as 1 in 500. A retinal scan, on the other hand, boasts an error rate of 1 in 10,000,000. Its close match, the iris scan, is slightly less precise, with an error rate of approximately 1 in 131,000. Retinal scan is almost exclusively used in high-end security applications.

The idea for retinal identification was first conceived by Dr. Carleton Simon and Dr. Isadore Goldstein. The idea was a little before its time, the first company to exploit the idea was Eydentify, founded by Robert Hill in 1976. He was an electrical engineer who first used the idea as a form of identification when he was helping his father, an ophthalmologist, detect eye disease through photographs. In 1978, specific means for a retinal scanner was patented, followed by a commercial model in 1981.

Diseases and disorders affecting the Retina

Retinitis pigmentosa (RP) is an inherited, degenerative eye disease that causes severe vision impairment and often blindness. In RP, the faulty genes cause the retinal cells to stop working and eventually die off. As there are many genes that can cause the retinal cells to stop working, there are many different types of RP, such as Leber's Congenital Amaurosis, cone and cone-rod dystrophies, and Choroideremia. In most cases, the peripheral rod cells are affected first and RP later affects the central cone cells. Sometime RP can occur alongside the hearing loss and is called Usher syndrome.

There are three types of RP inheritance. Autosomal dominant RP tends to be a known history of the condition in the family. Second type Autosomal recessive RP normally has no known history of condition in either family and cause more severe sight loss than dominant type. Both types affect men and women equally. The third type is known as an X-linked inheritance from which men are mostly affected. Female members of a family are carriers of the faulty gene. If a relative has passed on the faulty gene but have not developed symptoms themselves, then is not possible to establish which of the three types of inheritance have caused the RP.

The first symptom one usually notice is to find it difficult to see in poor light, such as outdoors at dawn/dusk, or in a dimly lit room. This is often referred to as "night blindness". A second symptom is the loss of some of your peripheral vision or peripheral visual field. This means that when you're looking straight ahead you become less able to see things either to the side, above or below. This is a sign of rod cells are being affected by RP.

All RP conditions are progressive, but the speed and pattern of deterioration of sight vary from one person to another. Mostly, the first effect is the gradual loss of peripheral vision. This means that you can start to miss things slightly to the side of you or trip over or bump into things you would have seen in the past. Eventually you have a very restricted visual field, leaving only a narrow tunnel of vision.

Treatment, there is no known cure or treatment for RP or associated retinal disorders. Many scientists are trying to understand genes causing RP, so a faulty gene could be replaced by gene therapy.

Macular degeneration (MD) usually affects older adults and results in a loss of vision in the centre of the visual field -the macula- because of damage to the retina. The macula is the central part of the retina and is responsible for detailed and colour vision, the vision we use to read, write, or recognize faces. The macula is a highly specialized part of the nervous system and the eye in which the photo-receptors that react to light stimulus and the neurons that interpret and transmit these signals are precisely organized and densely compacted. It is the macula that allows humans to see 20/20. Macular degeneration is a painless eye condition that leads to the gradual loss of the ability to see what is directly in front of you. Macular degeneration does not affect the peripheral vision.

There are two types of macular degeneration:

- *Dry macular degeneration* (also called non-neovascular) results from the gradual breakdown of cells in the macula, resulting in a gradual blurring of central vision. This results from atrophy of the retinal pigment epithelial layer below the retina, which causes vision loss through loss of photo-receptors (rods and cones) in the central part of the eye. No medical or surgical treatment is available for this condition; however, vitamins with high doses of antioxidants, lutein and zeaxanthin, improve visual acuity.
- *Wet* (exudative or neovascular) is a form of advanced AMD, newly created abnormal blood vessels grow under the centre of the retina. These blood vessels leak, bleed, and scar the retina, eventually cause irreversible damage to the photo-receptors and rapid vision loss if left untreated. Only about 10% of patients suffering from macular degeneration have the wet type.

Normal Vision

Figure 9.17 Same view with macular degeneration

Macular degeneration most commonly affects people who are over 50. Almost everyone affected will have enough peripheral vision to continue their daily activities. No one knows the aetiology, or cause, of age-related macular degeneration. Causes are likely to be genetically inherited, but environmental factors may also contribute. There is currently no treatment for dry macular degeneration, but wet AMD is more serious. There are several treatment options available that can slow the progression of wet AMD and, in some cases, restore some of the lost vision. The proliferation of abnormal blood vessels in the retina is stimulated by vascular endothelial growth factor (VEGF). When an anti-VEGF agent is injected directly into the vitreous humour can improve the vision. The first major anti-VEGF agent was *bevacizumab*.

Retinoblastoma is a rare type of eye cancer that usually develops in early childhood, typically before the age of 5. This form of cancer develops in the retina. In about two thirds of cases, only one eye is affected (unilateral retinoblastoma); in the other third, tumours develop in both eyes (bilateral retinoblastoma). The most common first sign of retinoblastoma is a visible whiteness in the pupil called "cat's eye reflex" or leukocoria.

There are two forms of the disease; a heritable form and non-heritable form. Mutations in the RB1 gene are responsible for most cases of retinoblastoma. RB1 is a tumour suppressor gene, which means that it normally regulates cell growth and keeps cells from dividing too rapidly or in an uncontrolled way. Most mutations in the RB1 gene prevent it from making any functional protein, so it is unable to regulate cell division effectively. As a result, certain cells in the retina can divide uncontrollably to form cancerous tumours. Researchers estimate that 40 percent of all retinoblastomas are germinal, which means that RB1 mutations occur in all of the body's cells, including reproductive cells (sperm or eggs).

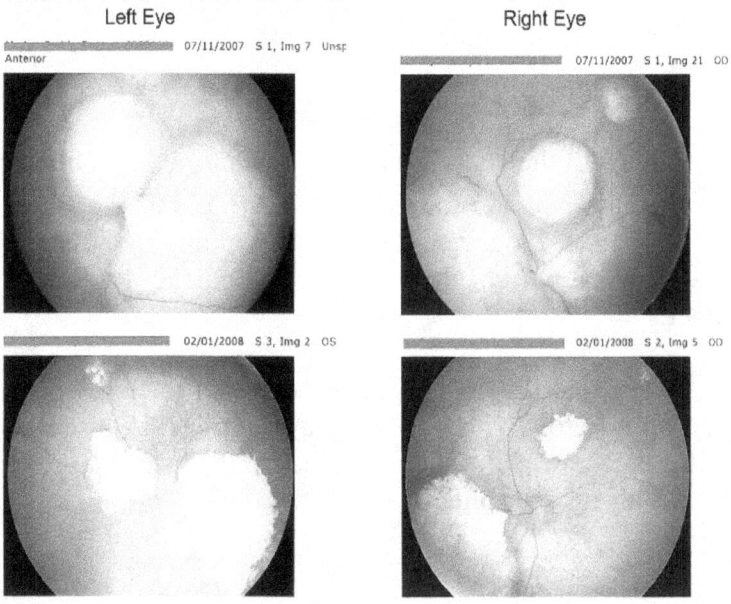

Figure 9.18 RB1 tumours with a retinoscan before and during chemotherapy.

Treatment will be decided by the ophthalmologist in discussion with the paediatric oncologist. Many treatment options exist, one of them is chemotherapy which is administered locally via a thin catheter threaded through the groin, through the aorta and the neck, into the optic vessels. The other treatment methods include, external beam radiotherapy, cryotherapy and final surgical removal of the eyeball.

11- Fovea is the centre most part of the macula. This tiny area is responsible for our central, sharpest vision. Unlike the peripheral retina, it has no blood vessels. Instead, it has a very high concentration of cones (photo-receptors responsible for colour vision), allowing us to appreciate colour. The cones in the foveal pit have a smaller diameter and can, therefore, be more densely packed. The high spatial density of cones accounts for the high visual acuity capability at the fovea. A healthy fovea is key for reading, watching television, driving, and other activities that require the ability to see detail. The human fovea has a diameter of about 1.0 mm. The centre of the fovea is the foveola – about 0.2 mm in diameter – where only cone photo-receptors are present and there are virtually no rods.

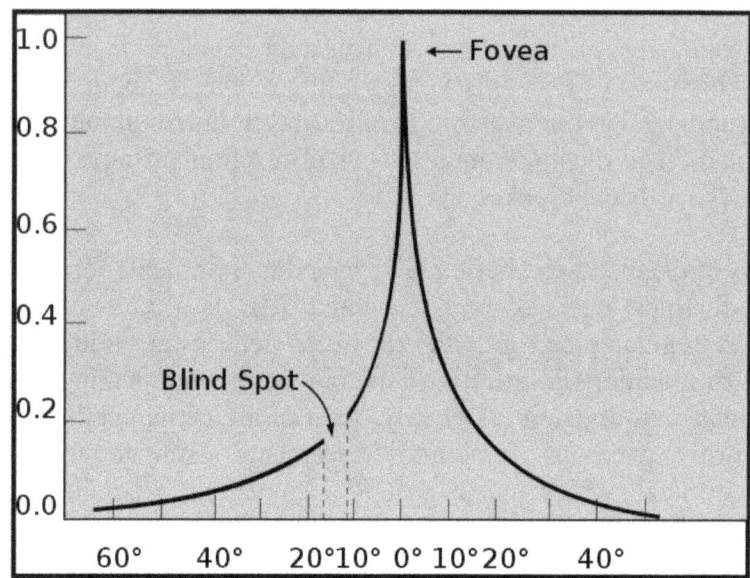

Figure 9.19 The diagram shows relative acuity of the left human eye in degree from the fovea.

Blind Spot is an obscuration of the visual field. The human eye has a blind spot in the area of the retina where the optic nerve leads back into the brain. There is a total absence of cones and rods in this area, and, consequently, each eye is completely blind in this spot. The blind spots in each eye are aligned symmetrically so that most of the time, one eye's field of vision will compensate for the loss of vision in the other.

In order to find the blind spot of the right eye, it is necessary to close your left eye, keeping your head motionless. Fixate on the cross with your right eye. This will cause the image of the cross to fall on your fovea. Adjust the viewing distance until the smiley face disappears. When this happens, the image of the spot is falling on your blind spot.

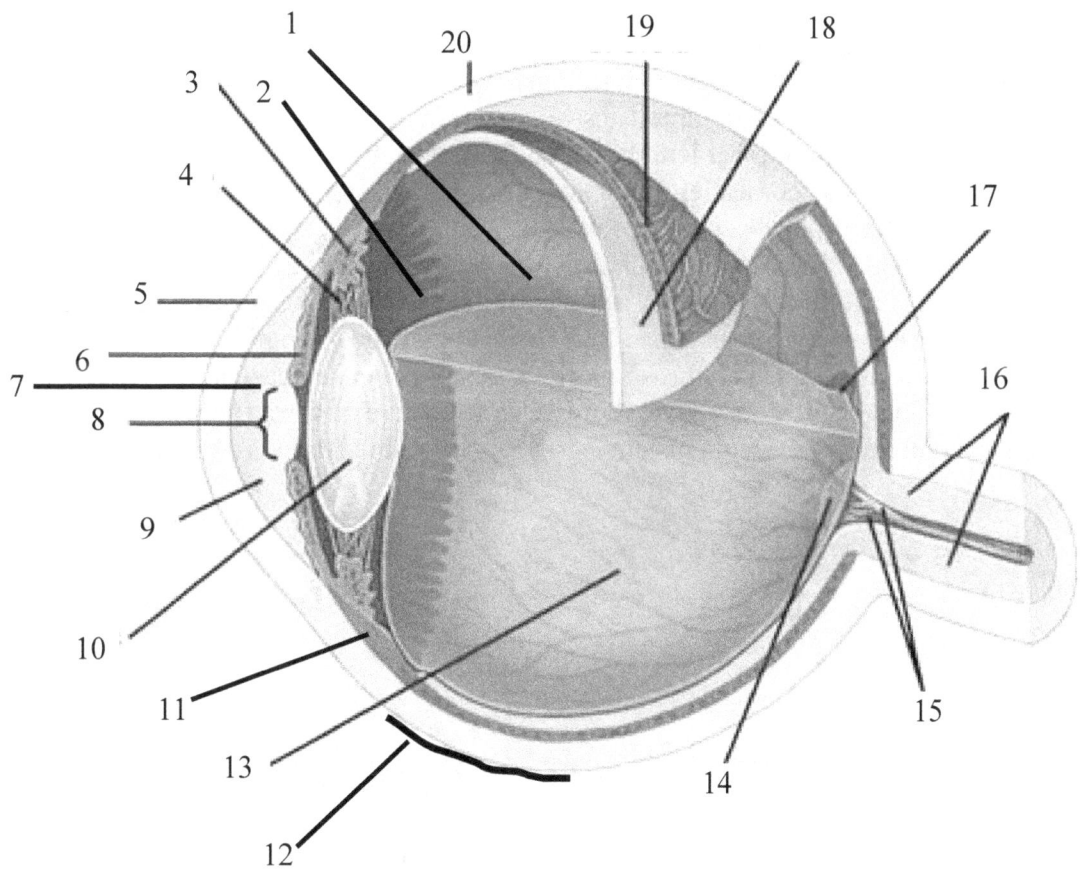

Figure 9.20 3-D view of Eye

1. posterior compartment   2. ora serrata   3. ciliary muscles

4. ciliary zonules or suspensory ligament   5. cornea

6. iris   7. anterior chamber   8. pupil

9. aqueous humour   10. lens   11. ciliary process

12. conjunctiva   13. retinal arteries and veins   14. blind spot – optic disk

15. central retinal artery and vein   16. optic nerve

17. fovea   18. retina   19. choroid

20. sclera

Accommodation of the Eye

Accommodate is the ability to increase the refractive power of the optical system of the eye. It produces the clearest image of near objects. Since a nearby object (small u) is typically focused at a further distance (large *v*), the eye accommodates by assuming a lens shape that has a shorter focal length (eq. 4.3). This reduction in focal length will cause more refraction of light and serve to bring the image back closer to the cornea/lens system and upon the retinal surface. To bring about accommodation , the ciliary body contracts, the lens zonules relax, and crystalline lens assumes more spherical shape, which increases the refractive power or shorten the focal length of the lens. On the other hand, a distant object (large u) is typically focused at a closer distance (small *v*). The eye accommodates by assuming a lens shape that has a longer focal length. So for distant objects the ciliary muscles relax and the lens returns to a flatter shape. This decrease in the curvature of the lens corresponds to a longer focal length or decrease in refractive power. The ability of the eye to accommodate is automatic and it occurs instantaneously. Accommodation is a remarkable design!

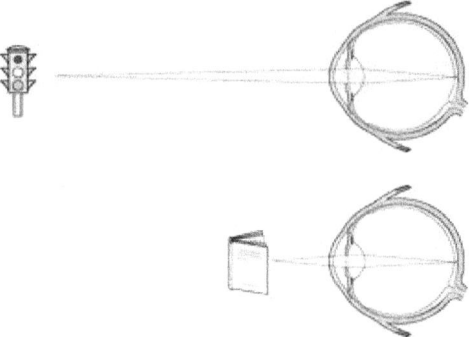

Figure 9.21 Accommodation the eye

Remember that the cornea provides 2/3 of the refractive power and the lens only provide 1/3 ? However, our eye changes the curvature of the lens, rather than the cornea. The curvature of the cornea cannot be changed. The young human eye can change focus from very far distance, normally considered as infinity, to 7 cm from the eye in 350 milliseconds. This dramatic change in focal power of the eye of approximately 13 dioptres.

When we grow old, our lens will turn hard. Our accommodation ability will decrease and it will get more and more difficult to focus, by the age of forty years. The accommodative amplitude has declined so the near point of the eye is more remote than the reading distance. The accommodative amplitude has declined so the near point of the eye is more remote than the reading distance. This defect is called *presbyopia* (is a condition when the eye exhibits a progressively diminished ability to focus on near objects). Once presbyopia occurs, those who are emmetropic (do not require spectacles for distance vision) will need spectacles for near vision; those who are myopic (near sighted and wear spectacles for distance vision), will find that they see better at near without their spectacles; and those who are hyperopic (far sighted) will find that they may need a correction for both distance and near vision.

The power of a lens is measured by opticians in a unit known as a dioptre. A *dioptre* is the reciprocal of the focal length.

$$\textbf{Dioptres (D)} = \frac{1}{focal\ length(m)} \quad (9.1)$$

Test

Hold your finger about 10cm in front of your nose. Now fixate at your finger for 10 seconds and then alternate between focusing on your finger and looking into the distance object past your finger. You will notice the change in your lens power.

To determine the near point of your eye's accommodation is to slowly bring this paragraph closer to your eye until the letters become blurry. Measure this distance and that is your near point.

Power of Accommodation

A healthy eye is able to bring both distant objects and nearby objects into focus without the need for corrective lenses. It would have the ability to view objects with a large variation in distance. The maximum variation in the power of the eye is called the *Power of Accommodation*. The healthy eye of a young adult has a Power of Accommodation of approximately 4 diopters.

The nearest point of the eye is the point near to the eye at which an object can be placed and still have a sharp image produced on the retina. In a normal eye, the near point is located 25cm from the eye. Whereas, the *far point* of the eye is the location of the farthest object on which the fully relaxed eye can focus. In a normal eye, the far point is located at infinity.

$1/f$ is the power of accommodation at a particular focal length of the eye lens. Max power of accommodation is at minimum focal length i.e. 25cm. Power is 1/0.25 i.e. 4. If the distance between eye lens and retina is taken as 2 cm. The table below shows the change in focal length and lens power as the object distance is varied.

Table1: Dependence of focal length (Power) of eye lens upon object distance.

Object distance (m)	Focal lengt (cm)	Power of lens (D)
0.25	1.85	54.00
1.00	1.96	51.00
3.00	1.98	50.33
100.00	1.99	50.01
infinity	2.00	50.00

Visual acuity

Visual acuity is an indication of clarity or clearness of one's vision, which is dependent on the sharpness of the retinal focus within the eye and the sensitivity of the interpretative faculty of the brain. Visual acuity is a measure of the spatial resolution of the visual processing system.

To test visual acuity, a person has to stand at a certain distance and identify characters on a chart. Chart characters are represented as black symbols against a white background for maximum contrast. Such a test is called "Snellen Test" after Dutch ophthalmologist Hermann Snellen (1834 – 1908) who created it. A Snellen test uses a chart with different sizes of letters or forms to evaluate your visual acuity, that is, the sharpness of your vision. The test shows how accurately you can see from a distance. It is viewed at 6 metres (20 feet). A visual acuity of 6/6 indicates that the chart was viewed at 6 metres, and the lowest line that could be read was labelled 6. See figure 9.22

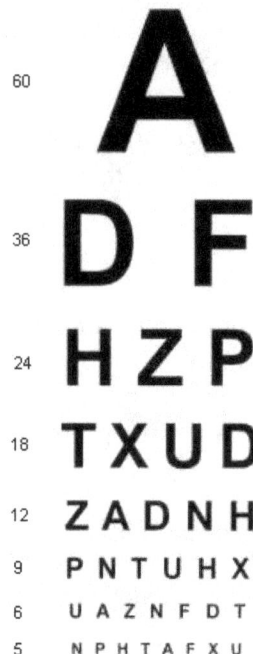

To have a UK driving license your visual acuity must be between 6/12 and 6/9. "Normal" vision is equivalent to 6/6, but many people who are well corrected by their spectacles are often able to see even better, perhaps 6/5 or 6/4. We normally refer to 20/20 vision and this is because in the United States, the standard length of the eye examination room is 20 feet. Most countries have adopted SI units so 6 meter is equal to 19.685 feet, which is close to 20 feet. Therefore, instead of using 20/20 for normal vision, a notation of 6/6 is now used in the UK. The less the bottom number in the acuity ratio, the better the acuity; and greater the bottom number, worse the acuity. What 20/20 or 6/6 actually means.

A person with 20/20 or 6/6 vision is just able to read a letter that subtends a visual angle of 5 minutes of arc (5'). 5' is equal to 5/60 of a degree, as there are 60' in in one minute. A person with 6/6 vision could stand at 9m (30 feet) away from a test chart and read a 6/9 (20/30) letter on the chart, since at a distance a 6/9 letter would subtend an angle of 5' at a person's eye. One can have better than 6/6 vision, for example 6/5 is better than 6/6 because a person with 6/5 acuity can stand 6 m away from the chart and see as well as person with 6/6 acuity moving up to 9 m away from the chart.

Example: If a person has a visual acuity of 6/5 looking at an object from 30 m away. Where would a person with 6/6 vision need to stand to see the object as well?

$$5/6 \times 30 = 25 \text{ m}$$

Let's now calculate the height of the letter for 6/6 vision. Where;
- the letter visual angle subtended at the eye is 5' of arc, and half the letter will subtend 2.5'
- d is the distance between the chart and the eye.
- h is half the height of the letter

$$\text{Let's take } d = 6m$$

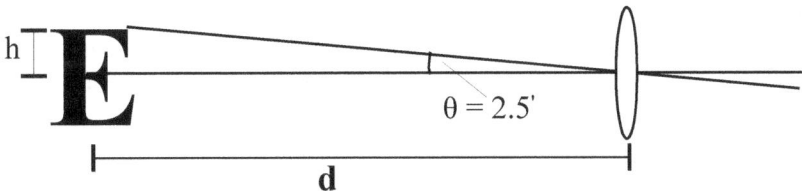

Figure 9.22 height of 6/6 vision letter

Lets convert 2.5 arc minutes into degree

$$2.5/60 = 0.04167^0$$

$$\tan \theta = h/d$$

or

$$h = d \times \tan \theta$$

$$h = 6 \tan 0.04167 = 4.363 \text{ mm}$$

The height of the letter is $2h = 2 \times 4.363 \approx 8.73$ mm

To calculate the standard distance for 6/6 vision test. First measure the height of the letter and use equation 9.2 to calculate the distance you should stand to view the chart.

Standing Distance in Meters = 6 x (Height of Letter in mm / 8.73)   (9.2)

The Resolution of the Human Eye

The absolute limit of optical resolution is determined by the wavelength of light that is used to illuminate the object. Our eye cannot see objects or detail that is smaller than a light wavelength. Human vision spans from 720 nanometres in the red wavelengths of light to 400 nanometres in the blue violet wavelengths. 560 nanometres is an average value for white light containing all colours of the rainbow.

Resolution of an optical system, including human eyes, is measured by the angular difference between two points that we can just resolve. At angles less than the resolution angle the points of light appear to be one bigger or brighter point. When any object is moved closer to the eye it appears larger with more detail because it is filling more of the light sensors in the eye's retina. The human eye has maximum resolution when an object is viewed as close to the eye as possible before it goes out of focus. This point is called the NEAR POINT. We have studied that this point is about 25 centimetres and the angular resolution of the eye at this point is about on the minutes (1/60 degree).

An object and its image can be considered to be composed of a large number of points of light which are called PIXELS (picture elements). The pixels can be magnified for viewing but the total number of pixels cannot be increased by magnification. This is similar to a normal television picture which is composed of 525 horizontal lines and about 250 vertical lines (525 x 250 = 131,250 pixels) and now High Definition TVs have 1920 horizontal lines and 1080 vertical lines (1920 x 1080 = 207,3600 pixels).

The standard definition of normal visual acuity (6/6 vision) is the ability to resolve a spatial pattern separated by a visual angle of one minute of arc. The average visual acuity of the human eye is one arc minute. The maximum possible is 0.4 arc minutes. The spatial resolution limit is derived from the fact that one degree of a scene is projected across 290μm of the retina by the eye's lens. See figure 9.23

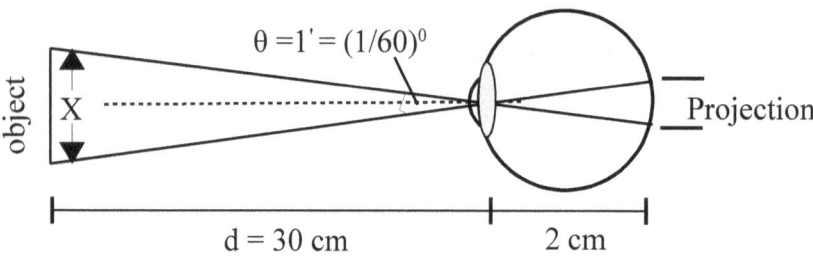

Figure 9.23 Eye resolution

Therefore, the average resolution of a good eye is between 0.4 and 1 arc minute. In the case of normal visual acuity the angle is 1/60 of a degree. By bisecting this angle we have a right triangle with angle θ as 1/120 of a degree. Using this right angle triangle it is easy to calculate the length X/2 for a given distance d. If we are inspecting an object at close detail the comfortable distance can be taken as 30 cm.

$$X/2 = d (\tan (\theta /2))$$
$$X/2 = (30 \times 10^{-3}) (\tan (1/120))$$
$$X/2 = (30 \times 10^{-3}) \times (1.45 \times 10^{-3}) = 4.36 \times 10^{-6} m$$

therefore, $\quad X = 2 \times (4.36 \times 10^{-3}) = 8.73 \times 10^{-6} m$

What this means is that if you had alternating black and white lines that were smaller than 8.73 μm wide, it would appear to most people as a mass of solid grey, or in other words any detail of the object smaller than this cannot be resolved by the eye at this distance.

To calculate the size of the projection on the retina, we again use the above equation;

$$P = (2 \times 10^{-2}) (\tan (1/120))$$
$$P = (2 \times 10^{-2}) \times (1.45 \times 10^{-3}) = 290 \times 10^{-6} m$$

Before these figures can be translated to pixels or displays, one needs to realize that the size of the pixel will vary with distance. Figure 9.24

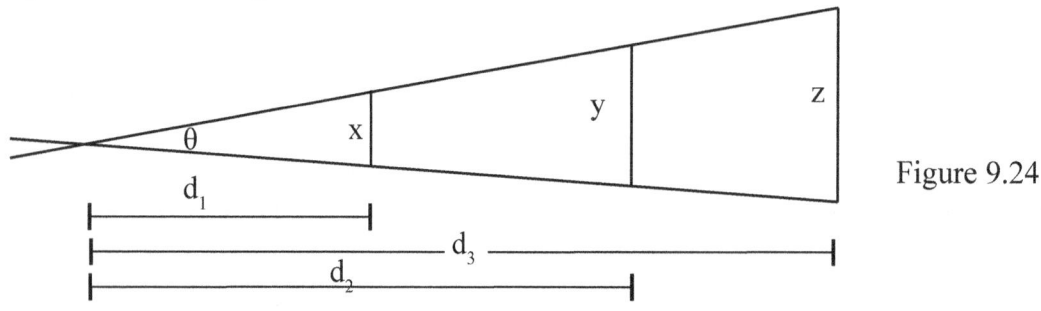

Figure 9.24

At 30 cm from the eye the pixel (sometimes called dot) size is 8.73 microns – for 1 arc minute and at 1m this size would be 291 microns. We also measure resolution as pixel (dot) per inch (ppi/dpi) An inch is 25.4mm. Therefore, 2910 pixels will fit into an inch at 30 cm and only 87 ppi at 1m.

Depth of field

Depth of field refers to the range of distance that appears acceptably sharp. Depth of field — the distance in front of, or behind, an object that still looks sharp — is affected by the amount of light hitting your retina and the magnifying power of your lens (focal length). The depth of field does not abruptly change from sharp to unsharpened, but instead occurs as a gradual transition. In fact, everything immediately in front of or in back of the focusing distance begins to lose sharpness — even if this is not perceived by our eyes. The speed of focus of your eye also means that you are not really aware that much of what you see is out-of-focus. If you look at something in the foreground, it appears sharp. If you shift your eye to look at the background, this snaps into focus.

To test this, hold your index finger up in front of your face, about 45cm away. Close one eye and look at your finger with the other. While concentrating on the finger, you will become aware that anything in the background is out of focus. Now look at the background and your finger will not look very sharp. This effect can be very useful in isolating the main subject from its background. However, it is also possible to make some of the out of focus areas appear sharper. The area of apparent sharpness in an image is called the 'depth-of-field', and it is one of the main creative controls you have.

Another way of defining the depth of field is that the range of object distances for which the circles of confusion are so small that the image is sharp enough to be considered 'in focus' is called the depth of field. The circle of confusion is the area of the retina over which the cells are stimulated by light from a point on an object. The bigger the circle, the more blurred the point becomes. The maximum acceptable circle of confusion determines the depth of field and focus of an eye.

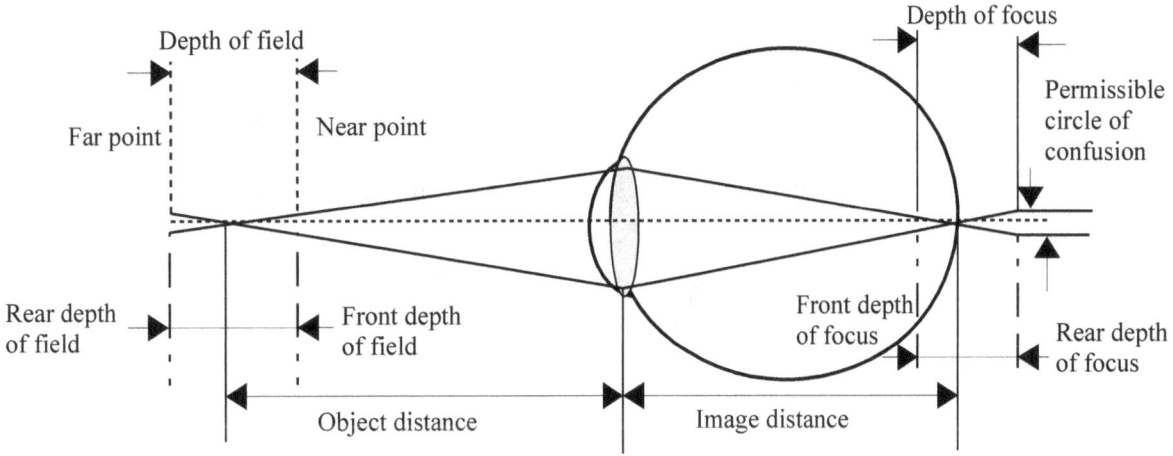

Figure 9.25 Depth of field and depth of focus

Depth of focus

The range of image distances over which the image of an improperly focused object is acceptably sharp is called the depth of focus. The depth of focus depends initially on the detecting element size and the focal length. In practice the depth of focused can be controlled. Decreasing the size of the pupil increases the depth of focus. By decreasing the size of the pupil means limiting the range of angles at which the rays may enter the lens, the pupil stop actually improves the depth of focus. As a trade-off, however, reducing the aperture stop limits the amount of light that can be collected by the retina, and therefore would require more sensitive detecting elements to achieve the same maximum detection range. This is why smaller apertures require longer exposure times in conventional photography.

Contrast

Contrast is the difference in brightness that makes an object distinguishable. When viewing a periodic object like bars or rectangle on a background. The maximum contrast of an image can be written mathematically as the contrast ratio.

Contrast is also the difference between the colour or shading of the printed material on a paper and the background on which it is printed. This is known as Weber contrast.

$$C = \frac{B_o - B_B}{B_B} \qquad (9.3)$$

where,
B_o = Brightness of the object
B_B = Brightness of the background

For eye to see well defined image, the black lines must appear black and white lines must appear white, Figure 9.26. Greater the brightness (intensity) difference between black and white better the contrast. So we can write this mathematically as percentage contrast. This is called Michelson contrast.

$$\text{Percentage contrast} = \frac{B_{max} - B_{min}}{B_{max} + B_{min}} \qquad (9.4)$$

where, B_{max} = maximum brightness, B_{min} = minimum brightness

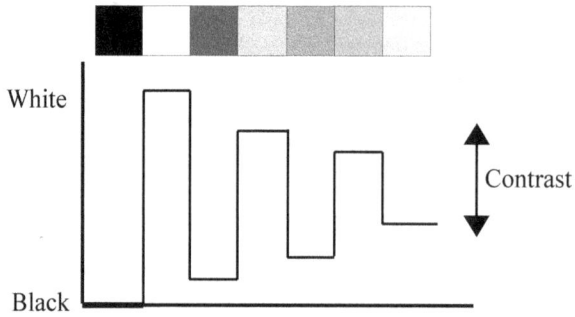

Figure 9.26 Contrast as percentage of brightness

The human eye, at any particular instant, can perceive contrast ratio over a range of 400:1 to 10,000:1. However, the human eye is a dynamic organ and can adjust, both chemically and via iris movement, over some 30 minutes in steady light conditions to perceive higher contrast ratios of up to between 1,000,000:1 and 10,000,000:1. The human eye is able to detect high contrast ratio for static images as compared to moving ones.

Chromatic Aberration

Chromatic aberration is a type of distortion which happens when a lens fails to bring all wavelengths of colour to a focus at the same point. Cause of chromatic aberration is that lenses have a different refractive index for different wavelengths of light. The refractive index is inversely proportional to wavelength. In chromatic aberration, objects can appear blurred, halos and streaks of colour appears in the image. There are two types of chromatic aberration: Longitudinal (*axial*) and Transverse (*lateral*).

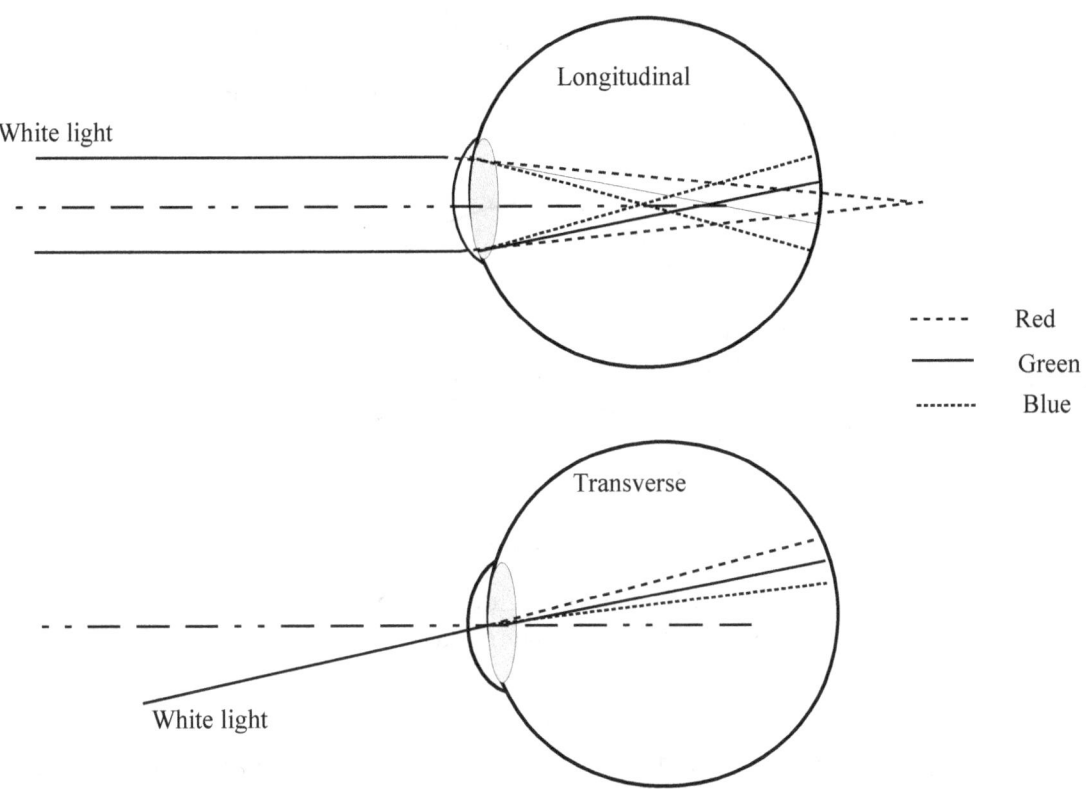

Figure 9.27 Chromatic Aberration

Transverse aberration occurs when different wavelengths are focused at different positions in the focal plane, whereas longitudinal aberration occurs when different wavelengths are focused at different distance from the lens on the optical axis.

In transverse aberration the edge of an object may appear fringed with colour and longitudinal chromatic aberrations can make it difficult to track objects and keep them in focus.

Optical defects of the eye

There are three common types of defects in image formation of the human eye. All these defects can be corrected by spectacles. Let's look at each defect in detail.

Myopia (Short sightedness)

Myopia is a medical term for short or near sightedness which occurs when light entering the eye is focused incorrectly, making distant objects appear blurred. A short sighted person sees close up objects clearly, but distant objects are blurred. Myopia is a type of refractive error of the eye and is due to when the physical length of the eye is greater than the optical length. This means that the eyeball is too long or the cornea, has too much curvature. This makes it more difficult for the eyes to focus light directly on the retina, so the image is formed in front of the retina so the images we see is blurry.

Near sightedness affects males and females equally. Most cases of myopia are caused by a combination of genetic and environmental factors that disrupt the normal growth of the eye. Most people with short sightedness have healthy eyes, but a small number of people with severe myopia develop a form of retinal degeneration.

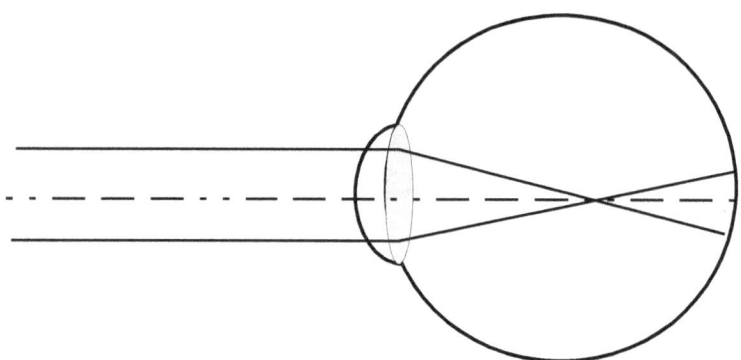

Figure 9.28 Short sightedness make image in front of the retina.

There is no way to prevent short sightedness. The other myth is also unproven that reading and watching television cause short sightedness. Short sightedness, which is measured in diopters by the strength of a corrective lens that focuses distant images on the retina, has also been classified by severity:

LOW: when myopia is of strength -3.00 diopters or less.

MEDIUM: when myopia is of strength -3.00 and -6.00 diopters. Most who suffer from this are more likely to have pigmentary glaucoma.

HIGH: when myopia has strength of more than -6.00 diopters. Most who suffer from this are more likely to have retinal detachment and also more likely to experience floaters.

Treating Myopia

Laser refractive surgery can reduce or even eliminate your dependence on glasses or contact lenses. This surgical procedure involves small alterations being made in your cornea using a laser, so that light rays are correctly focused onto your retina. The most common procedures for near sightedness include:

1- Photorefractive keratectomy (**PRK**) The first PRK procedure was performed in 1987 by Dr. Theo Seiler, then at the Free university medical centre in Berlin, Germany. In PRK the centre of the cornea called corneal stroma, is made flatter by removing more tissue from the centre than from the edge. To do this, the top layer on the surface of the cornea (just under the corneal epithelium) is removed using an excimer laser. The laser removes tissue from the cornea very accurately without damaging nearby tissues. There is a limit to how much tissue can safely be removed and therefore the amount of short sightedness that can be corrected. The corneal epithelium is removed and discarded, allowing the cells to regenerate after the surgery. The layer grows back during the healing process. PRK can be used to correct myopia and astigmatism at the same time.

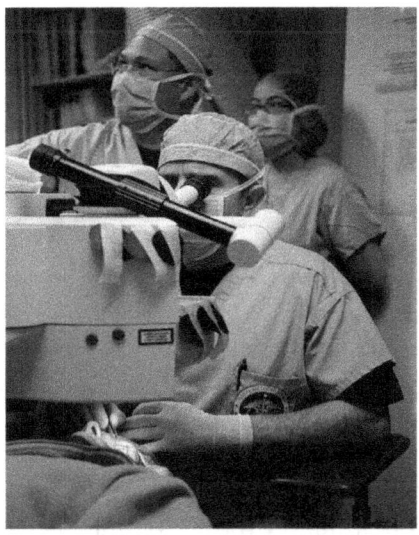

Figure 9.29 Photorefractive keratectomy at US Naval Medical Center San Diego.

After surgery everyone notice improvements in their vision after one of these surgeries. But not everyone gets perfect 6/6 vision. Most people with mild to moderate short sightedness will have uncorrected vision of 20/40 or better.

There are some risks associated with this surgery:

a) The most common problem is clouded vision, which is the result of a cornea healing process. This not permanent and will clear slowly.
b) People see halos and glare at night around bright lights.
c) New astigmatism. Some people will develop this which they did not have before surgery.
d) Some will develop double vision (diplopia)
e) As the cornea heals, cells may fill in the area that was shaped by the laser, causing at least some of the near sightedness to come back and is called regression and may even occur two years after the surgery.

***2-Laser-assisted** in **situ keratomileusis** (**LASIK**)* In this surgery tissue from the surface of the cornea is not removed rather its removed from its inner layer. This process is achieved with a mechanical micro-keratome using a metal blade, or a femto second laser and so a section of the outer corneal surface is cut. A hinge is left at one end of this flap. The flap is folded back, revealing the stroma, the middle section of the cornea.

Then an excimer laser (193 nm) is used to remove the precise amount of corneal stroma tissue needed to reshape the eye. The laser vaporises the tissue in a finely controlled manner (typical pulses are around 1 millijoule of pulse energy in 10 to 20 nanoseconds) without damaging the adjacent stroma. Performing the laser ablation in the deeper corneal stroma provides for more rapid visual recovery and less pain than the other technique of PRK. After the laser has reshaped the stromal layer, the flap is carefully repositioned to heal.

*3- **Corneal rings*** This surgery treat people who have keratoconus which is a degenerative disorder of the eye in which changes within the cornea cause it to thin and change to a more conical shape. Eye surgeon implant plastic rings, called Intacs, into your eye to help nearsightedness.

Figure 9.30 A simulation of the multiple images seen by a person with keratoconus.

This type of surgery treats mild myopia (-1.0 to -3.0 diopters), Intacs are two tiny, clear crescent-shaped pieces of a plastic polymer that are inserted into the the cornea to reshape the front surface of the eye called cornea. One advantage of the rings is that they may be left in place permanently, may be removed in case of a problem, or be adjusted should a prescription change be necessary.

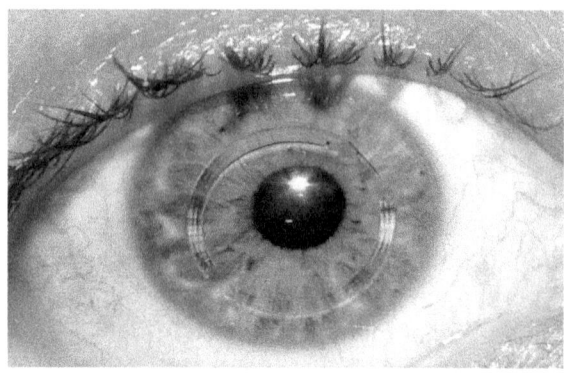

Figure 9.31 Intacs implant to correct keratoconus

4- Glasses The cheapest and safest way to correct myopia is glasses. To correct short sightedness we need a concave lens which has negative power. People who suffer from short sightedness their lens make image in front of the retina so concave lens bends light rays slightly outward before entering the cornea. The light rays now have a greater angle to bend back to focus when travelling through the cornea and lens. As a result, the light rays focus further back on the retina.

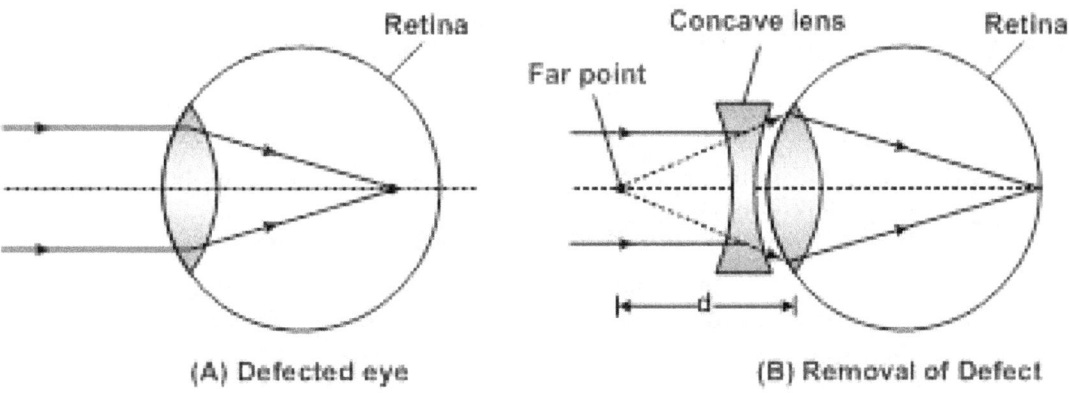

Figure 9.32 A) defective eye with myopia. B) correcting the eye vision by concave lens.

A person who can focus the distant object's image correctly on the retina has a normal vision. To calculate the power of the lens for such person we will use lens formula 4.3.

$$\frac{1}{f} = \frac{1}{u} + \frac{1}{v}$$

where u is the image distance that is the distance between lens and retina, which is approximately 2cm.
v is the distance of the distant object which we take at infinity and f is the focal length of the lens.

hence for normal eye,

$$\frac{1}{f} = \frac{1}{0.02} + \frac{1}{\infty} = \frac{1}{0.02}$$

therefore,

$$P_i = 50D$$

Now a person who is suffering from myopia may only be able to focus furthermost object at a distance of 70cm. The power of his eye lens when fully relaxed is:

$$\frac{1}{f} = \frac{1}{0.7} + \frac{1}{0.02}$$

therefore,
$$P_f = 1.43 + 50 = 51.43D$$

To calculates the power of the corrective lens, use equation 9.5

Power of corrective lens = Power of unaided eye − power of the defective eye   (9.5)

hence,

Power of corrective lens = 50 − 51.43 = −1.43D

The negative sign signifies that the corrective lens is concave lens with a power of -1.43D.

A person who is suffering from myopia will have his near point closer than the normal eye. If this person's eye has accommodation power of 5D, so

$$P_n - P_f = 5D$$

now, P_f is calculated as 51.43, therefore P_n is 56.43D.

To calculate his near point 'n' we again use formula 4.3

$$P_n = \frac{1}{n} + \frac{1}{u}$$

$$56.43D = \frac{1}{n} + \frac{1}{0.02} = \frac{1}{n} + 50D$$

$$\frac{1}{n} = 6.46$$

therefore,
$$n = 0.16 \text{m from the eye.}$$

Once this person starts wearing glasses his near point will change to n', which will be slightly further than the point he has without spectacles. The total power of his eye is now combined with the power of the lens.

$$P_{n'} = P_n + (-1.43) = 56.43 - 1.43 = 55D$$

the new near point can be calculated

$$P_{n'} = \frac{1}{n'} + \frac{1}{0.02} \quad \text{that is, } 55 = \frac{1}{n'} + 50$$

therefore,
$$n' = 0.2 \text{m from the eye.}$$

Hypermetropia (Long or Farsightedness)

Hypermetropia or Hyperopia is the medical term for long sightedness. Here a person sees a clearer vision when looking at objects in the distance, and have blurred vision when looking at closer objects. This could be because the eyeball is too short, or the cornea is not curved enough or lens is not thick enough so does not bend light sufficiently so it focuses at a point behind the retina. By placing a convex (plus powered) lens in front of the eye, the image is moved forward and focuses correctly on the retina.

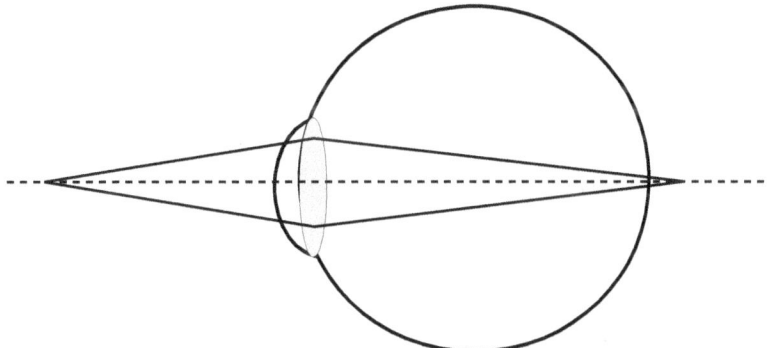

Figure 9.33a Long sightedness make image behind the retina.

There are four types of hypermetropia.

i. *Simple hypermetropia*: This one is caused by normal biological variation in the shape of the eye and the position of the lens and more often than not, is just a case of an eye that has not grown in length sufficiently. It is usually a continuation of the childhood hypermetropia which has not regressed with time. This is normally treated by spectacles or surgery.

ii. *Functional hypermetropia*: This is caused by the inability of the eye to accommodate, meaning that the muscles which focus the eye are not functioning correctly.

iii. *Pathological hypermetropia*: This is caused by abnormal eye anatomy which means the shape of the eye is not within the bounds of normal variation. This type of hyperopia is typically caused by disease within the eye, or abnormal development of the eye.

iv. *Acquired hypermetropia*: The one of the cause of this condition is removal of the crystalline lens after the operation of cataract. This can also occur after an injury or in certain disorders when the lens dislocates and slips backwards.

Hypermetropia can sometimes be divided into low, medium and high.
- *Low hypermetropia:* to + 3 D
- *Medium hypermetropia:* +3.25 to +5 D
- *High hypermetropia:* > +6 D

Treating Hypermetropia

The surgical procedure is the same as for myopia. However, larger amounts may be corrected with convex lens or contact lens. Convex lenses have a positive power, which causes the light to focus closer than its normal range.

To correct long sightedness we need a convex lens which has negative power.

Consider a person suffering from long sightedness has his near point at 3m from his eye. We can calculate the power of his eye.

$$P = \frac{1}{v} + \frac{1}{u} = \frac{1}{3} + \frac{1}{0.02} = 50.33D$$

to focus at 25 cm (unaided eye) he will need a lens power

$$P = \frac{1}{v} + \frac{1}{u} = \frac{1}{0.25} + \frac{1}{0.02} = 54D$$

To calculate the power of the corrective lens, use equation 9.5

Power of corrective lens = Power of unaided eye – power of the defective eye

Power of corrective lens = 54 – 50.33 = +3.67 D

The positive sign signifies that the corrective lens is convex lens with a power of +3.67 D.

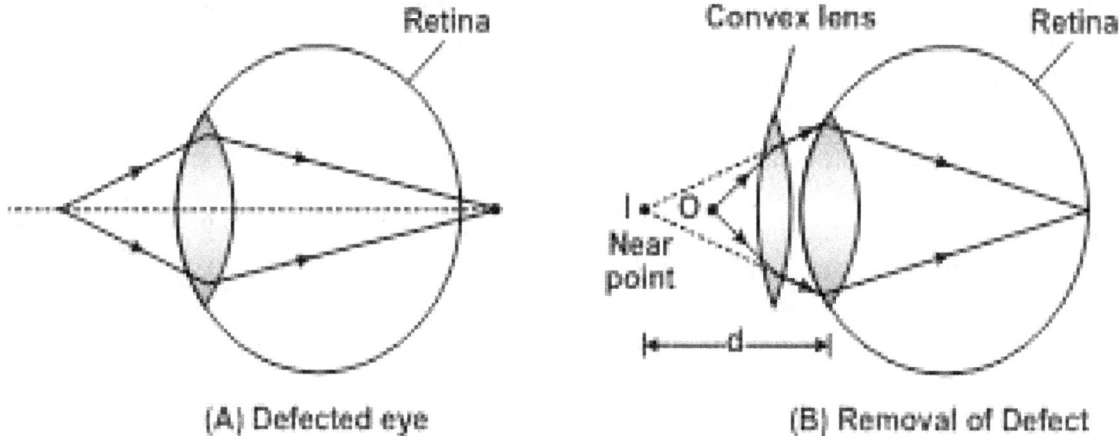

Figure 9.33b A) defective eye with hypermetropia. B) correcting the eye vision by convex lens.

Astigmatism

Astigmatism is a common (approximately 95% of the population have some degree of astigmatism) optical defect in which vision is blurred due to the inability of the optics of the eye to focus a point object into a sharply focused image on the retina. It occurs when a refractive error characterised by an irregular curvature of the cornea (toric curvature). Normal cornea should have a spherical shape like football, but for astigmatism it has an irregular curve (where one axis is steeper than the other) like a rugby ball. When light rays enters the eye will not focus properly, creating a blurred image. Astigmatism is usually congenital - is present at birth - and does not tend to change over time.

The two most common types of astigmatism are:

Corneal astigmatism - The cornea needs to have a perfect curve in order to bend (refract) light properly as it goes into the eye. Scientists are not sure why some people are born with a cornea that does not curve properly. Certain types of surgery or eye injuries that cause scarring of the cornea may cause astigmatism. Keratoconus, a degenerative disorder of the eye where the cornea gradually thins and changes to a more conical shape, can also cause astigmatism.

Lenticular astigmatism - The lens has variations in its curvature, rather than having a perfect curve, causing images to reach the back of the retina. Most patients with lenticular astigmatism have a normally shaped cornea - the defect is only in the curvature of the lens. People with diabetes commonly develop lenticular astigmatism because high blood sugar levels can cause the lens to change shape. The process usually develops slowly and is generally detected when the patient starts receiving treatment for the diabetes. When diabetes is controlled through treatment and blood sugar returns to more normal levels the lens' shape will also return to normal - many patients will notice the return as longsightedness.

If the astigmatism is very mild the optician suggest no treatment at all. The main treatment is corrective lenses bend the income light rays in a way that compensates for the error caused by faulty refraction so that images are properly received onto the retina. Whether the corrective lenses are in glasses or contact lenses is up to the patient - they are equally effective. There are three types of contact lenses. 1. Rigid contact lenses. 2. Gas permeable contact lenses. 3. Soft contact lenses.

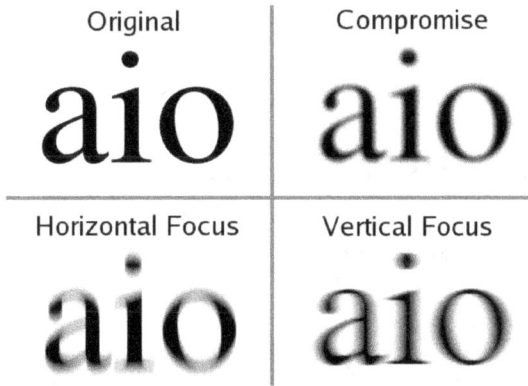

Figure 9.34. Blur from astigmatic lens at different distances

In most cases, astigmatism is present at birth. Astigmatism is often accompanies either myopia or hyperopia and affects people of all ages, however, it can develops after an eye injury or due to a complication of an eye operation. There are two types of astigmatism, one is called irregular astigmatism which is caused by a scar in the cornea or scattering in the crystalline lens. This type can not be corrected by wearing normal spectacles, but can be corrected by wearing contact lenses. The other type is called regular astigmatism and can be corrected by toric lens.

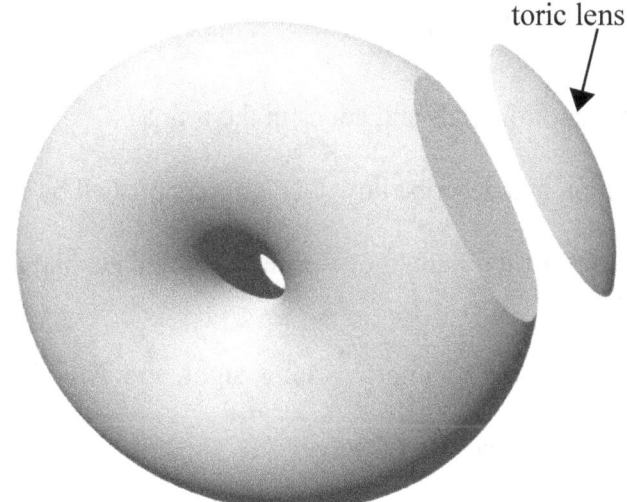

Figure 9.35 toric lens is a cap from the surface of torus

A toric lens has different optical power and focal length in two orientations perpendicular to each other. One of the lens surfaces is shaped like a "cap" from a tores (see figure 9.35), while the other one usually is spherical.

Exercise - 9

Q1- Correctly match the description with eye parts.

iris, retina, optic nerve, pupil, astigmatism, eye lens, rods, sclera, vitreous humour, fovea centralis, aqueous humour, suspensory ligament

a) the opening in the centre of the iris and changes size as the amount of light changes....................

b) sensory tissue that lines the back of the eye........................

c) the coloured part of the eye......................

d) a thick, transparent liquid that fills the centre of the eye...............................

e) the nerve that transmits electrical impulses from the retina to the brain....................

f) biconvex elastic structure that changes shape & focuses........................

g) contains a high density of cone receptors........................

h) night vision photo-receptors..................

I) transparent fluid that helps maintain eye shape and nourishes the cornea....................

J) elastic-like structures present in the eye that suspend the lens and pull it into shape for focusing distant objects onto the retina.......................

k) is the tough, white, outer layer (coat) of the eyeball, and is continuous with the cornea, it protects the entire eyeball........................

l) condition in which the surface of the cornea is not spherical, it causes a blurred or slightly twisted image to be received at the retina............................

Q2 - What are the functions of the rods and cones in the retina?

Q3 - Define accommodation and describe how this is accomplished by the eye.

Q4 - Explain what structure of the human eye primarily regulates the amount of light entering the eye and how it achieves this?

Q5 - What type of lens is used to improve long sight and why is this type of lens needed?

Q6 - a) What is meant by the **far point** of the eye?

b) A person has a far point of 1.20 m and a near point of 0.15 m. What type of lens must be used for her to view clearly objects at infinity?

c) Calculate the power of the corrective lens for this eye to allow objects at infinity to be viewed clearly.

d) This person has an uncorrected near point of 0.15 m. What would be her near point when using the corrective lens in **b(ii)**? (cea – GCE (AS) 3A 19 June 2009)

Q7 – a) Describe how a focused image is formed for objects at different distances from the eye.

b) Describe how the eye adapts over a period of time to an increase in the brightness of light.

c) A person has a normal distant vision, but a near point of 0.80 m. What name is given to this type of defective vision?

d) Suggest **two** possible causes of this condition and What type of lens is used to correct this condition?

e) Calculate the power of the spectacle lens required to correct the near point of the person in **(c)** to enable him to see clearly objects 0.25 m from his eyes.

f) What will his far point be when wearing these spectacles? (cea – GCE (AS) 3A 19 June 2008)

Q8 – a) Explain what is meant by the statement that two optical images can be **resolved** by the human eye.

b) State **one** factor upon which the resolving power of the human eye depends.

c) In the constellation of Lyra there is a double star called Epsilon. The two stars making up the double star are 2.41×10^{14} m apart. These stars are 2.37×10^{17} m from Earth. Calculate the angular separation of these stars as seen from Earth. Give your answer in degrees.

d) The resolving power of the human eye is 9.0×10^{-3} degrees. Explain whether or not these stars can be seen as separate by an unaided human eye. (cea – GCE (AS) 3A 22 June 2007)

Q9 - A student wears glasses to correct a defect in her vision. The prescription for each eye is the same and is stated as –2.75 D.

a) **(i)** State the name of the unit symbol D.

(ii) What type of lens is in the student's glasses?

(iii) State the physical property of the lens to which the prescription refers.

(b) (i) State the eye defect from which the student in **(a)** suffers.

(ii) Fig. 9.36 illustrates the student's eye with the uncorrected eye lens. The rays drawn come from an object at the normal near point. In Fig. 9.37 the rays come from a distant object. On Figs 9.36 and 9.37 sketches the rays between the eye lens and the retina.

(c) Determine the position of the student's uncorrected far point. Remember the correcting lens prescription is –2.75 D and a normal eye has a far point at infinity.

(cea – GCE (AS) 3A 22 June 2007)

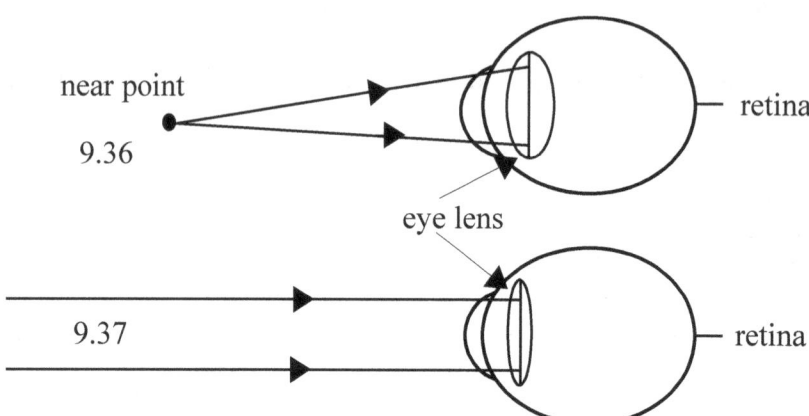

Q10 - **(a)** Define the terms

(i) near point,

(ii) far point.

(b) A person has a far point distance of 400 cm.

(i) Name the defect of vision from which this person suffers.

(ii) State the likely cause of this defect.

(iii) You are to use the lens equation to determine the power of the lens which would be used to correct the defect.

1. Write down the substitutions you will make for the quantities u and v in this equation. Include relevant signs and units.

2. Make these substitutions and calculate the power of the corrective lens. Give the magnitude of the power and state the appropriate unit.

(iv) The same person has a near point distance of 25.0 cm without spectacles. Calculate the near point distance when he is wearing the spectacles in **(b)(iii)**.

Q11 - A far sighted person person cannot see clearly objects that are closer to the eye than 73 cm. Determine the focal length of contact lenses that will enable this person to read a magazine at a distance of 25 cm.

Q12 - A person can see objects clearly only if they are between 80 cm and 500 cm from his eyes. His optometrist prescribes bifocals that enable him to see distant objects clearly through the top half of the lenses and read at a distance of 25 cm through the lower half. What are the powers of the top and bottom lenses.

10 – The Human Ear and Hearing System

Most of us take for granted the amazing gift of hearing. Without the ability to hear we can't communicate with others. Human hearing allows for the detection of sound and for 360° spatial detection and localization of sound sources. Our hearing system is sensitive to a limited range of sound intensities and frequencies. Our hearing system consists of two ears, located on the left and right sides of the head, the vestibulocochlear nerve, and the central auditory nervous system (CANS).

Transmission of sound in a medium

Sound can be transmitted through solid, liquid and gases. Sound cannot travel through a vacuum. Sound waves are created by a disturbance that can then propagate through a medium. For example, the transmission medium for sound received by the ears is usually air. It is this oscillation of air particles that triggers a cascade of mechanical and electrical events leading, ultimately, to the sensation of hearing.

Sound is a series of compression and rarefaction waves. It is produced by the vibration of the particles present in its medium. The presence of a medium is a must for the movement of sound waves. The speed and the physical characteristics of sound largely varies with the change in its ambient conditions. The speed of sound depends on the density of the medium through which sound is travelling as we have studied in chapter 1. If its density is fairly high, then the sound would travel at a faster speed. When sound travels through gaseous medium, its speed varies with respect to changes in temperature.

The frequency of a sound wave is the total number of oscillations that have been produced in one second. The length of sound wave varies according to its frequency as speed is fixed for given physical characteristic of the medium in which this sound travels. Sound waves with long wavelengths have low frequency or low pitch; and those with short wavelength have high frequency or high pitch. Sound that is heard by human ears has a frequency range of 20 Hz to 20 kHz.

The speed of sound varies from medium to medium. Sound travels almost 4.3 time faster in water and 17 times faster in steel at 20° C than in the air. Speed of sound in fluids depends on its compressibility and density. In solid the speed of sound depends on the shear deformation under shear stress and also the density of the medium. In gas, the speed of sound depends on temperature, molecular composition and heat capacity ratio. The ideal gas pressure has no effect on the speed of sound. Humidity has a very small effect on the speed of sound causing it to increase by about 0.1 to 0.6 percent as water replaces oxygen and nitrogen molecules.

The speed of sound in the air is given by Newton-Laplace equation:

$$v = \sqrt{\frac{E}{\rho}} \qquad (10.1)$$

where E = Bulk modulus, N m^{-2} or Pa and ρ = kg m^{-3}

The ratio of the change in pressure (Δ P) to the fractional volume ($\frac{\Delta V}{V}$) compression is called the bulk modulus of the solid material.

Hence,

$$E = \frac{\Delta P}{\frac{\Delta V}{V}} = -V \frac{dP}{dV} \qquad (10.2)$$

If we assume that when sound travels through the air the compressions and rarefaction made by sound waves are adiabatic. An adiabatic process is such a process which occurs without loss or gain of heat within a system. For ideal gases the bulk modulus E is simply the gas pressure multiplied by the adiabatic index, γ. The equation 10.1 can be written as:

$$v = \sqrt{\frac{\gamma P}{\rho}} \qquad (10.3)$$

For adiabatic pressure and volume obey the relationship:

$$V^{\gamma} P = \text{constant} = C \qquad (10.4)$$

rearranging for P

$$P = CV^{-\gamma}$$

and

$$\frac{dP}{dV} = -\gamma CV^{-\gamma-1}$$

Now, substituting in 10.2

$$E = -V(-\gamma CV^{-\gamma-1}) = -V\left(\frac{-\gamma C}{VV^{\gamma}}\right) = \frac{\gamma C}{V^{\gamma}} \qquad (10.5)$$

Also, the density of gas can be expressed as:

$$\rho = \frac{nM}{V} \qquad (10.6)$$

where n = number of moles, M = molecular mass of the gas and V = gas volume

substituting 10.5 and 10.6 into 10.1, gives

$$v = \sqrt{\frac{\gamma CV}{nMV^{\gamma}}} \qquad (10.7)$$

but

$$P = CV^{-\gamma}$$

$$v = \sqrt{\frac{\gamma PV}{nM}} \qquad (10.8)$$

If we consider air to behave like ideal gas, then pressure, volume and temperature is related:

$$nRT = PV \tag{10.9}$$

where R = universal gas constant = 8.314 J K^{-1} mol^{-1} and T = absolute temperature
hence;

$$v = \sqrt{\frac{\gamma RT}{M}} \tag{10.10}$$

The average molecular mass for dry air is 28.95 g mol^{-1}

The heat capacity ratio (γ) for an ideal gas can be related to the degree of freedom (f) of a molecule given by:

$$\gamma = 1 + \frac{2}{f} \tag{10.11}$$

Mon-atomic gas has three degrees of freedom (at room temperature these are three translational motions), while diatomic gas has five degrees of freedom (at room temperature these are three translational and two rotational). Thus

For monatomic gas γ = 5/3 and for diatomic gas γ = 7/5. The air is mostly consists of two diatomic gases, nitrogen (78%) and oxygen (21%). For air γ = 7/5 is the best approximation.

Example: Calculate the speed of sound in air at 0° C and at 30° C with pressure of 1 atmosphere.

$$v = \sqrt{\frac{\gamma RT}{M}}$$

where;
γ = 7/5, R = 8.314 J K^{-1} mol^{-1}, M = 28.95 g mol^{-1} and T = 273 K

$$v = \frac{\sqrt{1.4 \times 8.314 \times 273}}{0.02895} = 331.31 \text{ ms}^{-1}$$

at 30° C (303 K)

$$v = \frac{\sqrt{1.4 \times 8.314 \times 303}}{0.02895} = 349.03 \text{ ms}^{-1}$$

The approximate speed of sound in the air at different temperatures can be calculated from this formula:

$$v = 331.31 + (0.606 \, T) \text{ ms}^{-1} \tag{10.12}$$

where T is temperature in Celsius (°C)

Example: Calculate the speed of sound in water at $0\,°C$. The bulk modulus of water is 2.06 x 10^9 Nm^{-2} and density of water is 1000 kgm^{-3}.

$$V = \sqrt{\frac{E}{\rho}}$$

$$v = \sqrt{\frac{2.06 \times 10^9}{1000}} = 1435.27 \text{ ms}^{-1}$$

The following table gives the speed of sound in different mediums at 0°C.

Table 1: Speed of sound in different mediums

Medium	Speed (ms^{-1})
Rubber	60
Water Vapour 134 ^{0}C	494
Carbon dioxide	258
Hydrogen	1270
Lead	1230
Iron	5135
Copper	3560
Diamond	12000

Sound Intensity and Pressure

The amount of energy that is past a given area of the medium per unit of time is known as the intensity of the sound wave.

$$I = \frac{E}{At} \qquad (10.13)$$

Now, rate of energy is power.

$$I = \frac{P}{A} \qquad (10.14)$$

Since sound travels in every direction so the sound expand spherically from its origin. The area is thus is the surface area of sphere.

$$I = \frac{P}{4\pi r^2} \qquad (10.15)$$

Figure 10.1 Inverse square law

Sound intensity is a vector quantity. Also, Sound intensity is the time averaged product of pressure and particle velocity. The direction of the intensity is the average direction in which the energy is flowing.

$$I = Pv \qquad (10.16)$$

The sound pressure is the force of sound on a surface area perpendicular to the direction of sound. The lowest sound pressure human ears are sensitive to is 20 µPa (or in terms of sound intensity the threshold hearing is 1×10^{-12} Wm^{-2}). This is normally represented on a logarithmic scale called decibel (dB). So 20 µPa is 0 dB. The threshold of pain is about 140 dB.

The Sound Pressure Level can be expressed:

$$L_p = 10 \log_{10}(P^2/P_{th}^2) = 10 \log_{10}(P/P_{th})^2 = 20 \log_{10}(P/P_{th}) \quad (10.17)$$

where, L_p = sound pressure level (dB) or Loudness

P = sound pressure (Pa), $P_{th} = 2 \times 10^{-5}$ - threshold sound pressure (Pa)

Example: If you double the sound pressure, how much is this increases in dB. How many times power in dB must be increased from threshold of hearing to pain level. What this represents in the actual increase in power?

Double pressure = $2P$

$$L_p = 20 \log_{10}(2P/P_{th})$$
$$= 20 \times 0.3$$
$$= 6 \text{ dB}$$

Threshold to pain level power must be increased $\dfrac{140}{6} = 23\frac{1}{3}$ times

To calculate P we rearrange equation 10.17:

$$P = P_{th}(10^{\frac{L_p}{20}})$$
$$P = 2 \times 10^{-5}(10^{\frac{140}{20}}) = 200 \text{ Pa}$$

An increase of 10 million (10×10^6) in the pressure level. This is an incredible dynamic range of human hearing system. Scientists have found that at the threshold sound level the eardrum moves a distance smaller than the diameter of a hydrogen molecule. A normal speaking person has a sound level of 30 dB and an orchestra can hit sound pressure level of 110 dB. This gives a dynamic range of 80 dB. According to a 1970 article in the Journal of the Audio Engineering Society the maximum dynamic range of vinyl is about 90 dB where as for CD this is between 90 and 96 dB.

Example: What is the decibel reading corresponding to a pressure amplitude P = 0.4 N/m^2.

$$L_p = 20 \log_{10}(P/P_{th}) = 20 \log_{10}(0.4 / 2 \times 10^{-5}) = 66.02 \text{dB}$$

Example: Calculate the pressure amplitude corresponding to a decibel reading of 40 dB.

$$40\text{dB} = 20 \log_{10}(P/2 \times 10^{-5})$$
$$2 = \log_{10}(P/2 \times 10^{-5})$$

taking antilog on both sides $\quad 10^2 = (P/2 \times 10^{-5})$
thus

$$P = 2 \times 10^{-3} \text{ Nm}^{-2}$$

Our ears are most sensitive in a frequency band of 2-5 kHz. This is an important region where most of speech happens. When two tones are played at the same sound pressure level then one appears louder than the other.

The excessive sound pressure can damage our hearing so there is guideline how much and for how long we should expose ourselves to certain sound pressure. Many health organisations recommend that for every 3 dB sound pressure level over 85 dB, the permissible exposure time is cut in half – before damage to our hearing can occur. The table below shows a permissible by exposure time to sound pressure. When you hear music with earphone at full volume is equal to approximately 96 dB.

Table 2: Exposure time for sound pressure

Sound Pressure (dB)	Maximum exposure time
120	30 seconds
110	1.2 minutes
109	1.87 minutes
106	3.75 minutes
103	7.5 minutes
100	15 minutes
97	30 minutes
94	1 hour
91	2 hours
88	4 hours
85	8 hours
82	16 hours

National institute for occupational safety and health

This is obvious from equation 10.15 and 10.16 that the sound pressure decreases with distance from a point source with a relationship of $1/r$ and intensity decrease with a relationship of $1/r^2$. Since;

$$I \approx P^2 \approx 1/r^2 \implies P \approx 1/r$$

therefore,

$$I \alpha \frac{1}{r^2}$$

$$\frac{I_1}{I_2} = \frac{r_2^2}{r_1^2} \qquad (10.18)$$

$$\frac{P_1}{P_2} = \frac{r_2}{r_1} \qquad (10.19)$$

If we double the distance, the value of the sound pressure falls to a half of its initial value. If we double the distance, the value of the sound intensity falls to a quarter of its initial value. For a sound level change we can use this expression.

$$\Delta L = 10 \log_{10}\left(\frac{r_2^2}{r_1^2}\right) = 20 \log_{10}\left(\frac{r_2}{r_1}\right) \quad (10.20)$$

where $\Delta L = L_2 - L_1$ the value of the sound at distance r_2 and at r_1 respectively.
Hence,

$$L_2 = L_1 - 10 \log_{10}\left(\frac{r_2^2}{r_1^2}\right) \quad (10.21)$$

The sound intensity level is a logarithmic measure of the sound intensity in comparison to reference intensity.

$$L_I = 10 \log_{10}\left(\frac{I}{I_0}\right) \quad (10.22)$$

where, $I_0 = 1 \times 10^{-12}$ W/m².

The human audible range is then $\log_{10}(1/10^{-12}) = 12$ bel

What is decibel, Phons and Sones?

The decibel (dB) is a logarithmic unit to describe a ratio, which could be power, voltage, intensity and pressure. A decibel is defined as 10 times the logarithm of the ratio of the quantity being measured. The decibel is 1/10 of Bel, named in the honour of Alexander Graham Bell. Decibel provides the relative measure of sound intensity.

A phone is a unit of *loudness* for pure tones perceived by individual's. Sounds with equal intensities but different frequencies are perceived by the same person to have unequal loudness. For example, a 50 dB sound with a frequency of 2 kHz sound may be louder than a 50 dB sound with a frequency of 1kHz. Therefore, 1 phon is equivalent to 1 dB at 1 kHz.

The phone scale is not directly proportional to loudness. The sones scale was invented to provide a linear measure of loudness. The sone scale is based on the observation that a 10 phon increase in a sound level is most often perceived as a doubling of loudness. If we assign the lower end of the loudness of one sone equal to 40 phons, then 50 phons would have a loudness of 2 sones. Figure 10. 2 illustrates that.

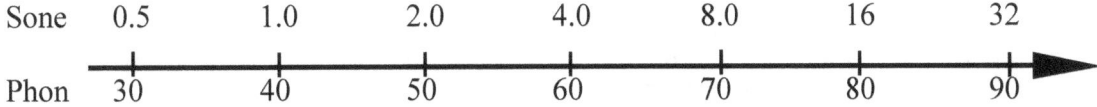

Figure 10.2 Loudness scale with units of sone and phon

Human hearing does not have a flat frequency response. We do not perceive high and low sound as well as sound near 2 kHz. This is due to the change of frequency response with amplitude. Figure 10.3 shows several equal loudness curves. They were first measured by Harvey Fletcher and Wilden A. Munson in 1933 and the produced these curves in 1937 now known as Fletcher-Munson curves or equal-loudness curves.

When listeners were presented with a pure tone at 1kHz at 70 dB; this has a loudness of 70 phon. They were then presented with different frequency and listener then adjusted the decibel level until it was perceived to have the same loudness as it had at 1 kHz. Then the decibel against frequency was plotted for 70 phon, which is shown as 70 phone curve, Then other loudness levels were presented. The curves shown in figure 10.3 are the average of many volunteers.

As our ears are more sensitive between 2-5 kHz, the sound at 20 Hz has to be at least 70 dB more powerful than at 3 kHz. The curve at 100 phons is a loud sound level. To equally loud at this level the sound at 20 Hz must be 40 dB more powerful.

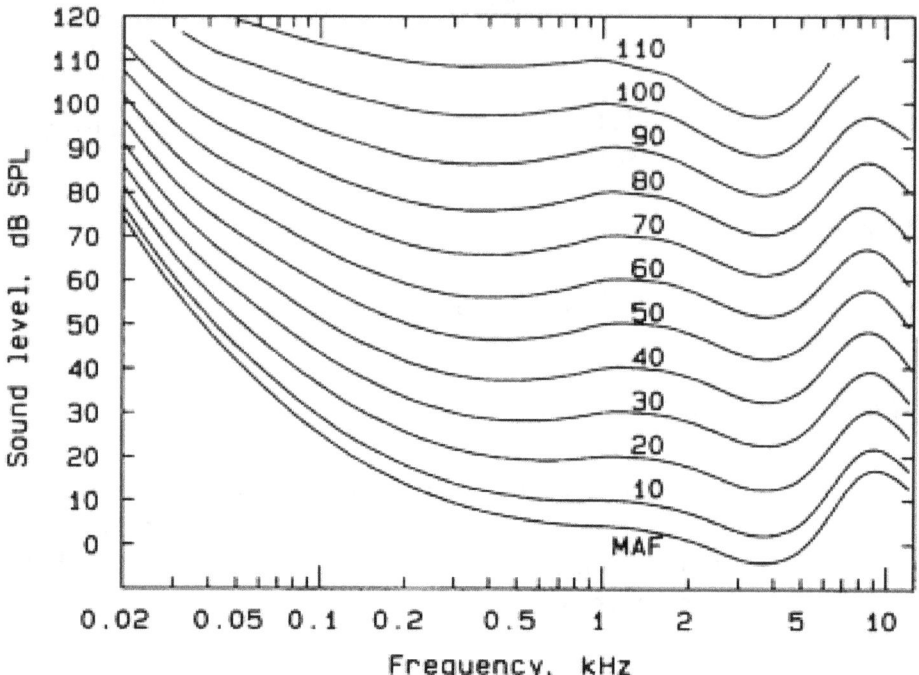

Figure 10.3 Equal loudness contours

Acoustic impedance

Acoustic impedance indicates how much sound pressure is generated by the vibration of molecules of a particular acoustic medium at a given frequency. Acoustic impedance Z is frequency dependent. Mathematically, it is the sound pressure p divided by the particle velocity v_p and the surface area S, through which an acoustic wave of frequency f propagates. Acoustic impedance can be expressed in either its constituent units (pressure per velocity per area) or in rayls per square meter.

$$Z = \frac{P}{v_p S} \qquad (10.23)$$

The *specific acoustic impedance* z is the ratio of sound pressure p to particle velocity v_p at a single frequency and is expressed in rayls.

$$z = \frac{P}{v_p} = ZS = \rho v_s \qquad (10.24)$$

where, v_s is the speed of sound.

The *characteristic acoustic ice* Z_0 of a medium the the inherent property of a medium.

$$Z_0 = \rho_0 c_0 \qquad (10.25)$$

here Z_0 is the characteristic impedance, measured in Rayls, and ρ_0 and c_0 are the density and speed of sound in the unperturbed medium (i.e. when there are no sound waves travelling in it). In a viscous medium, there will be a phase difference between the pressure and velocity, so the specific acoustic impedance Z will be different from the characteristic acoustic impedance Z_0.

The characteristic impedance of air at room temperature is about 415 Pa·s/m or N·s/m³. By comparison the sound speed and density of water are much higher, resulting in an impedance of 1.5 MPa·s/m, about 3400 times higher.

Sound Power

Sound waves, like any other waves, transport energy. The amount of energy transported is proportional to the square of the amplitude of the wave. From conservation of energy law we can say that this energy must come from the source which produce it and how loud the sound is depends on how fast the object is converting this energy into sound. Sound power can also be computed as the sound intensity times area:

$$P = IA \qquad (10.26)$$

Sound power belongs strictly to the sound source. Sound pressure is a measurement at a point in space near the source, while sound power is the total power produced by the source in all directions.

As we have seen before that the difference between two intensities or pressure can be expressed logarithmically. Sound power can more practically be expressed as a relation to the threshold of hearing that is 1×10^{-12} W. Sound power level can be expressed:

$$L_W = 10 \log_{10} \left(\frac{P}{P_0} \right) \qquad (10.27)$$

where,

L_w = Sound Power Level in Decibel (dB),

P = *sound power (W)*

P_o = *reference sound power - 10^{-12} (W)*.

Example: An omnidirectional loudspeaker radiates two hundred milliwatts (100 m W). Calculate the sound intensity level at a distance of 1 m, 2 m and 4 m from the loudspeaker?

Compare the two levels.

Using 10.24

$L_W = 10 \log_{10} (P_{actual}/P_0) = 10 \log_{10} (200 mW/(1 \times 10^{-12} W)$

$= 10 \log_{10}(2 \times 10^{11}) = 113$ d

The sound intensity at a given distance can be calculated using Equations 10.15 and 10.22 as:

$$L_I = 10 \log_{10} (I_{actual}/I_0) = 10 \log_{10} ((P_{source}/4\pi r^2)/I_0)$$

this can be simplified to give:

$$L_I = 10 \log_{10} (P_{source}/P_0) - 10 \log_{10}(4\pi) - 10 \log_{10}(r^2)$$

which can be simplified further to:

$$L_I = 10 \log_{10} (P_{source}/P_0) - 20 \log_{10}(r) - 11 \text{ dB}$$

We can use this final equation to calculate the intensity level at the three distances as:

$$L_I = 10 \log_{10} (200 \text{ mW}/10^{-12}\text{W}) - 20 \log_{10}(1) - 11 \text{ dB} = 113 \text{ dB} - 0 \text{ dB} - 11 \text{ dB} = 102 \text{ dB at 1m}$$

$$L_I = 10 \log_{10} (200 \text{ mW}/10^{-12}\text{W}) - 20 \log_{10}(2) - 11 \text{ dB} = 113 \text{ dB} - 6 \text{ dB} - 11 \text{ dB} = 96 \text{ dB at 2m}$$

$$L_I = 10 \log_{10} (200 \text{ mW}/10^{-12}\text{W}) - 20 \log_{10}(4) - 11 \text{ dB} = 113 \text{ dB} - 12 \text{ dB} - 11 \text{ dB} = 90 \text{ dB at 4m}$$

We can see that the sound intensity level reduces by 6 dB every time we double the distance; we have seen before that this is a direct consequence of the inverse square law. Strictly speaking this is only true when the sound source is well away from any surface which may act as a sound reflector.

Table 3: Sound pressure level

Source of sound	Sound pressure (dB)
Shock waves	190
Rocket launching	180
Jet engine at 3m	140
Threshold of pain	130
Rock concert	120
Non-electric chain saw	110
Pneumatic hammer at 2m	100
Heavy city traffic	90
Vacuum cleaner	80
Noisy room	70
Conversational speech	60
Washing machine at full speed	50
Normal conversation	40
Soft whisper	30
Rustling leaf	20
Human breathing at 3m	10
Thresholds of hearing	0

The Anatomy of the Ear

The ears are paired sensory organs for the auditory system. They not only receive sound but also maintain body balance and position. Physiologically and anatomically ear is divided into three sections: the *external ear,* the *middle ear,* and the *inner ear.*

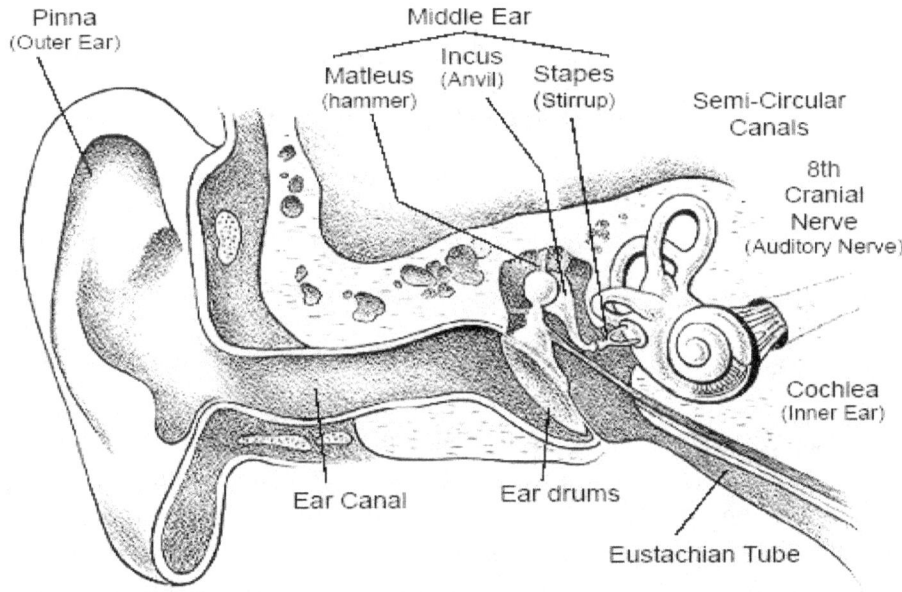

Figure 10.4 Anatomy of the ear

Hearing is the ability to perceive sound by detecting vibration in the air. The outer part of the ear collects the sound which is then amplified by the middle ear and finally passed into the hollow part of the inner ear which contain sensory epithelium studded with hair cells and the hollow channel is filled with liquid. When these hair cells are stimulated with sound they release chemical neurotransmitter. This way sound vibration is transformed into nerve impulse. These impulses from both ears then travel through the eighth cranial nerve to both sides of the brain stem and the up to the part of the cerebral cortex which is dedicated to sound reception.

Outer Ear:

Mainly consists of Pinna, ear canal and ear drum.

Pinna: This is the visible part of the ear that resides outside the ear is also called auricle. The pinna is made with cartilage and skin are attached to the skull by ligaments and muscles. Pinna collects, amplify and direct the sound to the ear canal. It also acts like a filter so it enhances the frequencies which our hearing system requires. Pinna also adds directional information to the sound.

For low frequencies, Pinna behaves similarly to a reflector dish, directing sounds toward the ear canal and for high frequencies, some of the sounds that enter the ear travel directly to the canal, others reflect off the contours of the pinna before entering the ear canal. This delay eliminates the frequency components whose wave period are twice the delay period.

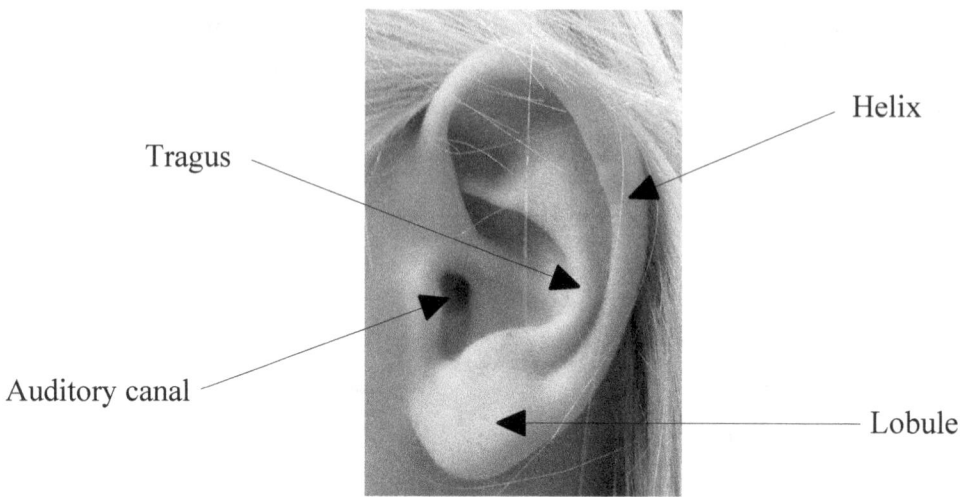

Figure 10.5 Pinna

The Auditory Canal: Once sound pass through the pinna, the sound waves enter the head through the auditory canal or more commonly known as the ear canal. The auditory canal is just like a pipe open on one side to the environment and closed on the other end by the eardrum, also known as tympanic membrane. These three parts make outer ear. The auditory canal has two main functions: one is to provide a passageway for sound travelling from the pinna to the eardrum, and the second is to protect eardrum and middle ear from infection as well as to protect it from foreign objects. The canal contains small hairs that filter out debris. The ear canal also produces wax to trap dust particles. There are also fibres called cerumen strands which are electro-statically charged, allowing them to catch small dust particles.

As the canal's structure is like a pipe open on one side, standing waves are formed inside. The standing waves will give rises to resonance. The question arise than what is the fundamental frequency?

The average length of the auditory canal is approximately 25mm. The minimum wavelength which will resonate in one open ended pipe is ¼ of wavelength.

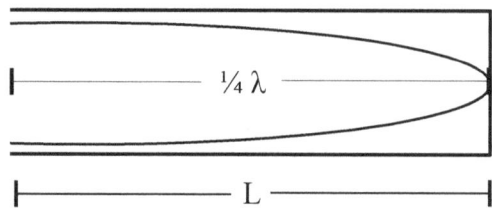

10.6 The resonance in auditory canal

$\frac{1}{4} \lambda = L = 0.025$ m

$\lambda = 0.1$ m

Lets take the speed of the sound in the air as 343 ms^{-1} at 20 °C.
Using equation
$$v = \lambda f$$
hence,
$$f = \frac{v}{\lambda} = \frac{343}{0.1} = 3430 \text{ Hz}$$

Most of the sound energy is absorbed and some is reflected by the tympanic membrane and this cause damping which results in broader range of resonant frequency, thus the fundamental frequency range from 3 to 4 kHz.

Eardrum: The eardrum is a thin membrane which is stretched with adjustable tension and also separates the outer ear from the middle ear. It acts like a transducer and vibrates in response to sound energy and transmits the resulting mechanical vibrations to the structures of the middle ear, which amplifies the vibrations and pass them on to the inner ear.

The diameter of the eardrum is about 1 cm and thickness of 0.5 mm. The eardrum consists of three layers of different materials. The first layer from auditory canal is composed of skin; the middle layer is mainly made with fibrous and elastic material and the one on the inner ear side is a mucus producing lining layer.

The eardrum can be perforated (a tear or hole) due to loud noise, infection of the middle ear, sudden excessive pressure change, or injury to the ear. In most cases this will heel over a period of around two months. You may feel uncomfortable and experience some hearing loss and in some cases earache.

During second world war pilots needed to dive quickly in aerial combat and that manoeuvre usually ruptured their eardrum, some German Luftwaffe pilots have said to pierce their eardrums in order to prevent air pressure issue. What height drop will rupture the pilot's eardrum?

A pressure difference of 8×10^3 Pa, can cause a rupture in the eardrum. We first have to find the pressure on the air depends on the height over sea level h in the gravitational field of the Earth. If we take a volume of air in the auditory canal with area A in contact with the eardrum, then the weight of this column of air is given by:

$$\text{weight} = m \cdot g = \rho \cdot V \cdot g = \rho \cdot h \cdot A \cdot g$$

where ρ is the air density = 1.29 kgm^{-3}

Now,
$$P = \frac{F}{A} = \frac{\rho g h A}{A} = \rho g h \qquad (10.28)$$

therefore, change in height will bring change in pressure.

$$\Delta P = \rho g \, \Delta h$$

$$\Delta h = \frac{\Delta P}{\rho g} = \frac{8 \times 10^3}{1.29 \times 9.81} = 632.17 \text{m}$$

When sound waves travel along the auditory canal and hit the eardrum, some of the energy carried by the waves will be absorbed and some will reflected by the eardrum. For optimum hearing sensitivity more energy should be absorbed. We have studied above that speed of sound is different in materials with different density. The product of velocity and density is called the characteristic impedance of the material.

$$Z_0 = \rho\, v \qquad (10.29)$$

If the acoustic impedance of the two mediums are different most energy will be reflected. When a sound wave travels in a medium with impedance Z_1, hit the medium with impedance Z_2. Some of the waves will pass through the material and some will reflect back. If the amplitude of the incoming wave is A_I, the amplitude of the reflected wave is A_R and transmitted wave amplitude is A_T, then their ratios can be written as:

$$\frac{A_R}{A_I} = \frac{Z_2 - Z_1}{Z_2 + Z_1} \qquad (10.30)$$

$$\frac{A_T}{A_I} = \frac{2 Z_2}{Z_2 + Z_1} \qquad (10.31)$$

But our ears are only sensitive to intensity and we have studied that $A \propto I^2$ therefore,

$$\frac{I_R}{I_I} = \frac{[Z_2 - Z_1]^2}{[Z_2 + Z_1]^2} \qquad (10.32)$$

$$\frac{I_T}{I_I} = \left(\frac{Z_1}{Z_2}\right) \cdot \left(\frac{2 Z_2}{Z_1 + Z_2}\right)^2 \qquad (10.33)$$

Table 4: Effect of temperature on characteristic impedance

Temperature (°C)	Acoustic impedance (N s m^{-3})
25	409.40
20	413.30
15	417.00
10	420.50
5	424.35
0	429.00
-5	432.60
-10	436.50
-15	440.50
-20	444.70
-25	450.00

The primary function of the middle ear is to offset the decrease in acoustic energy that would occur if the low impedance auditory canal was directly contacted the high-impedance cochlear fluid. When a sound wave is transferred from a low-impedance medium (e.g., air) to one of high impedance (e.g., water), a considerable amount of its energy is reflected and fails to enter the liquid. If no middle ear were present, only 0.1% of the acoustic wave energy travelling through air would enter the fluid of the cochlea and 99.9% would be reflected. See question in exercise 10.

Middle Ear: The middle ear is between the eardrum and the oval widow of the inner ear. The middle ear contains a small grouping of three tiny bones. They are essentially the tiniest bones in our body. They are commonly called *ossicles*. These three bones form a connection from the eardrum to the oval window of the inner ear. These three bones are:

Malleus or Hammer: One side malleus called the manubrium ("handle"), which adheres to the tympanic membrane. Whenever eardrum vibrates it make malleus to move to and fro. On the other side of the malleus is the *incus*.

Incus or Anvil: This middle bone connects malleus to the third bone of the middle ear, *stapes*. The incus transmits sound vibrations from the malleus to the stapes.

Stapes or Stirrup: One side of the stapes is connected to incus and the other side is connected to the outer side of the oval window.

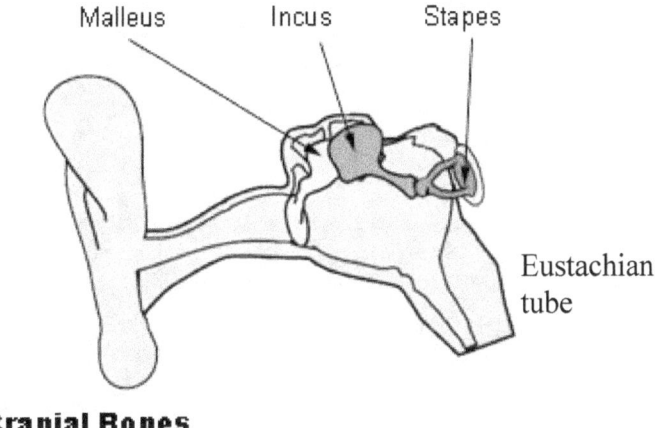

Cranial Bones

10.7 Three bones of ossicles of the middle ear.

The function of the three bones is to transfer sound vibrations from the outer ear into the inner ear. Then why do we need three bones, why simply one bone connects the eardrum to the oval window. Outer ear receives sound via air and thus a minuscule amount of force will make the eardrum to move. Now the oval window is connected to the *cochlea* of the inner ear. This sends sound through the fluid. To move fluid require a lot more force than moving air.

The human auditory system respond to a vast range of pressure change from as small as 10^{-5} Pa to 10 Pa. What would be the force on the eardrum at the loudest sound? The area of the eardrum is approximately 78 μm^2, but the actual area connected to the malleus is about 43 μm^2. The total force on the malleus is:

$$F_1 = P_{sound} A_{malleus} = 10 (43 \times 10^{-6}) = 0.43 \text{ mN}$$

and
on the eardrum,

$$F_1 = P_{sound} A_{eardrum} = 10 (78 \times 10^{-6}) = 0.78 \text{ mN}$$

The oscillations of the eardrum are transferred to the ossicles, which consists of a linkage of three bones and they function as a lever, therefore, increase the force on the eardrum to be 50 % larger when it reaches to the third bone of ossicles which is connected to the oval window.

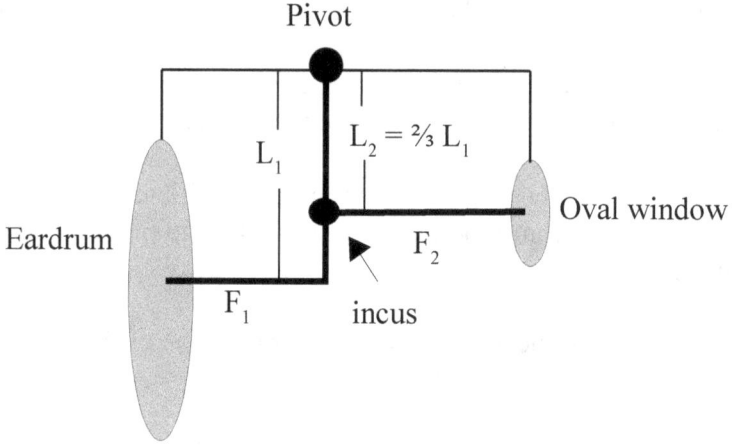

10.8 The lever system of ossicles of the middle ear.

The area of oval window is approximately twenty times smaller than the area of the eardrum. The force on the eardrum cause a torque τ_1 at the incus. This torque in turn transmits a force F_2 on the oval window.

$$\tau_1 = F_1 A_{mallus} = F_2 A_{oval\ window}$$

$$P_1 A_{mallus} L_1 = P_2 A_{oval\ window} L_2$$

$$P_2 = \frac{L_1}{L_2} \cdot \frac{A_{mallus}}{A_{ovalwindow}} \cdot P_1 \qquad \text{also } P_1 = P_{sound}$$

$$P_2 = 1.5 (20) P_{sound}$$

$$P_2 = 30 P_{sound}$$

This shows us that sound pressure has been amplified by 30 time. This represent an increase on dB scale

$$dB = 20 \log_{10} 30 = 29.5 \text{ dB}$$

Eustachian Tube: An eustachian tube connects the middle ear to the throat. It is vital for the tympanic membrane to have maximal mobility, the air pressure within the middle ear must equal that of the external environment. It is important in maintaining equal air pressure on both sides of the eardrum, which is important for normal hearing. Under normal condition, the eustachian tube is closed, but it can open to let a small amount of air through to prevent damage by equalizing pressure between the middle ear and the atmosphere.

The importance of the eustachian tube is experienced in a plane during rapid altitude change i.e. when descending. In this case the air pressure on the outside of the eardrum becomes greater and the eardrum will be pushed inward. This will not only cause discomfort but also cause temporary conductive hearing loss. Swallowing, yawning, or chewing may aid in opening the eustachian valves and equalize pressure thus bringing eardrum back to its normal position. When this happen, we normally hear a popping sound

Inner Ear: This is the most complicated part of the auditory system located in a tiny space of the hardest temporal bone and not accessible to direct examination, The inner ear comprise of three closely related structures - the cochlea (spiral tube), three semicircular canals and the vestibule (labyrinth).

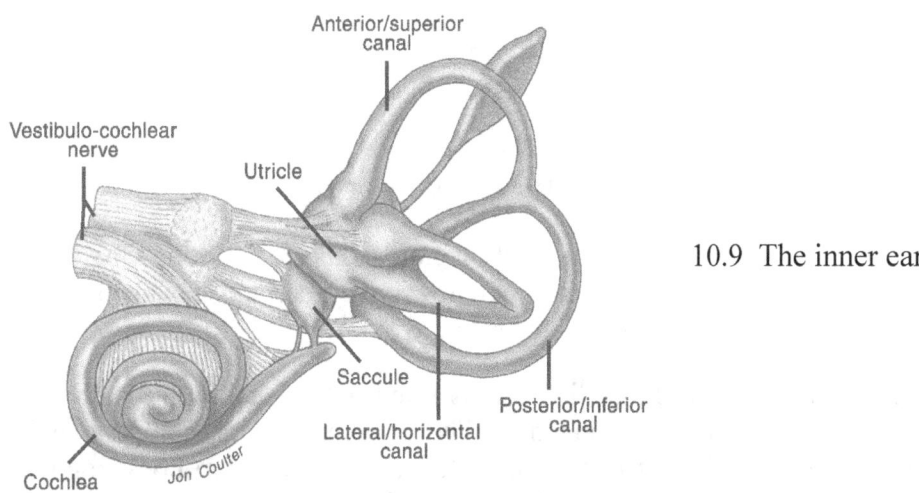

10.9 The inner ear

Cochlea: It is a spiral shape cavity making two and a half turns around its axis. The sound waves propagates from the oval window, also called base the round window also called the apex. Cochlea contains a bony core (modiolus) and a thin bony shelf (spiral lamina) that winds around the core like threads of a screw. The shelf divides the bony labyrinth of the cochlea into upper and lower compartments. The upper compartments, called the scala vestibulli (vestibular canal), leads from the oval window to the apex of the spiral. The lower compartment, the scala tympani (tympanic canal), extends from the apex of the cochlea to a membrane-covered opening in the wall of the inner ear called the round window. See figure 10.10.

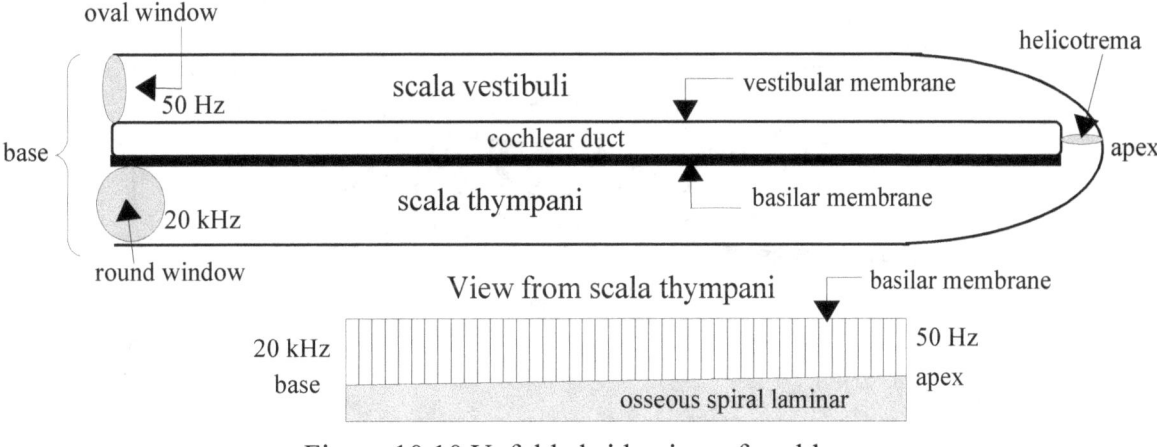

Figure 10.10 Unfolded side view of cochlea

Two canals connected at the end of the cochlea are filled with fluid called perilymph. Perilymph can flow from one canals to the other through an opening called the helicotrema at the apex of the cochlea. The principal ion of perilymph is Na^+.

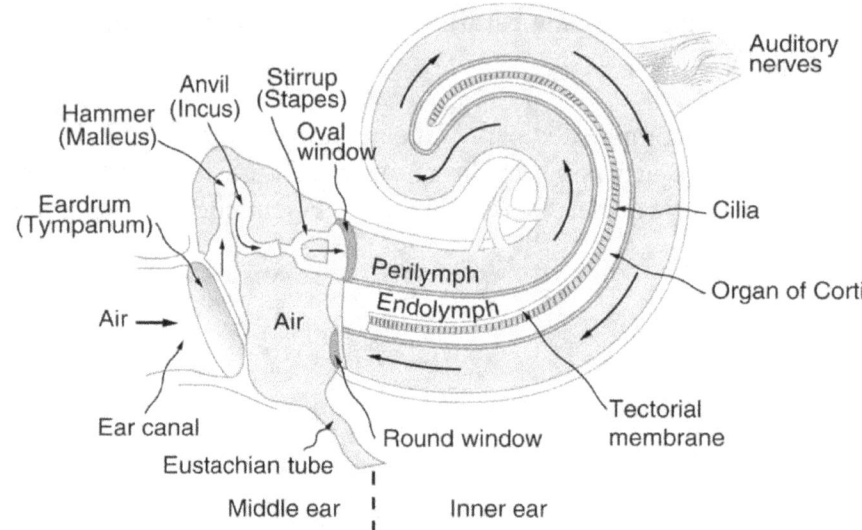

Figure 10.11 Structural diagram of the cochlea showing how fluid pushed in at the oval window moves, deflects the cochlear partition, and bulges back out at the round window.

The cochlea duct or scala media. This is filled with viscous fluid called endolymph, this makes the cochlear duct acoustically inactive. The principal ion is K^+ but it also contains a very low concentration of Ca^{2+} ions. Cochlear duct lies between the two bony compartments and ends as a closed sac at the apex of the cochlea. The cochlear duct is separated from the scala vestibuli by a vestibular membrane and from the scala tympani by a basilar membrane. The vibration of the basilar membrane is amplified by the outer hair cells situated in the organ of Corti. This process is called active amplification. Higher frequencies tend to vibrate the front end of the basilar membrane and low frequencies vibrate the back end.

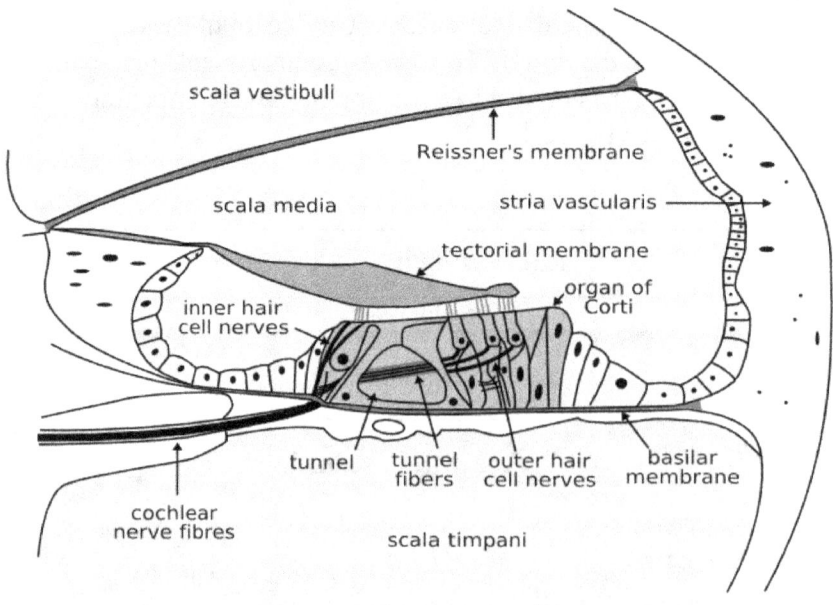

Figure 10.12 A cross section of the cochlea illustrating the organ of corti.

The organ of corti (named after Italian anatomist Alfonso Corti (1822- 1888) who discovered it) has a very specialised structure that responds to vibration in the fluid caused by the stapes footplate on the oval window and it runs the length of the cochlear tube. The sensory cell also know as hair cells (approx. 30,000) are located along the basilar membrane within the organ of corti. The length of these hairs varies becoming progressively longer from the base of the cochlea to its apex.

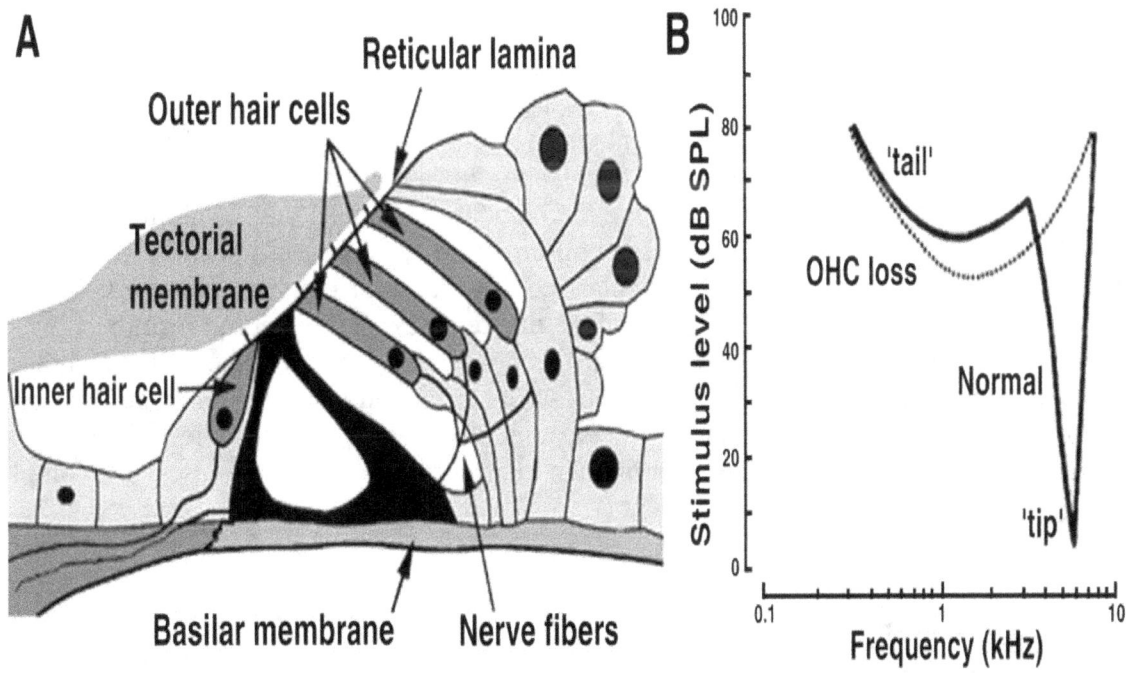

Figure 10.13 Section through the spiral organ of corti.

The hair cells are tuned to a certain frequency by way of their location on the cochlea (fig. 10.10). This is due to the stiffness in the basilar membrane. Stiffness is higher near the base, it allows only high frequency vibrations to move the basilar membrane and the hair cell present at that location. The stiffness decrease toward the apex and lower frequency vibration moves down the basilar membrane.

There are two anatomically distinct types of hair: the inner hairs and outer hairs. Our cochlea contains approximately 3500 inner cells and about 12,000 outer cells. The other important element in the organ of corti is a cellular membrane of collagen and specialised protein fibrils, known as *tectorial membrane*. This membrane makes mechanical contact with the tip of the hair cell. Along the basilar membrane hair cell are arranged as one row of inner cells followed by three to five rows of the outer cells. Outer hair cells have a special function within the cochlea. They are shaped cylindrically, like a can, and have *stereocilia* at the top of the cell, and a nucleus at the bottom. (fig. 10.13)

When the stapes rocks in & out, causing the membrane of the oval window to produce pressure waves within perilymph of the scala vestibuli and pass through the vestibular membrane to enter the endolymph of the cochlear duct, where they cause the basilar membrane to vibrate up and down. This creates a shearing force between the basilar membrane and the tectorial membrane, causing the hair cell stereocilia to bend back and forth. This leads to internal changes within the hair cells that creates electrical signals thus stimulate the receptor cells. Various receptor cells, however, have slightly different sensitivities to such deformation of the hairs. Thus, a sound that produces a particular frequency of vibration will excite certain receptor cells, while a sound involving another frequency will stimulate a different set of cells. This means the cell changes in length. So, with every sound wave, the cell shortens and then elongates. This pushes against the tectorial membrane, selectively amplifying the vibration of the basilar membrane. The pressure wave displaces the basilar membrane, transmitting pressure to the scala tympani and displacing the membrane of the round window.

The tips of the stereocilia of the outer hair cells are embedded in the tectorial membrane, and the bodies of hair cells rest on the basilar membrane (Fig. 10.13). An upward displacement of the basilar membrane creates a shearing force that results in lateral displacement of the stereocilia. Mechanical displacement of the stereocilia and the kinocilium (a special type of cilium on the apex of hair cells located in the epithelium) in a lateral direction causes an influx of K^+ potassium ions through their membranes, the hair cell is depolarized. The receptor potential opens a voltage gated calcium channel. The influx of Ca^{2+} triggers the release of the neurotransmitter (probably glutamate) containing vesicles in the cytoplasm near its base, fuse with the cell membrane and release neurotransmitter substance into the outside. This neurotransmitter simulates the sensory nerve fibres, and in turn they transmit nerve impulses along the cochlear branch of the *vestibulocochlear nerve* to the brain. The brain then interprets these nerve impulses, and completing the hearing process.

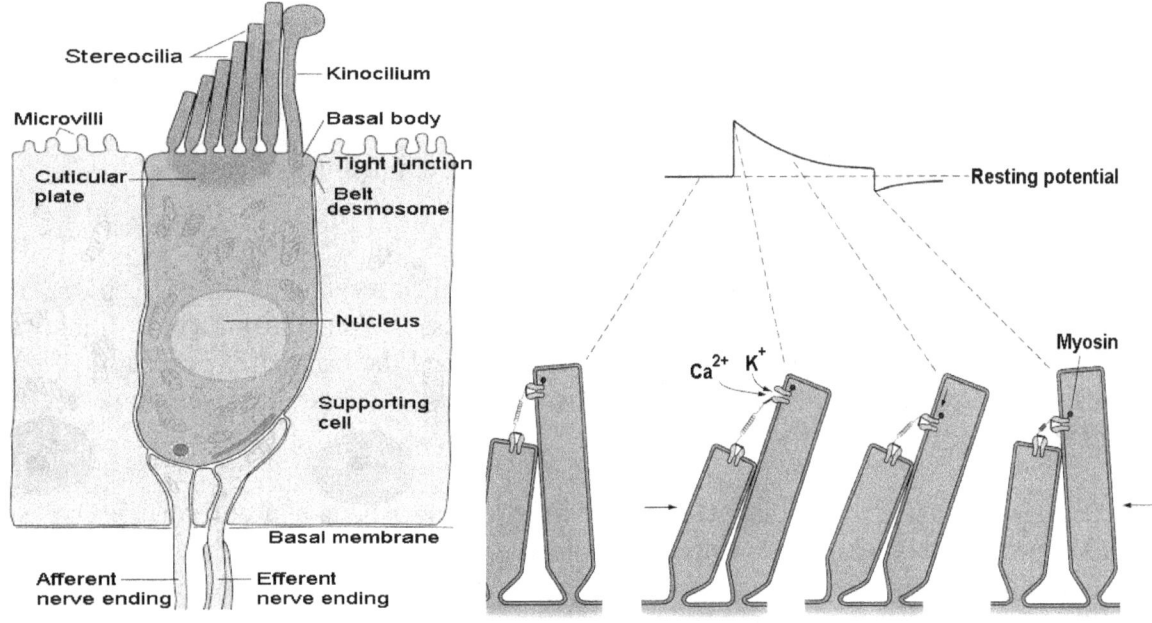

Figure 10.14 The changes in the receptor potential in the haircell. The resulting emission of neurotransmittors can elicit an action potential (AP) in the post-synaptic cell

Hearing Loss

Hearing loss occurs when the hearing system diminishes in sensitivity of normal sound. Following are the risk factors for hearing loss:

Early Age Factors:
- genetically inherited – both dominant and recessive genes can cause mild to profound impairment. A first gene mapped to non-syndromic deafness is DFNA1.
- prenatal disease – meningitis and measles may cause auditory nerve damage, rubella mumps
- premature birth – cause sensorineural hearing in 5% cases
- medications – irreversible effects are due to two important groups, one is the aminoglycosides (gentamicin) and the other is platinum based chemotherapeutics (cisplatin)
- chronic otitis media
- neurological disorder – multiple sclerosis and strokes.
- Chemicals – *metals* (lead, mercury), *solvent* (styrene, xylene, toluene, ethyl benzene, white spirit), *pesti/ herbicides* (paraquat, organophosophates)

Late Age Factors:
- noise – cochlear damage
- physical trauma
- ageing – presbycusis
- drugs – aspirin, antineoplastic, lasix, diuretics

Degree of hearing loss

- 0 – 25 dB   - none
- 26 – 40 dB - mild
- 41 – 55 dB - moderate
- 56 – 70 dB - moderate to severe
- 71 – 90 dB - severe
- 90 +   - profound

Types of hearing loss:

There are three main types of hearing loss:

Conductive hearing loss: is the result of sounds not being able to pass freely to the outer ear and the tiny bones (ossicles) of the middle ear. With this loss we feel a reduction in sound level or the ability to hear faint sounds.

Some possible causes of conductive hearing loss in the outer ear:
- blockage due to earwax (cerumen)
- build up of fluid due to ear infection (otitis externa)
- abnormality in the outer ear or ear canal
- tumour of the ear canal
- exostoses (formation of a new bone on the surface of a bone)

- tympanic membrane perforation
- tympanic retraction (part of the eardrum lies deeper within the ear than its normal position)

Some possible causes of conductive hearing loss in the middle ear:
- fluid in the middle ear from cold
- ear infection (otitis media)
- Allergies (serous otitis media)
- otosclerosis (abnormal growth of bone in the middle ear, the excess bone prevent the ossicles from moving freely)
- cholesteatoma (destructive and expanding growth consisting of keratinizing squamous epithelium in the middle ear)
- temporal bone tumour

Some possible causes of conductive hearing loss in the inner ear:
- severe otosclerosis – occurs when there is obstruction in either the oval or round window
- superior canal dehiscence syndrome (The symptoms are caused by a thinning or complete absence of the part of the temporal bone overlying the superior semicircular canal of the vestibular system)

Depending on its cause, a conductive hearing loss can either be temporary or permanent. Conductive hearing losses can often be corrected with medical treatment, or minor surgery.

Sensorineural hearing loss: also called *SNHL,* occurs when there is damage to the inner ear (cochlea) due to the sensitive hair cells either inside the cochlea or the auditory nerve are damaged, either naturally through ageing, or as a result of injury. This is the most common type of permanent hearing loss. The sufferer will be unable to hear faint sounds, even loud sound may still be unclear.

Some possible causes of SNHL:
- genetic
- toxic drugs
- severe head trauma
- illness

Sensorineural hearing loss may be congenital or acquired

congenital

- congenital cholesteatoma – the squamous epithelium hyperplasia in the middle ear becomes a invasive tumour.
- aplasia of the cochlea
- congenital rubella syndrome

acquired

> - inflammatory – viral, syphilis, suppurative labyrinthitis
> - ototoxic drugs – aminoglycosides, loop diuretics, antimetabolites
> - prolong exposure to noise > 90dB
> - perilymph fistula – when either oval or round window rupture and perilymph leaks into the middle ear. The sufferer will also experience imbalance.
> - Autoimmune – This occurs due to IgE (immunoglobulin E) and IgG food allergies

A person suffering with this type of hearing loss will often struggle to understand speech. When the cochlea hair cells become damaged, they will remain so for the rest of a person's life. Therefore sensorineural hearing loss is irreversible and cannot be cured. Now there is a hope for treating patients and that is using stem cells and gene therapy.

Hearing Aid

Those who suffer from mild conductive hearing or have a hearing impaired can use hearing aids. This is an electro acoustic device. There are different types of hearing aid offering different advantages, depending on size, levels of amplification and design. They are all battery operated and the main types are 'in the ear' (which sit in the outer ear), 'behind the ear' and 'in the canal' (which sit in the ear canal). A hearing aid is normally consists of:

- a microphone
- an amplifier
- a loudspeaker
- a battery

A microphone in a hearing aid device will picks up sound signals and these signals are amplified and fed to a speaker. The amplifier circuits can distinguish between background noise and foreground noise, so to make hearing pleasant. Now digital hearing aids are much better and can distinguish a quiet room from noisy room. There are many types of hearing aids.

Behind the ear (BTE) hearing aid consists of a case and an ear mould which sits inside the ear. The case contains the electronic circuit, controls and battery and usually sit behind the pinna. The sound from the case is connected to the tinny speaker in the ear mould. BTE can be used for mild to profound hearing loss.

In-the-ear (ITE) hearing aid fits perfectly in the outer ear bowl. All the working components are either located in a small compartment that is attached to the ear mould or inside the ear mould itself.

In-the-canal (ITC) hearing aid is same as ITC but it fills the outer part of the ear canal and are just visible.

Completely in-the-canal (CIC) hearing aid is even smaller and less visible than ITE hearing aids this gives a more natural experience of sound but they may not be recommended if someone has severe hearing loss.

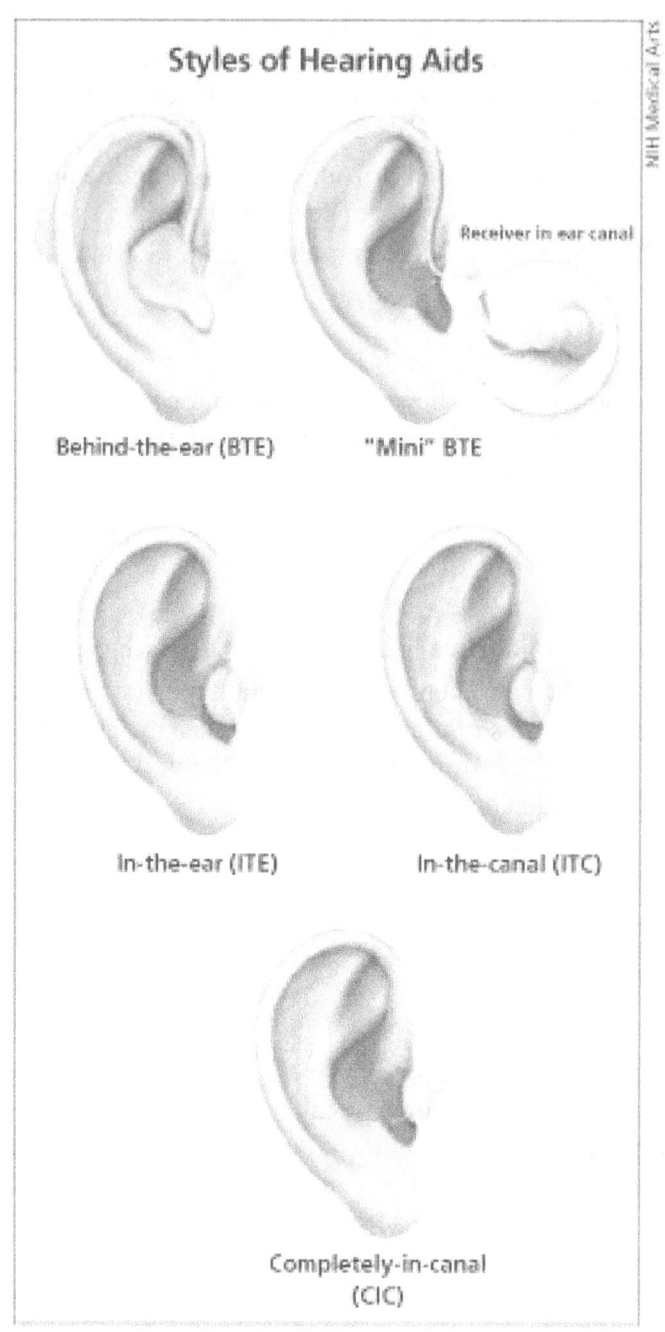

Figure 10.15 Hearing aid types

Cochlear implant

A cochlear implant is a surgically implanted device to help patients with damage to their sensory hair cells in the cochlea and gives them a sensation of hearing. Unlike hearing aids, cochlear implants bypass the damaged hair cells in the inner ear and stimulate the hearing nerve directly.

Cochlear implants are electronic devices made up of two parts:

An internal part: an implanted receiver and stimulator components that are inserted during a surgical operation in the mastoid bone behind the ear. They convert sound signals into electrical impulses through an internal wire to electrodes in the cochlea. There are around 22 electrodes connected to the cochlea. They send the impulses received from the receiver directly (bypassing hair cells) to the nerves in the scala tympani which then send signals to the brain through auditory nerves.

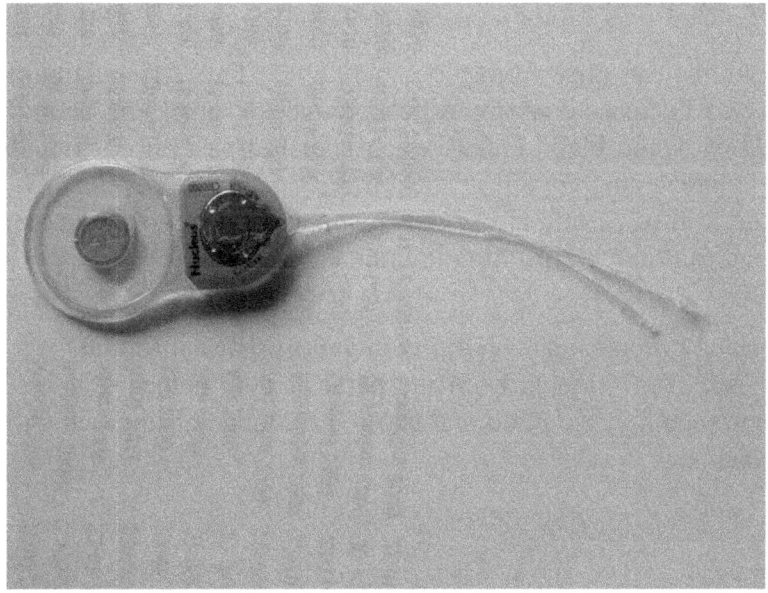

Figure 10.16 The internal part of a cochlear implant (model Cochlear Freedom 24 RE)

An external part: external components worn on the head or body like a hearing aid. This part consists of the following components.

- A microphone, which picks up sound from the environment.
- A speech processor, which selects using filter circuitry and arranges sounds into channels picked up by the microphone and send them to the transmitter.
- A transmitter which is coil held behind the external ear by a magnet, which receive signals from the speech processor and convert them into electric impulses and send them to internal part.

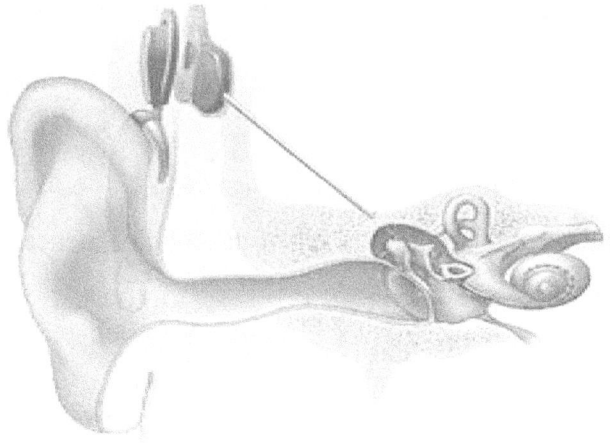

Figure 10.17 cochlear implant

Tinnitus: Tinnitus is not a disease, but a condition that results from a range of causes. Tinnitus is the medical name for the perception of noise in one ear, both ears or the head. Tinnitus is a sound that's heard in one or both ears for which there's no external source.

People with tinnitus can experience different types of sound. Most people describe it as a ringing sound but other sounds may include, whistling, roaring, hissing and humming. The sound is not produced by any abnormality in the hearing system, it is a misinterpretation by the brain. Tinnitus can be permanent or intermittent.

In some cases tinnitus can arise from the patient's ear. This type of tinnitus is called *objective tinnitus* can be caused by muscle spasm which produces a crackling sound in the middle ear or sometime is caused by sound blood makes when it rushes past an obstruction. Sometime tinnitus could be sign of life threatening condition such as a *carotid artery aneurysm*. This is a blood filled balloon shape of the wall of carotid artery which supply oxygen to the brain. When the size of an aneurysm increases it can rupture, resulting in brain haemorrhage.

If part of the cochlea becomes damaged, it will stop sending information to the brain. The brain will then actively "seek out" signals from the parts of the cochlea that are still working. These signals are over-represented in the brain and cause the sounds of tinnitus. Other possible causes of tinnitus. These include:

- build up of earwax
- exposure to noise
- stress
- ear infections
- méniére's disease
- antiviral drugs
- head injury

Presbycusis: is the loss of hearing that gradually occurs in most individuals as they grow older. Presbycusis is a complex and multi-factorial disorder, characterized by symmetrical progressive loss of hearing over many years. Presbycusis involves bilateral high-frequency hearing loss associated with difficulty in speech discrimination and central auditory processing of information. The hearing loss is most marked at high frequencies. This occurs due to epithelial atrophy with loss of sensory hair cells in the organ of Corti. This process starts at the base of the cochlea and slowly progresses toward the apex.

There are many causes of presbycusis:

- hereditary
- diabetes – proliferation in the blood vessels of the cochlea, hence depriving of oxygen
- hypertension
- atherosclerosis – diminishes vascularity of the cochlea

Exercise – 10

Q1 – Name the structure that connects the ear to the part of the throat located behind the nose?

Q2 – What is the audible range of human ears?

Q3 – The human ear operates by sensing pressure variations above and below atmospheric pressure. True or false.

Q4 – What does equal-loudness contour shows?

Q5 – Distinguish between the terms *sound intensity* and *sound intensity level*.

Q6 – During examinations, the sound intensity level in the college hall was 35 dB. At the end of the examination when candidates were leaving, the intensity level rises to 75.0 dB.
Calculate the numerical factor by which the sound intensity increases.

Q7 – State *two* reasons for using the decibel scale to measure sound intensity levels.

Q8 – A student has plotted a graph of $\log_{10}(I/I_0)$ against sound intensity level measured in dB (fig. 10.17). Find the gradient of this graph. (Hint: Use the defining equation for sound intensity level).

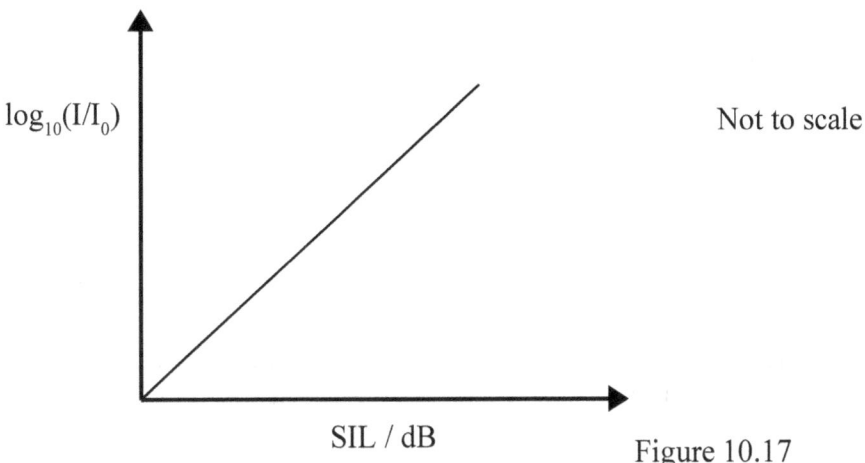

Figure 10.17

Q9 – A high frequency sound that is audible only to young people. The purpose of this device is to stop groups of teenagers congregating in certain areas by creating an irritating sound. At a distance of 2 m from it, the intensity level is 90 dB.

Calculate the intensity of the sound at this location. If the device is mounted on a wall, the sound energy is spread uniformly over the area of a hemisphere of radius 2 m. Calculate the energy emitted every second by the device.
The surface area of a hemisphere = $2\pi r^2$.

Q10 – Calculate the frequency of the sound we can best localize it. Take the average distance between each ear as 15 cm?

Q11 – Which statement accurately describes the mathematical relationship between the intensity of a sound and the decibel rating?

a. The decibel rating is the intensity value multiplied by 10.
b. The decibel rating doubles every time the intensity rating increases by 1000.
c. The decibel rating increases by 10 dB for every 10-fold increase in the intensity level.
d. The decibel rating is 10 times the absolute power of the exponent on 10 of the intensity level.

Q12 – How is the decibel scale different than the phon scale?

a. All sounds have a different decibel rating but the same phon rating.
b. The decibel scale is an objective measure of a sound; the phon scale is more subjective.
c. There is no limit on the decibel scale; the phon scale is limited to a smaller range of numbers.
d. The decibel scale measures the loudness of sound; the phon scale measures a person's perception.

Q13 – What would be the phon rating of a 100 Hz sound that has a decibel rating of 60 dB? (see figure (10.3)

a. 30 phons
b. 35 phons
c. 40 phons

d. 60 phons

Q14 – Based on the information in Figure 10.2, how many times louder would a typical person perceives a 70 phon sound to be compared to a 40 phon sound?
a. 0.5 times louder
b. 3 times louder
c. 8 times louder

d. 30 times louder

Q15 – A sound has an intensity level of 1 x 10^{-2} W/m² and a frequency of 200 Hz. Use Table 1, Figure 10.3, and Figure 10.2 to determine its rating on the Sone Scale.
a. Approximately 10 sones
b. Approximately 64 sones
c. Approximately 100 sones
d. Approximately 110 sones

Q16 – Sound vibrations reach the cochlea via...

a. the footplate of the stapes in the oval window

b. the three bones of the middle ear

c. the footplate of the stapes in the round window.

d. the basilar membrane.

Q17 – Short wavelengths of high pitched sounds cause displacement of the basilar membrane...

a. far from the oval window

b. near the oval window

c. near the round window

d. far from the round window

Q18 – The function of the round window is to dampen sounds after they stimulate the hair cells.

a. true

b. false

Q19 – Calculate the difference in intensity (loudness) levels for two uncorrelated sounds which differ by a factor of 4 in intensity, $I_2 = 4 I_1$ (Hint: use equation 10.22)

Q20 – Two uncorrelated sounds at a frequency of 1 kHz with intensity (loudness) $L_1 = 60$ dB and $L_2 = 80$ dB are emitted. Show that $I_2 = 100 I_1$ (Hint: use equation 10.22)

Q21 – What would be the total loudness when both sounds are emitted simultaneously at the same spot in Q20 with only on the difference that they are correlated i.e. have no phase difference? (Hint: If sounds are correlated then must add intensities).

Q22 – A chef is working with Blander in a restaurant kitchen at an intensity level of 80dB. Another person in the kitchen is also using dough making machine at an intensity level of 78 dB. Calculate the intensity level experienced by the chef when both machines are switched on.

Q23 – The area of a typical eardrum is about 5.0×10^{-5} m^2. Calculate the sound power (the energy per second) incident on an eardrum at (a) the threshold of hearing and (b) the threshold of pain.
Threshold of hearing: $I = 10^{-12}$ W/m^2
Threshold of pain: $I = 1$ W/m^2

Q24 – The human ear canal is about 2.6 cm long. If it is regarded as a tube open at one end and closed at the eardrum, what is the fundamental frequency around which we would expect hearing to be better? Speed of sound, use: $v = 340$ m/s

Q25 – Calculate the speed of sound in the substances given in the table below and fill the table.

Substances	Bulk Modulus Elasticity $\times 10^9$ (N/m^2)	Density (kg/m^3)	Speed (m/s^{-1})
Ethyl Alcohol	1.06	810	
Mercury	2.85	13595	
Iridium	320	22400	
Blood	2.4	1027	

Q26 – During a fireworks display, a rocket explodes high in the air. Assume that the sound spreads out uniformly in all directions and that reflections from the ground can be ignored. When the sound reaches listener 2, who is r_2 = 800 m away from the explosion, the sound has an intensity of I_2 = 1.1 W/m^2. What is the sound intensity detected by listener 1, who is r_1 = 200 m away from the explosion?

Q27 – The smallest change in loudness that an average person can detect is about 1 dB. What percentage increase in intensity does this represent?

Q28 – When you are playing a violin produces a sound level of 60 dB at your ears.

a.) What is the intensity of the sound energy in your ear?
b.) If a second violin plays, producing the same sound level at your ear, what is the sound level from both violins playing?

Q29 – Four singers are singing a song as a group. They all produce sound level of the same magnitude at 2m. Their combined loudness is 80 dB.

What is the sound level from one person?

Q30 – Which of the following is true about cochlea?

1. It helps maintain dynamic equilibrium or balance.
2. It is a tube that helps maintain normal pressure in the ear.
3. It contains the organ of corti, the sensory organ of hearing.
4. It is a nerve which brings signals from the inner ear to the brain.

11 – Biopotential and measurements

An electric potential that is measured between points in living cells, tissues and organisms is called biopotentials. This arises from the electrochemical gradients established across cell membranes. In most living cells, potassium ions K^+ are 30 to 40 times more concentrated internally than externally whereas, sodium Na^+ ions are 10 timeless concentrated internally than externally. Also, chloride ions Cl^- are in less concentration inside cells than outside cells, even though there are abundant intracellular fixed negative charges. While calcium ion Ca^{2+} concentration is relatively low in body fluids external to the cells, the concentration of ionized calcium internally is much lower than that found external to the cells.

The electrical activity of neuron cell occurs due to ion exchange through cell membranes. For inactive cell this membrane potential is called *resting potential*. As described above that K^+ have a higher concentration internally, therefore, a diffusion gradient occurs towards the exterior of the cell which will make the interior more negative with respect to the exterior. This will result in building up of electric field towards the interior of the cell. In addition to these ion channels, there is a pump that moves out three Na^+ ions for every two K^+ entering the cell. Normally, the diffusion gradient of K^+ ion is balanced by electric field until equilibrium is reached at -70mV. Whenever a cell is electrically stimulated by central nervous system, the diffusion of Na^+ ions increases towards the interior of the cell. This will raise the internal potential of the cell. When this potential reaches to +40mV, the diffusion gradient of Na^+ decreases and the diffusion gradient to K^+ increases, resulting in a sharp decrease in membrane potential towards a steady state.

Action Potential

The resting potential occurs when a neuron is at rest. An action potential is caused when different ions cross the neuron membrane. The action potential is an electrical activity that is caused by a *depolarizing current*. An action potential occurs in the types of cells called *excitable cells*, which include neurons, muscle cells and endocrine cells.

The action potential sequence is essential for neural communication and many such action potentials are required for communication. For modelling the action potential for a human nerve cell, a nominal rest potential of -70mV will be used. The process involves six steps. To understand these steps you should read each step and also studying the diagram 11.1a-e and 11.2.

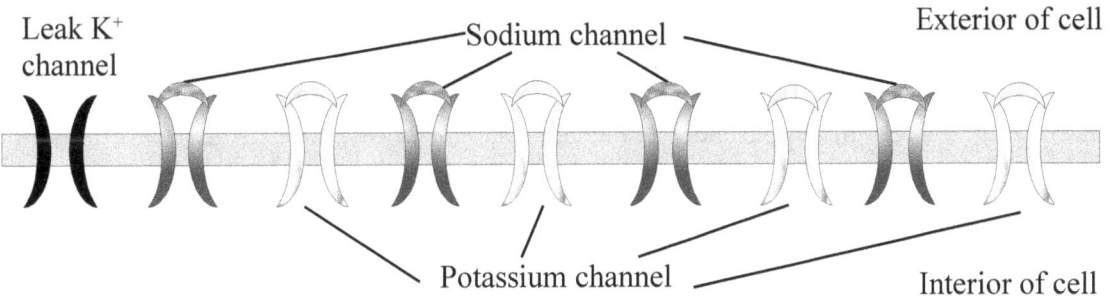

Figure 11.1a resting phase of neuron

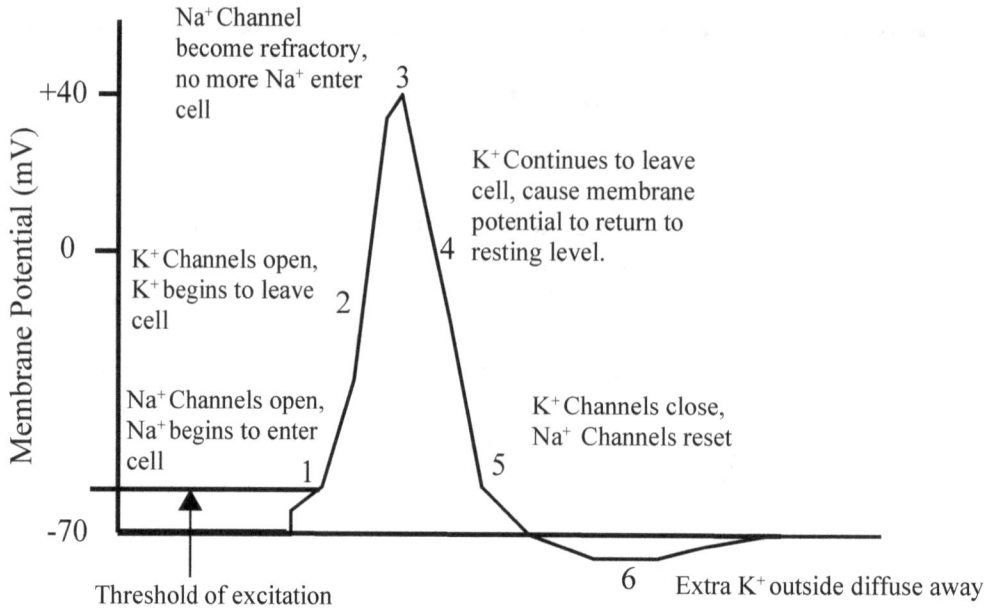

Figure 11.2 A view of action potential shows its various phases as the action potential passes a point on a cell membrane.

1- Resting Phase: When the neuron is inactive and polarised it said to be in a steady state or resting phase. At this stage only the so called "leak" K^+ channels are open thus establishing the resting potential. These channels are separate from other K^+ channels as they remain open.

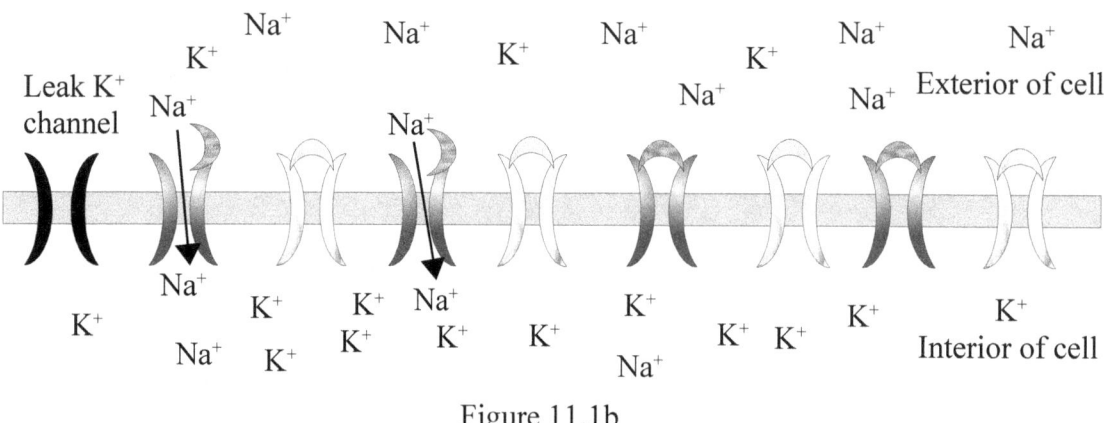

Figure 11.1b

2- Rising Phase: A stimulus is received by the dendrites of a nerve cell. This causes the Na^+ channels to open. If the opening is sufficient to drive the interior potential from -70 mV up to -55 mV, the process continues. The increasing positive shift in membrane potential is driven by the opening of progressively more and more Na^+ channels. This allows the Na^+ that was on the outside of the membrane to go rushing into the cell. The cell goes from being polarized to being depolarized state. As more positive ions go inside the cell making it positive (remember at resting phase interior of the cell was more negative) and the threshold is reached. After this more gated ion channel open and allow more Na^+ ions inside the cell. An action potential is created. The neuron continues to open Na^+ channel all along the membrane.

3- **Overshoot Phase:** Inside of the cell is now flooded with Na^+ and the membrane potential is at its most positive state (40 mV) which has overshot 0 mV. At this stage two processes occur simultaneously. First, voltage-gated sodium channel which were active during the rising phase began to close, so sodium conductance declines. Secondly, K^+ channels begin to open, driving the membrane potential back toward the equilibrium potential for K^+. Since the K^+ channels are much slower to open, the depolarization has time to be completed. Having both Na^+ and K^+ channels open at the same time would drive the system toward neutrality and prevent the creation of the action potential.

4- These voltage gated potassium channels differ from leak potassium channels because they are normally closed at the resting potential but open in response to depolarisation.

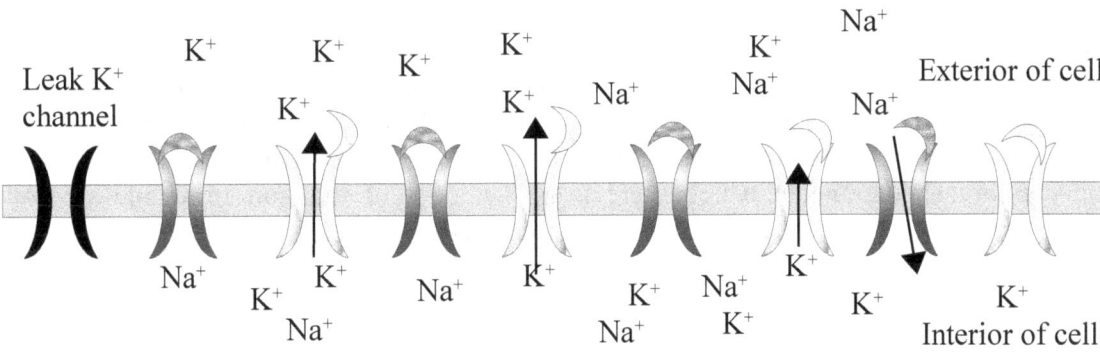

Figure 11.1c

Now the action potential is in re-polarizing phase in which membrane potential very quickly returning to the resting potential. During this phase activation of voltage-gated potassium is at the maximum level and the number of open sodium channels are rapidly declining.

5- **Hyper-polarization:** The action potential re-polarize beyond the resting membrane voltage. The undershoot to about -90mV because most voltage-gated K^+ channels are still open which results in K^+ conductance of the membrane being greater than when the membrane is at its resting potential. Hyper-polarization prevents the neuron from receiving another stimulus during this time, or at least raises the threshold for any new stimulus. Part of the importance of hyper-polarization is in preventing any stimulus already sent up an axon from triggering another action potential in the opposite direction. This period doesn't last long, though.

6- Recovery Phase: The Na^+/K^+ pump eventually brings the membrane back to its resting state of -70mV so membrane potential returns to the original steady state resting potential.

Quantitative model of the action potential.

Two types of mathematical models of the action potential have been developed. First type model the data qualitatively. *Hodgkin-Huxley model* is of this type. This model has limitations as it only considers two ions (K^+ and Na^+) each with only one voltage-sensitive channel. Therefore, this model cannot describe all types of an excitable cell membrane, like *cardiac action potential* which is produced by voltage-sensitive calcium channel and different types of potassium/sodium channels. The second type is a simplified version of the first type. This model tries to understand qualitatively the role of action potential in neural circuit. The *Fitzhugh-Nagumo model* is an example of this type.

Hodgkin-Huxley model

Two British Physiologist and Biophysicist Alan Lloyd Hodgkin (5 February 1914 – 20 December 1998) and Andrew Huxley (22 November 1917 – 30 May 1912) developed a model in 1952 to explain their experimental voltage-clamp data on the axon from Loligo squid. They received the Nobel prize in 1963 in recognition of their work. The model simulates an electrical signal called an action potential that passes through the axon of a neuron. Hodgkin-Huxley describe their work by saying " *Our object here is to find equations which describe the conductances with reasonable accuracy and are sufficiently simple for theoretical calculation of the action potential and refractory period. For sake of illustration we shall try to provide a physical basis for the equations, but must emphasize that the interpretation given is unlikely to provide a correct picture of the membrane.*" *(Hodgkin and Huxley, 1952d, p. 506)* Hodgkin-Huxley model is shown in Figure 11.3. Each component of an excitable cell is analogue to biophysical properties. The lipid bilayer is represented by constant capacitance C_m (F/cm^2). Voltage-gated ion channels are represented by a non-linear electrical conductance g_n (where n is the specific ion channel). Conductance is voltage-time dependent and mediated by voltage-gated cation channel proteins. Leak channels are represented by linear conductance g_L. The *emf* is represented by batteries E_n. The value of these *emf* is determined by *Nernst Potential* (the reverse potential of an ion is the membrane potential at which there is no net flow of that ion from one side of the membrane to the other) of the ion. Ion pumps are represented by current sources I_p.

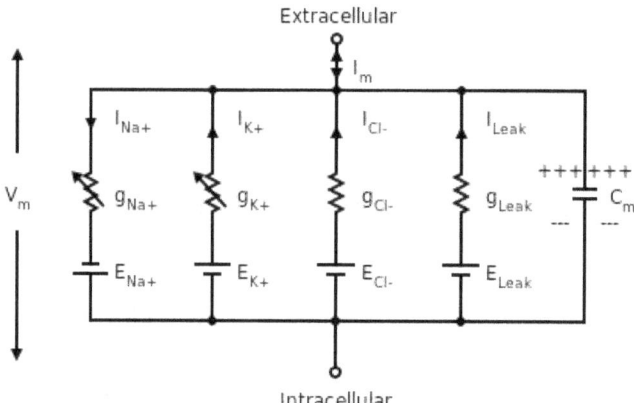

Figure 11.3 Equivalent electrical circuit for the Hodgkin-Huxley model of the action potential.

The voltage V_m changes with the total transmembrane current I_m according to this equation:

$$C_m \frac{dV}{dt} = I_m = I_{Na} + I_K + I_L + I_{ext} \qquad (11.1)$$

where I_{Na}, I_K, and I_L are currents per unit area (mA/cm^2) through the sodium channels, potassium channels, and "leakage" channels respectively. The current I_{ext} represents the current arriving from external sources, such as excitatory post synaptic potential from the dendrites or a scientist's electrode.

The ion channels are either fully closed or open. When the channel is closed its conductance is zero and when open then it is some constant value g. The net current through any channel depends on two variables. One is the probability of channel being open and the other is the difference in voltage from ion's equilibrium voltage. According to Kirchhoff's voltage laws, current is dependent on voltage and resistance, or voltage and conductance.

$$I = \frac{V}{R} \quad \text{Or} \quad I = Vg \tag{11.2}$$

The current flowing through the ion channel can be written as:

$$I_i = g_i (V_m - V_i) \tag{11.3}$$

where V_m (mV) is the membrane potential and V_i (mV) represent reverse potential and g_i is the ion's membrane conductance which is dependent on the average number of open ion channels at resting membrane potential. We can represent the membrane current, I_m (mA/cm^2) in terms of equation 11.3.

$$I_m = C_m \frac{dV}{dt} + g_k(V_m - V_k) + g_{Na}(V_m - V_{Na}) + g_l(V_M - V_l) \tag{11.4}$$

Based on ohm's law:

$$g_k = \frac{I_k}{V_m - V_k} \quad \text{membrane conductance per unit area for potassium, S/cm}^2 \tag{11.5}$$

$$g_{Na} = \frac{I_{Na}}{V_m - V_{Na}} \quad \text{membrane conductance per unit area for sodium, S/cm}^2 \tag{11.6}$$

$$g_l = \frac{I_l}{V_m - V_l} \quad \text{leakage conductance per unit area for other ions, S/cm}^2 \tag{11.6}$$

In equation 11.4 V_k, V_{Na}, and V_l are Nernst volts and are define by the Nernst equation. This equation defines the relation between the concentrations of an ion on either side of a cell membrane that it perfectly selective for that ion and the potential difference (voltage) that will be measured across that membrane under equilibrium conditions. The Nernst equation can give a value for the voltage that must exist across the membrane in order to balance a chemical gradient that exists for that ion.

The membrane potential can also be expressed in terms of ionic concentrations inside the cell and outside the cell and ion membrane permeability, illustrated by the Goldman, Hodgkin, Katz (GHK) equation. When Chloride (Cl$^-$) is taken into account, its part is flipped to account for the negative charge.

$$V_m = \frac{RT}{F} \ln\left(\frac{P_k[k]_o + P_{Na}[Na]_o + P_{cl}[cl]_i}{P_k[k]_i + P_{Na}[N]_i + P_{cl}[cl]_o}\right) \tag{11.7}$$

V_m is the membrane potential. This equation is used to determine the resting membrane potential in real cells, in which K^+, Na^+, and Cl^- are the major contributors to the membrane potential. Note that the unit of V_m is the Volt. However, the membrane potential is typically reported in millivolts (mV). If the channels for a given ion (Na^+, K^+, or Cl^-) are closed, then the corresponding relative permeability values can be set to zero. For example, if all Na^+ channels are closed, $p_{Na} = 0$.

- **R** is the universal gas constant (8.314 J.K^{-1}.mol^{-1}).
- **T** is the temperature in Kelvin (K = °C + 273.15).
- **F** is the Faraday constant (96485 C.mol^{-1}).
- p_K is the membrane permeability for K^+. Normally, permeability values are reported as relative permeabilities with p_K having the reference value of one (because in most cells at rest p_K is larger than p_{Na} and p_{Cl}). For a typical neuron at rest, $p_K : p_{Na} : p_{Cl} = 1 : 0.05 : 0.45$. Note that because relative permeability values are reported, permeability values are unit less.
- p_{Na} is the relative membrane permeability for Na^+.
- p_{Cl} is the relative membrane permeability for Cl^-.
- $[K^+]_o$ is the concentration of K^+ in the extracellular fluid.
- $[K^+]_i$ is the concentration of K^+ in the intracellular fluid.
- $[Na^+]_o$ is the concentration of Na^+ in the extracellular fluid.
- $[Na^+]_i$ is the concentration of Na^+ in the intracellular fluid.
- $[Cl^-]_o$ is the concentration of Cl^- in the extracellular fluid.
- $[Cl^-]_i$ is the concentration of Cl^- in the intracellular fluid.

Example: Assume the following concentrations of Na^+ and K^+ ions: $[Na^+]_o = 130$ mM, $[Na^+]_i = 8$ mM, $[K^+]_o = 4$ mM, and $[K^+]_i = 160$ mM. Assume further that at the peak of the action potential, $p_K : p_{Na}$ is 1 : 12 and temperature is 300K. Calculate V_m at the peak of the action potential.

Using equation 11.7

$$V_m = \frac{8.314 \times 300}{96485} \ln\left(\frac{1 \times 4 + 12 \times 130 + 0 \times 0}{1 \times 160 + 12 \times 8 + 0 \times 0}\right)$$

$V_m = 0.0259 \ln 6.109$

$V_m = 46.79$ mV

We can derive the Nernst equation. Recall that the electrical energy of an ion with unit charge e, placed in a region with V_1 is eV_1. For ion with valence z will have the energy zeV_1. In a cell there is a concentration, c_1 of one ion with potential V_1 and on the other side with a concentration, c_2 of another ion with potential V_2. In equilibrium, the probability of finding a positive ion in either side can be represented by applying Boltzmann distribution.

$$P = \frac{1}{Z} e^{-zeV/kT} \tag{11.8}$$

where k Boltzmann constant ($1.3806488 \times 10^{-23}$ kgm^{2}s^{-2}K^{-1}) and T is temperature.

The ratio of the two probabilities is equal to the ratio of the number of ions in each region, which in turn are proportional to the ion concentrations in the two regions. This can be written as;

$$\frac{c_1}{c_2} = \frac{p_1}{p_2} = \frac{e^{-zeV_1/kT}}{e^{-zeV_2/kT}} \qquad (11.9)$$

Taking the logarithm of both side of the equation, give the Nernst equation:

$$V_2 - V_1 = \frac{kT}{ze} \ln \frac{c_1}{c_2} \qquad (11.10)$$

Now, $k = \frac{R}{N_A}$ and $N_A = \frac{F}{e}$ where F is a Faraday constant. The above equation is normally written in terms of R and F.

$$V_2 - V_1 = \frac{RT}{zF} \ln \frac{c_1}{c_2} \qquad (11.11)$$

This is the starting point for understanding the way cells move charge in order to set up membrane potentials. Now we can write equation for each ion.

$$V_{Na} = V_{Na}^\ominus - \frac{RT}{zF} \ln \frac{C_{iNa}}{C_{oNa}} \qquad (11.12)$$

$$V_k = V_k^\ominus - \frac{RT}{zF} \ln \frac{C_{ik}}{C_{ok}} \qquad (11.13)$$

$$V_l = V_l^\ominus - \frac{RT}{zF} \ln \frac{C_{il}}{C_{ol}} \qquad (11.14)$$

where,

$V_{Na}^\ominus, V_k^\ominus, V_l^\ominus$ is the standard cell potential at temperature of interest

R is the universal gas constant: $R = 8.314$ JK^{-1}mol^{-1}

T is the absolute temperature.

F is the Faraday constant, the umber of coulomb per mole of electron: $F = 9.648 \times 10^4$ Cmol^{-1}

z is the number of moles of electrons transferred in the cell reaction.

C_i and C_o are the chemical activity of ions inside and outside the cell respectively.

For one mole at 20 ^{0}C

$$\frac{RT}{zF} = 0.0253$$

Conversion factor from ln to log$_{10}$ is 2.303 $\qquad \frac{\ln a}{\log_{10} a} = 2.303$

So equation 11.12 – 11.14 can be written in terms of $\log_{10}$.

$$V_{Na} = V_{Na}^{\ominus} - 58.15 \log_{10} \frac{C_{iNa}}{C_{oNa}} \quad, \quad V_k = V_k^{\ominus} - 58.15 \log_{10} \frac{C_{ik}}{C_{ok}} \quad, \quad V_l = V_l^{\ominus} - 58.15 \log_{10} \frac{C_{il}}{C_{ol}}$$

Example: Consider two beakers containing silver nitrate solutions of different concentrations, 0.010 M and 0.50 M. Calculate Nernst Volts for these ions?

Two half reactions are

$$Ag_{(S)} \longrightarrow Ag^+_{(Aq)} \qquad Ag^+_{(Aq)} \longrightarrow Ag_{(S)}$$

$$V_{Ag} = 0.00 - 58.15 \log_{10} \frac{0.01\,M}{0.5\,M} = 0.00 - (-98.79) = 98.79\,mV$$

Table 1: Typical values of mammalian cell

ion	Concentration (mM)		Nernst potential (mV)
	Intercellular	Extracellular	
K^+	140	5	-98
Na^+	05-15	145	67
Cl^-	4	110	-90
Ca^{2+}	0.0001	02-05	130

When two or more ions contribute to the membrane potential across the plasma membrane (V_m) of a cell, it is likely that the membrane potential would not be at the equilibrium potential ($V_{eq.}$) for any of the contributing ions. Thus, no ion would be at its equilibrium (i.e., $V_{eq.} \neq V_m$). When an ion is not at its equilibrium, an *electrochemical driving force* (V_{DF}) acts on the ion, causing the net movement of the ion across the membrane down its electrochemical gradient. The driving force is quantified by the difference between the membrane potential and the ion equilibrium potential:

$$V_{DF} = V_m - V_{eq} \qquad (11.15)$$

Table 2: Sign of driving force and direction of ion flow

Type of ion	Sign of driving force	Direction of ion flow
Cation	+	Outward
	0	No net flow
	-	Inward
Anion	+	Inward
	0	No net flow
	-	Outward

Example: At the peak of the action potential, V_m is approximately +40 mV. Assuming normal intracellular and extracellular K⁺ concentrations at 300K (1) calculate the driving force (in mV) that acts on K⁺ ions, and (2) use the information obtained in part 1 to determine the direction in which K⁺ ions will flow (i.e., into the cell or out of cell).

Using Nernst equation 11.11 and from table 1 $K^+_i = 140$ and $K^+_o = 5$

$$V = \frac{RT}{zF} \ln \frac{c_1}{c_2} = 0.0259 \ln \frac{140}{5} = -86.14 \text{ mV}$$

$V_{DF} = V_m - V_{eq} = 40 \text{ mV} - (-86.14 \text{ mV}) = 126.14 \text{ mV}$

Looking at table 2 show that ion's movement is out of the cell across the plasma membrane

The ions cross the membrane only through channels that are specific for that ion. Each individual ion channel can be thought of as containing a small number of physical *gates* that regulate the flow of ions through the channel. Hodgkin and Huxley supposed that the opening and closing of these channels are controlled by electrically charged particles called n-particles. An individual gate can be in one of two states, *permissive* (open) or *non-permissive* (closed). When all the gates for a particular channel are in the permissive state, ions can pass through the channel if any of the gates are in the non-permissive state, ions cannot flow. So they move between these states with first-order kinetics. The probability of an n-particle being in the open position is described by the parameter *n*, and in the closed position by (1 – n).

Mathematically, the voltage and time dependent transitions of the n-particles between the open and closed positions are described by the changes in the parameter *n* with the voltage-dependent transfer rate coefficients α_i and β_i. This can be written like this.

$$\text{Shut} \underset{\beta}{\overset{\alpha}{\rightleftarrows}} \text{Open}$$

For a whole population of gates, let us say a proportion P_i (i means for specific gate) are in the open state, where P_i varies between 0 and 1. That represents the *probability* of an individual gate being in the permissive state. This means that a proportion 1- P_i will be in the *non-permissive* state. The fraction of the total population which open in a given time is dependent on the proportion of gates which are shut, and the rate at which shut gates open:

fraction of open gates = $\alpha_i (V) (1 - P_i)$ (11.15)

fraction of close gates = $\beta_i (V) P_i$ (11.16)

In equilibrium, the fraction of gates closing must be equal to fraction of gates opening.

Hence, $\alpha_i (V) (1 - P_i) = \beta_i (V) P_i$

rearranging,
$$P_{i,t\to\infty(V)} = \frac{\alpha_i(V)}{\alpha_i(V)+\beta_i(V)} \quad (11.17)$$

The infinity subscript is used for P because the system only achieves equilibrium if α and β remain stable for a relatively long period of time. If the membrane voltage V_m is clamped at some fixed value V, then the fraction of gates in the permissive state will eventually reach a steady-state value (i.e., $dP_i/dt = 0$) as t → ∞

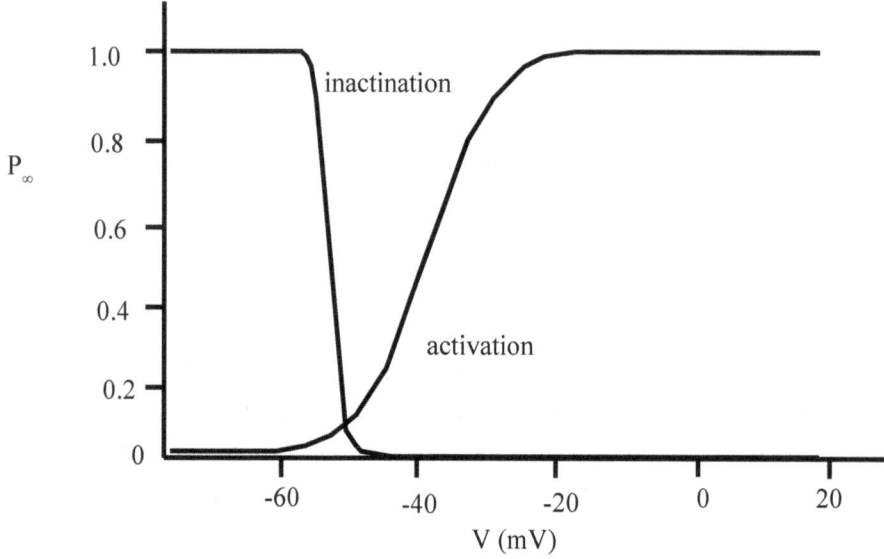

Figure 11.4 P_∞ is related to the fraction of channels in the open state.

The voltage dependency of P arises because the fundamental transition rate constants α_i and β_i are themselves voltage dependent. Clearly, if the membrane potential changes, and consequently the values of α_i and β_i for a particular class of gate change, then the open probability P for that class of gate must also change. For activation gates the voltage dependency of α_i and β_i is such that a depolarising shift in membrane potential causes P to *increase*, while for inactivation gates the change in α_i and β_i causes P to *decrease*.

The HH model assumes that α_i and β_i change instantly with a change in voltage. However, this does not lead to an instantaneous change in the value of P. The rate at which P achieves its new value following a change in α_i and/or β_i is equal to the difference in the rate of shutting and the rate of opening can be calculated by solving the differential equation:

$$\frac{dP}{dt} = \alpha_i(V)(1-P) - \beta_i(V)P_i \quad (11.18)$$

Thus, following a change in voltage, the *rate of change* of P, as well as the direction and size of change, is dependent on the values of α_i and β_i. Depending on the values of α_i and β_i, some classes of gates will respond more rapidly to changes in voltage than others.

The differential equation (6) has a solution:

$$P = P_\infty - (P_\infty - P_{start})e^{-t/\tau} \quad (11.19)$$

where,
$$\tau = \frac{1}{\alpha_i + \beta_i}$$
(11.20)

These equations can be understood as follows. We start with assuming that the system has been at a fixed constant voltage for a long period of time, and therefore P is at a starting equilibrium value P_{start} defined in equation (11.17). The voltage is then changed suddenly, and α_i and β_i immediately switch to new values appropriate to the new voltage. P then starts to change, and approaches its new equilibrium value P_∞ (also defined in equation 11.17, but with the new values for α_i and β_i) with an exponential time course with a time constant of τ. If either α_i or β_i are large, then the time constant is short and P arrives at its new value rapidly. If both are small, then the time constant is long and it takes longer for P to reach equilibrium.

By combining equations (11.17) and (11.20) it is possible to express α_i and β_i in terms of P_∞ and τ:

$$\alpha_i = \frac{P_\infty}{\tau}$$
(11.21)

$$\beta_i = \frac{1 - P_\infty}{\tau}$$
(11.22)

When an individual ion channel is open (i.e., when all the gates are in the permissive state), it contributes some small, fixed value to the total conductance and zero otherwise. The macroscopic conductance for a large population of channels is thus proportional to the number of channels in the open state which is, in turn, proportional to the probability that the associated gates are in their permissive state. Thus the macroscopic conductance g_k due to channels of type k, with constituent gates of type i, is proportional to the *product* of the individual gate probabilities P_i.

$$g_k = \bar{g}_k \prod_i P_i$$
(11.23)

where $\bar{g}_k$ is a normalization constant that determines the maximum possible conductance when all the channels are open.

For each class of gate in each type of channel, α_i and β_i have values appropriate for the voltage, and P (the probability of a gate being open) is at its steady-state equilibrium value given in equation (11.17). If a channel contains several (say n) gates of that class within it, the probability of the whole channel being open is P raised to the power of the number of gates within the channel (i.e. P^n). This is because *all* the gates have to be open for the channel to be open.

Hodgkin and Huxley proposed that each K^+ channel has 4 identical activation gates (n = 4). Thus, to make this concrete, if at a particular voltage the probability of an x-gate being open is one half (x = 0.5), then the probability of an individual K^+ channel being open is 0.5x0.5x0.5x0.5, or 1 in16. That's 1 out of every 16 in the whole population of K^+ channels will be open, and thus the actual K conductance (g_k) will be 1/16 of the maximum possible K^+ conductance, i.e.

$$g_k = \bar{g}_k n^4$$
(11.24)

To conform to the standard notation of the Hodgkin and Huxley model, the probability variable P_i in Equation 11.17 is replaced by a convenient notation in which the variable name is the same as the gate type. For example, Hodgkin and Huxley modelled the sodium conductance using three gates of a type labelled m and one gate of type h. Applying Equation 11.24 to the sodium channel using both the generalized notation and the standard notation yields:

$$g_{Na} = \bar{g}_{Na} P^3_m P_h \equiv \bar{g}_{Na} m^3 h \tag{11.25}$$

Summarizing the ionic currents in the HH model Eq. 11.4 in standard notation, we have:

$$I_{ion} = C_m \frac{dV}{dt} + \bar{g}_{Na} m^3 h (V_m - V_{na}) + \bar{g}_k n^4 (V_m - V_k) + \bar{g}_l (V_m - V_l) \tag{11.26}$$

The gating variable satisfy the non-linear ordinary differential equations

$$\frac{dm}{dt} = \alpha_m(V)(1-m) - \beta_m(V)m = \frac{m_\infty(V) - m}{\tau_m(V)} \tag{11.27}$$

$$\frac{dh}{dt} = \alpha_h(V)(1-h) - \beta_h(V)h = \frac{h_\infty(V) - h}{\tau_h(V)} \tag{11.28}$$

$$\frac{dn}{dt} = \alpha_n(V)(1-n) - \beta_n(V)n = \frac{n_\infty(V) - n}{\tau_n(V)} \tag{11.29}$$

Hodgkin and Huxley were able to determine the steady-state conductance values n_∞ ((Vc) = *clamped voltage*) and time constants τ_∞ (Vc) as a function of command voltage. Once values for n_∞ (Vc) and τ_∞ (Vc) have been determined by fitting the conductance data, values for α_i (Vc) and β_i (Vc) can be found from the following relationships:

$$\alpha_i(V) = \frac{n_\infty(V)}{\tau_i(V)} \tag{11.30}$$

$$\beta_i(V) = \frac{1 - i_\infty(V)}{\tau_i(V)} \tag{11.31}$$

Hodgkin and Huxley then found smooth curves that went through these data points. The empirically determined expressions for the rate constants α_i and β_i are:

K$^+$ activation

$$\alpha_n(V) = \frac{0.01(V + 10)}{\exp\left(\frac{V - 10}{10}\right) - 1} \tag{11.32}$$

$$\beta_n(V) = 0.125 \exp\left(\frac{V}{80}\right) \tag{11.33}$$

Na⁺ activation

$$\alpha_m = \frac{0.1(25+V)}{\exp\left(\frac{25+V}{10}\right)-1} \qquad (11.34)$$

$$\beta_m = 4\exp\left(\frac{V}{18}\right) \qquad (11.35)$$

Na⁺ inactivation

$$\alpha_h = 0.07\exp\left(\frac{V}{20}\right) \qquad (11.36)$$

$$\beta_h = \frac{1}{\exp\left(\frac{V+30}{10}\right)+1} \qquad (11.37)$$

Although Hodgkin–Huxley model is regarded as one of the great achievements in biophysics. Nevertheless, this models has been extended in several important ways:
- Additional ion channel populations have been incorporated based on experimental data.
- Models often incorporate highly complex geometries of dendrites and axons, often based on microscopy data.

The Cardiovascular System

The main components of of the human cardiovascular system are the heart, blood and blood vessels (arteries , venules, capillaries , venules, veins). It includes: the pulmonary circulation a "loop" through the lungs where blood is oxygenated; and the systemic circulation a "loop" through the rest of the body to provide oxygenated blood. An average adult contains five to six quarts (roughly 4.7 to 5.7 liters) of blood, accounting for approximately 7% of their total body weight. Blood consists of plasma, red blood cells, white blood cells and platelets. Also, the digestive system works with the circulatory system to provide the nutrients the system needs to keep the heart pumping.

The heart connects the two major portions of the circulation's continuous circuit, the *systemic circulation* and the *pulmonary circulation*. The blood vessels in the pulmonary circulation carry the blood through the lungs to pick up oxygen and get rid of carbon dioxide, while the blood vessels in the systemic circulation carry the blood throughout the rest of our body.

Pulmonary circulation first described by Ibn al-Nafis (1213 -1288) is the half portion of the cadiovascular system which carries oxygen depleted blood away from the heart. De-oxygenated blood leaves the heart, enter the lungs through the right ventricle (function like the pump) through the pulmonary artery to the capillaries where carbon dioxide diffuses out of the blood cell into the alveoli in the lungs, and oxygen diffuses out of the alveoli into the blood and then re-enters the heart. Blood leaves the capillaries to the pulmonary vein to the heart, where it re-enters at the left atrium (function like a reservoir).

A systemic circulation refers to the part of the circulatory system in which the blood leaves the heart, services the body's cells, and then re-enters the heart. Blood leaves through the left ventricle to the aorta, the body's largest artery. The aorta leads to smaller arteries, arterioles, and finally capillaries. Waste and carbon dioxide diffuse out of the cell into the blood, and oxygen in the blood diffuses into the cell. Blood then moves to venious capillaries, and then the venae cavae: the lower inferior vena cava and the upper superior vena cava, through which the blood re-enters the heart at the right atrium. Figure 11.5 shows a schematic diagram of blood flow system and heart.

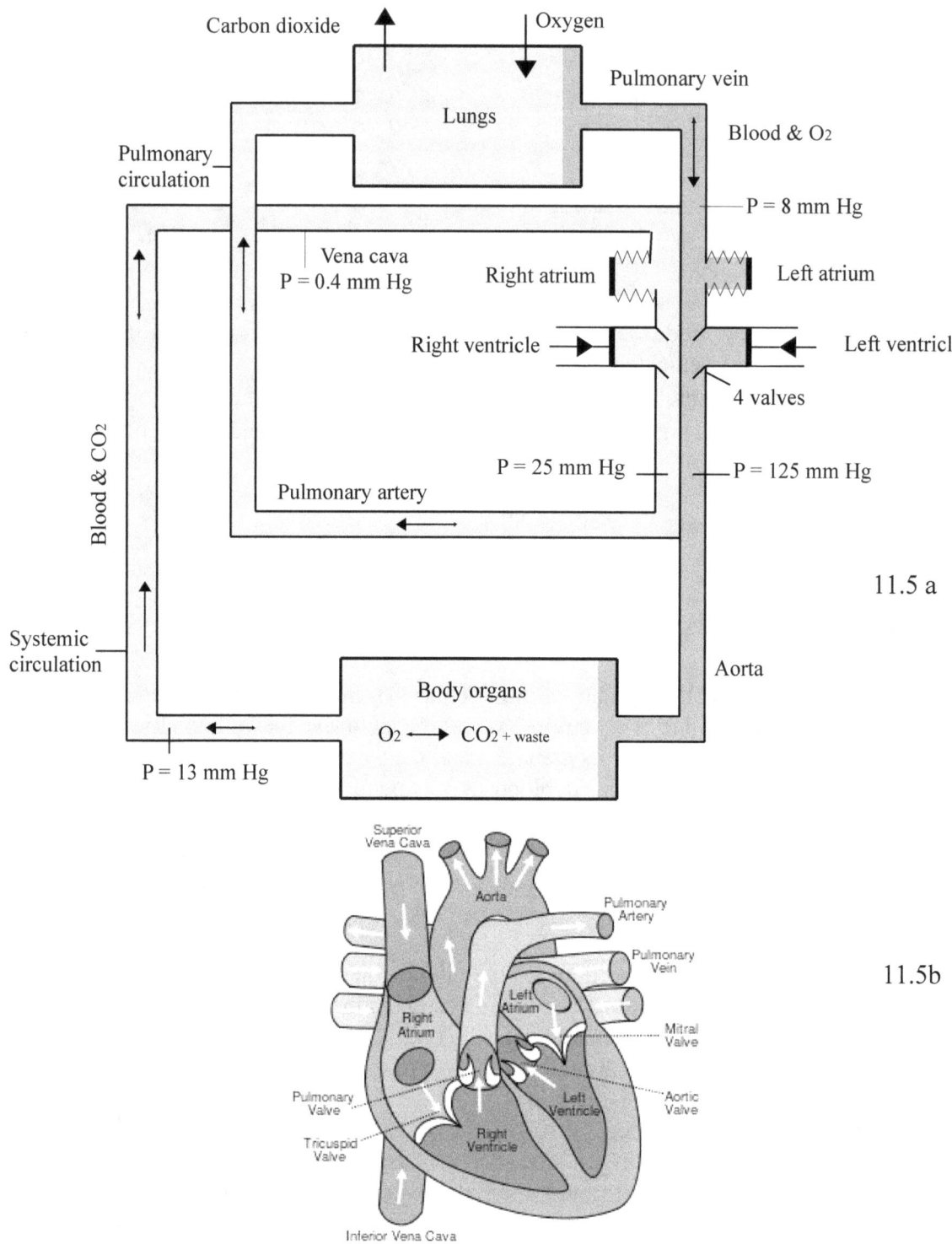

Figure 11.5 Schematic diagram of the circulatory system and view from the heart

The coronary circulatory system refers to the movement of blood through the tissues of the heart. From the left atrium, blood flows through the mitral valve (also known as the bicuspid valve) into the left ventricle. Contraction of the ventricle closes the mitral valve and opens the aortic valve at the entrance to the aorta. The first branches from the aorta occur just beyond the aortic valve still within the heart. Two openings lead to the right and left coronary arteries, which supply blood to the heart itself. Although the coronary arteries arise within the heart, they pass directly out to the surface of the heart and extend down across it. They supply blood to the network of capillaries that penetrate every portion of the heart. The capillaries drain into two coronary veins that empty into the right atrium.

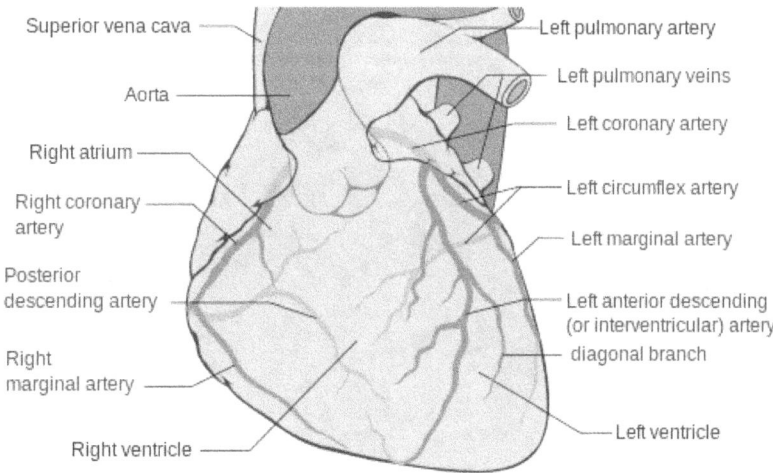

Figure 11.6 Coronary arteries

Cardiac Action potential

Action potentials are quick, transient fluctuations in a cell's membrane that often signal cell activation. The membrane potential is the electrical voltage created between the inside and outside of the cell and is a property of the concentration of charged particles on either side of the cell's membrane. Action potential which is crucial to the functioning of nerve cells, is also seen in muscle and certain endocrine cells. The cardiac action potential differs from skeletal muscle action potentials in three ways:

i- some muscle cells are self-excitable, all cardiac muscle cells are electrically connected by gap junctions and so contract together as a unit.

ii- the cardiac action potential has a much longer absolute refractory period (the period of time following a contraction during which the muscle cannot contract again) than skeletal muscle cells.

iii- contrary to skeletal muscle cells, about 1% of cardiac fibres are auto-rhythmic, meaning they have the special ability to depolarise spontaneously and control the pace of the heart. These auto-rhythmic cells initiate the cardiac action potential.

The cardiac action potential spans 5 phases, numbered 0-4. Once the cell is electrically stimulated it begins a sequence of actions involving the influx and efflux of multiple cations and anions that together produce the action potential of the cell, propagating the electrical stimulation to the cells that lie adjacent to it. In this fashion, an electrical stimulation is conducted from one cell to all the cells that are adjacent to it, to all the cells of the heart.

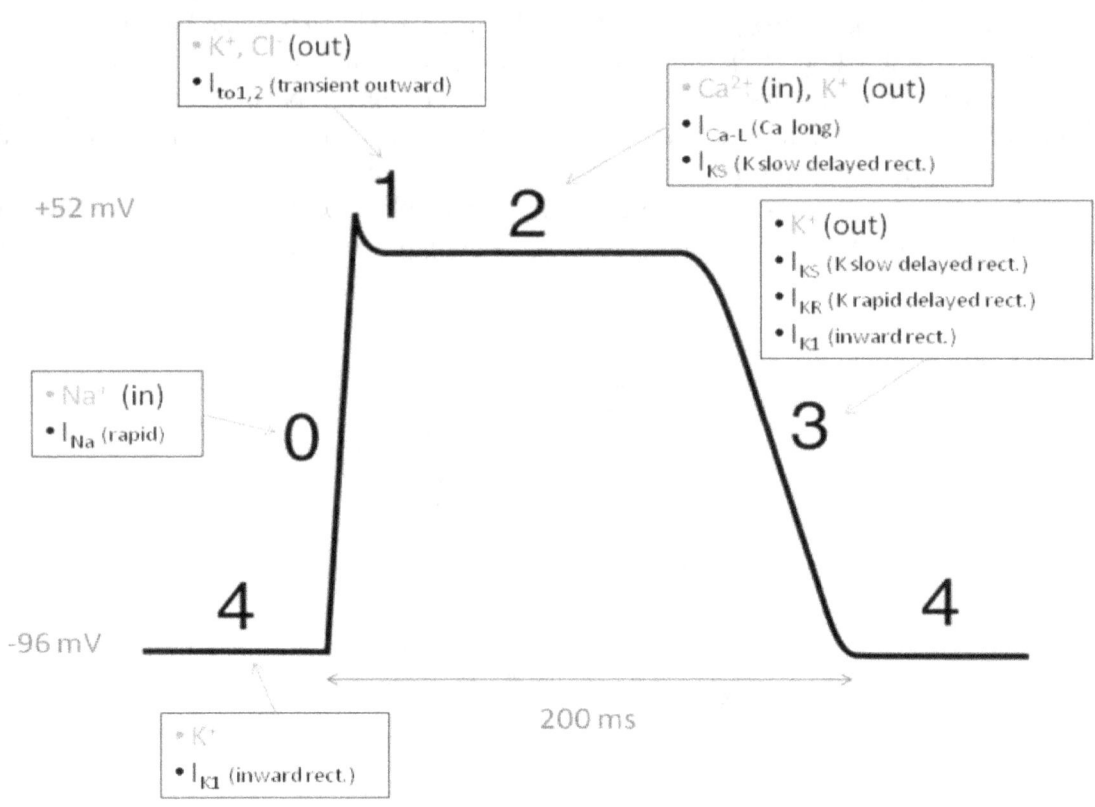

Figure 11.7 Five phases of cardiac action potential

Phase 4: - *increased* K^+ and *decreased* Na^+ and Ca^{2+} conductances.

At phase 4 the cell has returned to resting membrane potential that remains near the equilibrium potential for K^+ (V_K). The resting membrane potential during phase 4 is about -96 mV. This is because potassium channels are open duo to K^+ conductance $[gK^+]$ and K^+ currents $[I_K]$ being high. This phase is associated with K^+ currents, in which positive potassium ions are leaving the cell and thereby making the membrane potential more negative inside. At the same time, fast sodium channels and slow calcium channels are closed.

The cardiac muscle cells do not require a nerve action potential stimulation like skeletal muscle cells do. These cardiac cells (the sinoatrial (SA) node especially) are self activating. Though they do receive some input from the autonomic nervous system they need no stimulus to fire. This phase of the action potential is associated with diastole (is the period of time when the heart refills with blood after contraction) of the chamber of the heart.

Phase 0: - *increased* Na^+ and *decreased* K^+ conductance.

This is the rapid depolarisation phase because the opening of the fast Na^+ channel, causing a rapid increase in the membrane conductance g_{Na} and this result in a repid influx of Na^+ ions into the cell so high I_{Na}. At the same time sodium channels open, g_K and outward directed K^+ currents fall as potassium channels close. These two conductance changes move the membrane potential away from V_K (which is negative) and closer toward the equilibrium potential for sodium V_{Na}, which is positive.

Phase 1: - *decreased* Na⁺ and *increased* K⁺ conductance (initial re-polarisation)

The action potential of phase 1 begins as the sodium ion channels begin to inactivate. The transient net outward current causing the small downward deflection of the action potential is due the cell continues to lose potassium and chloride ions, carried by the I_{to1} and I_{to2} currents respectively.

Phase 2: - *increased* Ca²⁺ conductance (Plateau Phase)

This phase is sustained by a balance between inward movement of Ca^{2+} (I_{Ca}) through L-type Potassium channels (I_K). When the threshold of approximately -40 mV is achieved, calcium ion channels open up. Calcium ions rush into the cell and begin the rising phase of the action potential, reversing the membrane potential.

Phase 3: - *increased* K⁺ and *decreased* Ca²⁺ conductances (Rapid repolarisation)

During this phase the L-type Ca^{2+} channels close, while the slow delayed rectifier K⁺ channels are still open causing a net outward current, corresponding to negative change in membrane potential, thus allowing more types of K⁺ channels to open. This net outward, positive current causes the cell to repolarize. The delayed rectifier K⁺ channels close when the membrane potential is restored to about -85 mV, while I_{K1} remains conducting throughout phase 4, contributing to set the resting membrane potential.

To calculate the membrane potential we can use this expression.

$$Vm = g'K^+ (-96 \text{ mV}) + g'Na^+ (+50 \text{ mV}) + g'Ca^{2+} (+134 \text{ mV}) \qquad 11.38$$

Electrocardiogram

An electrocardiogram or ECG - is a simple and useful test which records the rhythm and electrical activity of the heart. The heart produces tiny electrical impulses which spread through the heart muscle to make the heart contract. These impulses can be detected by the ECG machine. Electrode leads on the chest are able to detect electrical impulses that are generated by the heart. Multiple leads provide many different electrical signals from the heart. By interpreting the tracing, the physician can learn about the heart rate and rhythm as well as blood flow to the ventricles.

Heart rate shows how fast the heart is beating. SA (SinoAtrial) node generates an electrical impulse 60-100 times per minute. If the heart beat is less than 60 bpm is called Bradycardia and if heart beat is more than 100bpm is called Tachycardia. Rhythm refers to the type of heartbeat. Normally, the heart beats in a sinus rhythm with each electrical impulse generated by the SA node resulting in a ventricular contraction, or heartbeat. There are a variety of abnormal electrical rhythms, some are normal and others are potentially dangerous.

An ECG can diagnose abnormal rhythms of the heart, caused by damage to the conductive tissue that carries electrical signals, In a myocardial infarction the ECG can identify if the heart muscle has been damaged in specific areas. The ECG cannot reliably measure the pumping ability of the heart, for which ultrasound-based called *echocardiography* is used. The ECG device detects and amplifies the small electrical changes on the human skin that are caused when the heart muscles depolarise during each heartbeat. At rest, each heart muscle cell has a negative charge. When this decreases towards zero, via the influx of the positive cations, Na^+ and Ca^{++}, is called depolarization, occurs in the sinoatrial (SA) node; current travels through internodal tracts of the atria to the atrioventricular (AV) node; which divides into right and left bundle branches; left bundle branch divides into left anterior and posterior fascicles.

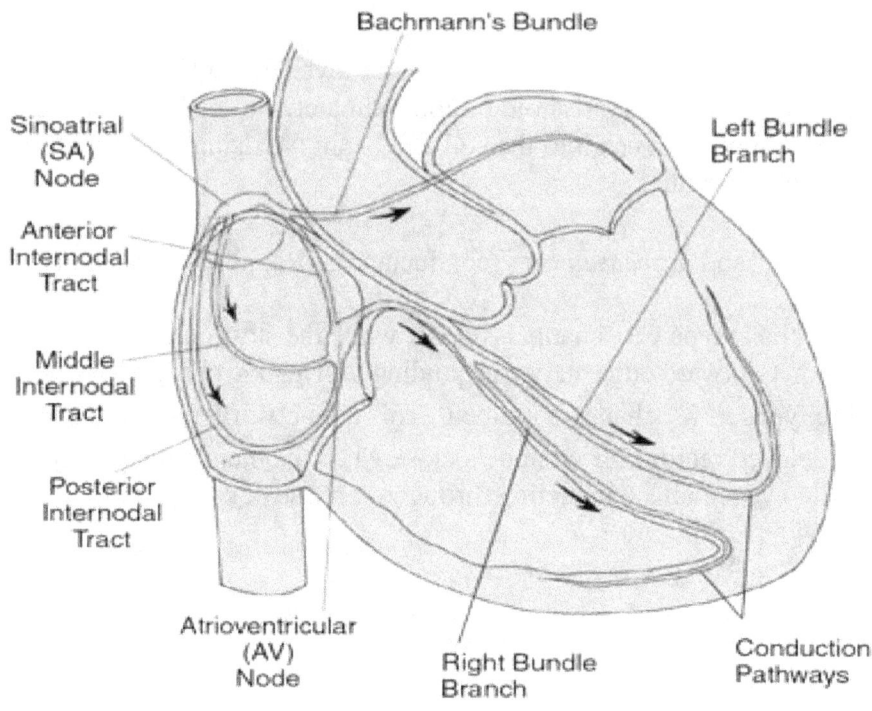

Figure 11.8

Usually, more than two electrodes are used, and they can be combined into a number of pairs (For example: left arm (LA), right arm (RA) and left leg (LL) electrodes form the three pairs LA+RA, LA+LL, and RA+LL). The output from each pair is known as a *lead*. Each lead looks at the heart from a different angle. Different types of ECGs can be referred to by the number of leads that are recorded, for example 3-lead, 5-lead or 12-lead ECGs. A 12-lead ECG is one in which 12 different electrical signals are recorded at approximately the same time and will often be used as a one-off recording of an ECG, traditionally printed out as a paper copy. Three- and 5-lead ECGs tend to be monitored continuously and viewed only on the screen of an appropriate monitoring device, for example during an operation.

The output of an ECG recorder on a moving graph paper which move at 25mm per second during real-time recording. This means that when looking at the printed ECG a distance of 25 mm along the horizontal axis represents 1 second in time. ECG paper is marked with a grid of small and large squares. Each small square represents 40 milliseconds (ms) in time along the horizontal axis and each larger square contains 5 small squares, thus representing 200 ms. Standard paper speeds and square markings allow easy measurement of cardiac timing intervals. This enables the calculation of heart rates and identification of abnormal electrical conduction within the heart. A calibration signal may be included with a record. A standard signal of 1 mV must move the stylus vertically 1 cm, that is, two large squares on ECG paper(see Figure 11.8).

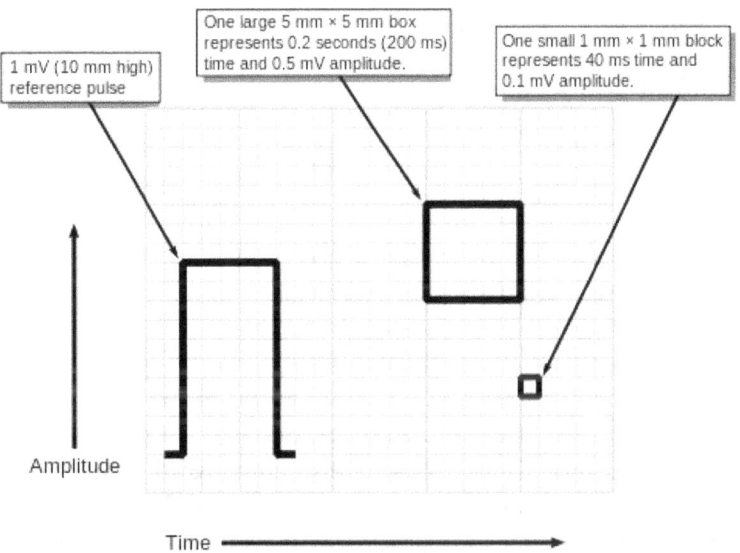

Figure 11.9 ECG one second graph paper

Augustus Désiré Waller measured the human electrocardiogram in 1887. He selected five electrode locations: the four extremities and the mouth. These points produced 10 leds as shown in Figure 11.9a. In 1908 Willem Einthoven showed the first clinically important ECG measuring system. The Einthoven lead system is illustrated in Figure 11.10b.

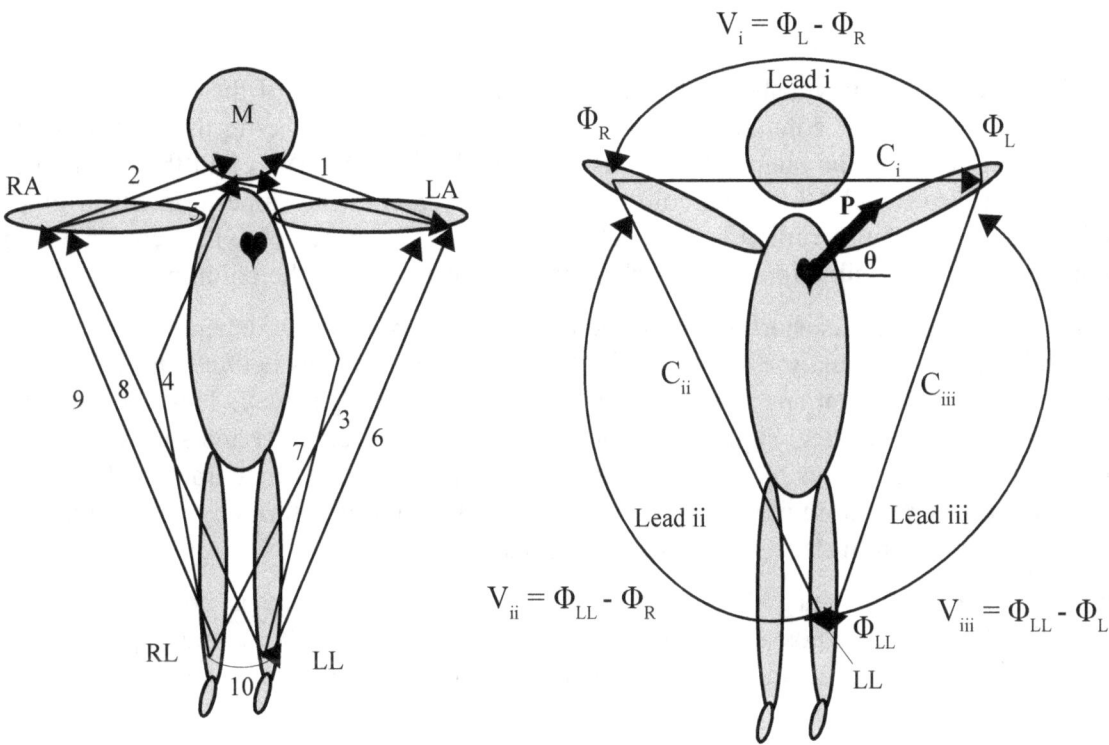

Figure 11.10 (A) The 10 ECG leads of Waller. (B) Einthoven limb leads and triangle.

$$\text{Lead 1:} \quad V_i = \Phi_L - \Phi_R \tag{11.40}$$

$$\text{Lead 2:} \quad V_{ii} = \Phi_{LL} - \Phi_R \tag{11.41}$$

$$\text{Lead 3:} \quad V_{iii} = \Phi_{LL} - \Phi_L \tag{11.43}$$

Where V is the lead voltage and Φ is potential at given point. According to Kirchhoff's law these lead voltages have the following relationship:

$$V_i + V_{iii} = V_{ii} \tag{11.44}$$

If the position of the right arm, left arm, and left leg are at the vertices of an equilateral triangle, having the heart located at its centre, then the lead vectors also form an equilateral triangle. Also if cardiac sources are represented by a dipole located at the centre of a sphere representing the torso, which can also be considered the centre of the equilateral triangle. The voltages measured by the three limb leads are proportional to the projections of the electric heart vector on the sides of the lead vector triangle as shown in Figure 11.10.

The voltages of the limb leads are obtained as:

$$V_i = P \cos \theta = P_Y \tag{11.45}$$

$$V_{ii} = \frac{P}{2}\cos\theta - \frac{\sqrt{3}}{2} P \sin\theta = \frac{1}{2} P_y - \frac{\sqrt{3}}{2} P_z = 0.5 P_y - 0.87 P_z \tag{11.46}$$

$$V_{iii} = \frac{-P}{2}\cos\theta - \frac{\sqrt{3}}{2} P \sin\theta = \frac{-1}{2} P_y - \frac{\sqrt{3}}{2} P_z = -0.5 P_y - 0.87 P_z \tag{11.47}$$

A 12-lead ECG will show a short segment of the recording of each of the 12-leads. This is often arranged in a grid of four columns by three rows, the first columns being the limb leads (I, II and III), the second column the augmented limb leads (aVR, aVL and aVF) and the last two columns being the chest leads (V1-V6). Each column will usually record the same moment in time for the three leads and then the recording will switch to the next column, which will record the heart beats after that point. It is possible for the heart rhythm to change between the columns of leads.

Each of these segments is short, perhaps one to three heart beats only, depending on the heart rate, and it can be difficult to analyse any heart rhythm that shows changes between heart beats. To help with the analysis, it is common to print one or two "rhythm strips", as well. This will usually be lead II (which shows the electrical signal from the atrium, the P-wave, well) and shows the rhythm for the whole time the ECG was recorded (usually 5–6 sec). Some ECG machines will print a second lead II along the very bottom of the paper in addition to the output described above. This printing of lead II is continuous from start to finish of the process.

Bipolar Lead: One in which the electrical activity at one electrode is compared with that of another. By convention, a positive electrode is one in which the ECG records a positive (upward) deflection when the electrical impulse flows toward it and a negative (downward) deflection when it flows away from it.

Unipolar Lead: One in which the electrical potential at an exploring electrode is compared to a reference point that averages electrical activity, rather than to that of another electrode. This single electrode, termed the *exploring electrode*, is the positive electrode.

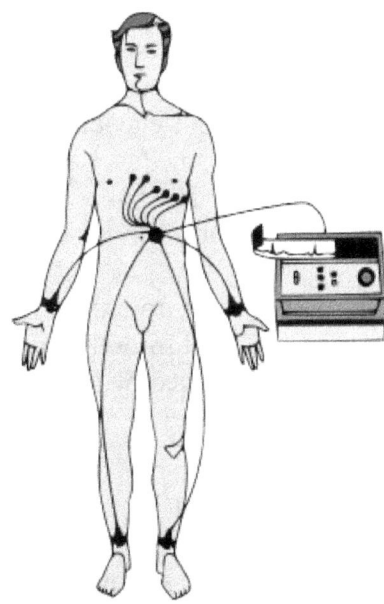

Figure 11.11 Image showing a patient connected to the 10 electrodes necessary for a 12-lead ECG

Standard Limb Leads: I, II, III; bipolar, form a set of axes 60° apart
Lead I: Composed of negative electrode on the right arm and positive electrode on the left arm.
Lead II: Composed of negative electrode on the right arm and positive electrode on the left leg.
Lead III: Composed of negative electrode on the left arm and positive electrode on the left leg.

Augmented Voltage Leads: aVR, aVL, aVF; unipolar ; form a set of axes 60° apart but are rotated 30° from the axes of the standard limb leads.
Reference Point for Augmented Leads: The opposing standard limb lead; i.e., that standard limb lead whose axis is perpendicular to the particular augmented lead.

aVR: Exploring electrode located at the right shoulder.
aVL: Exploring electrode located at the left shoulder.
aVF: Exploring electrode located at the left foot.

Chest Leads: Vl, V2, V3, V4, V5, V6; unipolar; explore the electrical activity of the heart in the horizontal plane.

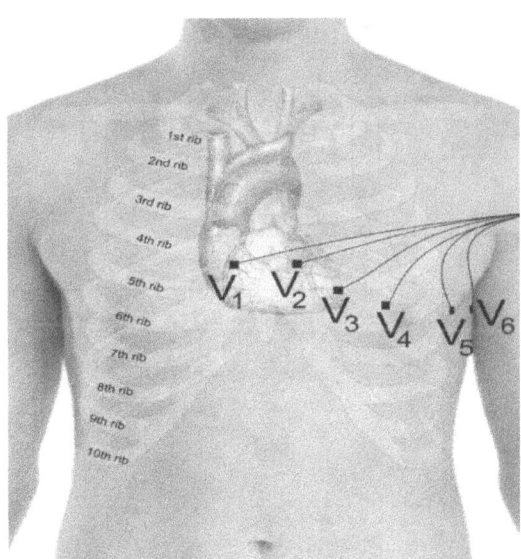

Figure 11.12. Placement of the precordial leads

V1: Positioned in the 4th intercostal space just to the right of the sternum.
V2: Positioned in the 4th intercostal space just to the left of the sternum.
V3: Positioned halfway between V2 and V4.
V4: Positioned at the 5th intercostal space in the mid-clavicular line.
V5: Positioned in the anterior axillary line at the same level as V4.
V6: Positioned in the mid axillary line at the same level as V4 and V5.
V1 and V2*: Monitor electrical activity of the heart from the anterior aspect, septum, and right ventricle.
V3 and V4*: Monitor electrical activity of the heart from the anterior aspect.
V5 and V6*: Monitor electrical activity of the heart from the left ventricle and lateral aspect.
Reference Point for Chest Leads: The point obtained by connecting the left arm, right arm, and left leg electrodes together.

Frank Norman Wilson (1890-1952) investigated how electrocardiographic *unipolar* potentials could be defined. Ideally, those are measured with respect to a remote reference (infinity). But how is one to achieve this in the volume conductor of the size of the human body with electrodes already placed at the extremities? Wilson suggested the use of the *central terminal* as this reference. This was formed by connecting a 5 kΩ resistor from each terminal of the limb leads to a common point called the central terminal, as shown in Figure 11.13 Wilson suggested that unipolar potentials should be measured with respect to this terminal which approximates the potential at infinity. The Wilson central terminal represents the average of the limb potentials. Because no current flows through a high-impedance voltmeter, Kirchhoff's law requires that $I_R + I_L + I_F = 0$

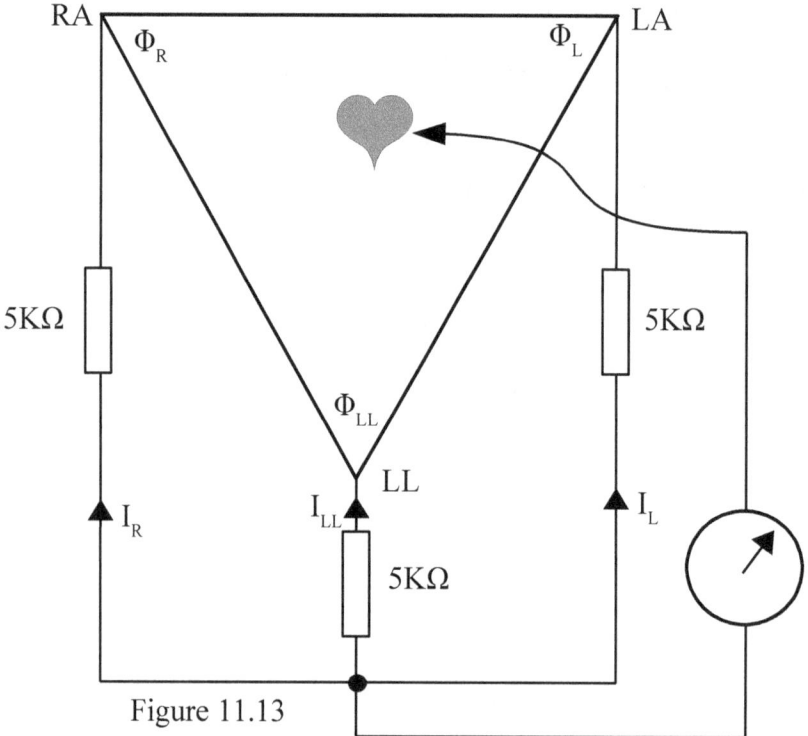

Figure 11.13

$$I_R + I_L + I_{LL} = \frac{V_W - \Phi_R}{5k\Omega} + \frac{V_W - \Phi_L}{5k\Omega} + \frac{V_W - \Phi_{L.L}}{5k\Omega} = 0 \qquad (11.48)$$

hence,

$$V_W = \tfrac{1}{3}(\Phi_R + \Phi_L + \Phi_{LL}) \qquad (11.49)$$

Leads aVR, aVL, and aVF are augmented limb leads, they are derived from the same three electrodes as leads I, II, and III. However, they view the heart from different angles because the negative electrode for these leads is a modification of Wilson's central terminal. This zeroes out the negative electrode and allows the positive electrode to become the "exploring electrode". This is possible because the Einthoven's Law states that I + (−II) + III = 0. Einthoven reversed the polarity of lead II in Einthoven's triangle, because he wanted to view upright QRS. Wilson's central terminal paved the way for the development of the augmented limb leads aVR, aVL, aVF and the precordial leads V_1, V_2, V_3, V_4, V_5 and V_6.

Lead augmented vector right (aVR) has the positive electrode on the right arm. The negative electrode is a combination of the left arm electrode and the left leg electrode, which "augments" the signal strength of the positive electrode on the right arm. See figure 11.13

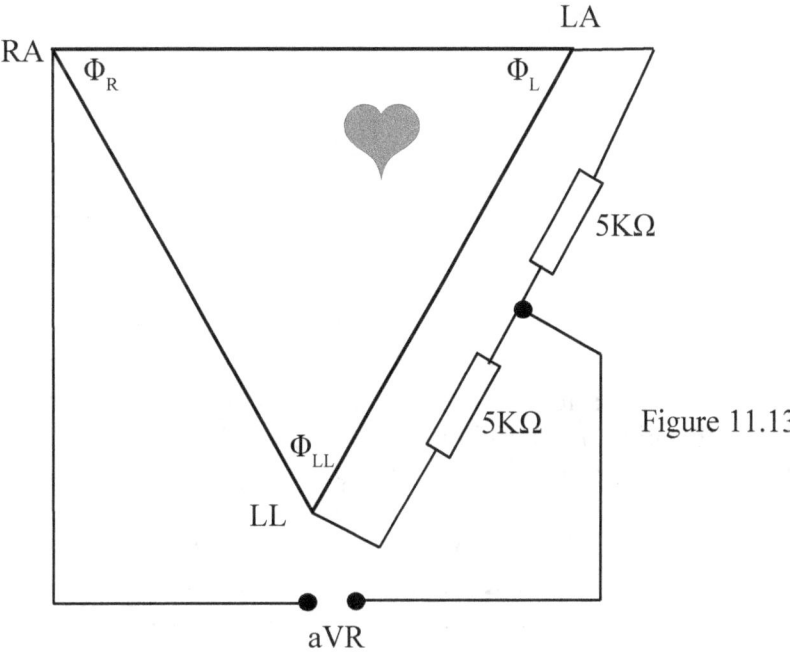

Figure 11.13

$$aVR = \Phi_R - 1/2\,(\Phi_L + \Phi_{LL}) = 3/2\,(\Phi_R - V_w) \tag{11.50}$$

Example: Show that $aVR = -\left(\dfrac{V_i + V_{ii}}{2}\right)$

Substituting for V_i and V_{ii} from equation 11.40 and 11.41 respectively in the above expression.

$$aVR = -\left(\frac{(\Phi_L - \Phi_R) + (\Phi_{LL} - \Phi_R)}{2}\right)$$

$$= \left(\frac{-\Phi_L + \Phi_R - \Phi_{LL} + \Phi_R}{2}\right) = \left(\frac{2\Phi_R}{2} - \frac{\Phi_L}{2} - \frac{\Phi_{LL}}{2}\right)$$

$$= \Phi_R - 1/2\,(\Phi_L + \Phi_{LL})$$

Lead augmented vector left (aVL) has the positive electrode on the left arm. The negative electrode is a combination of the right arm electrode and the left leg electrode, which "augments" the signal strength of the positive electrode on the left arm. See Figure 11.14

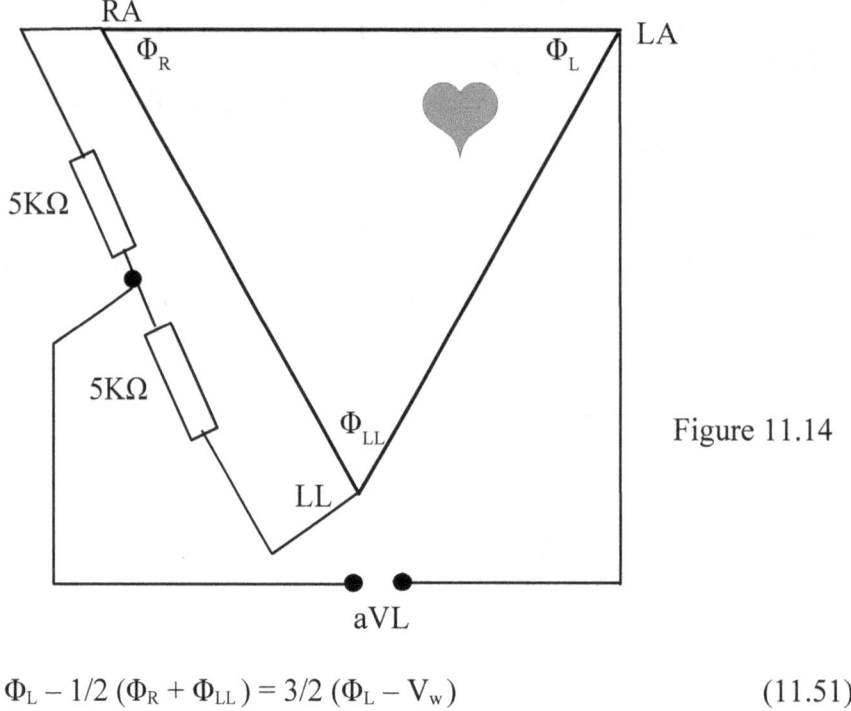

Figure 11.14

$$aVL = \Phi_L - 1/2\,(\Phi_R + \Phi_{LL}) = 3/2\,(\Phi_L - V_w) \qquad (11.51)$$

Lead augmented vector foot (aVF) has the positive electrode on the left leg. The negative electrode is a combination of the right arm electrode and the left arm electrode, which "augments" the signal of the positive electrode on the left leg. See figure 11.15.

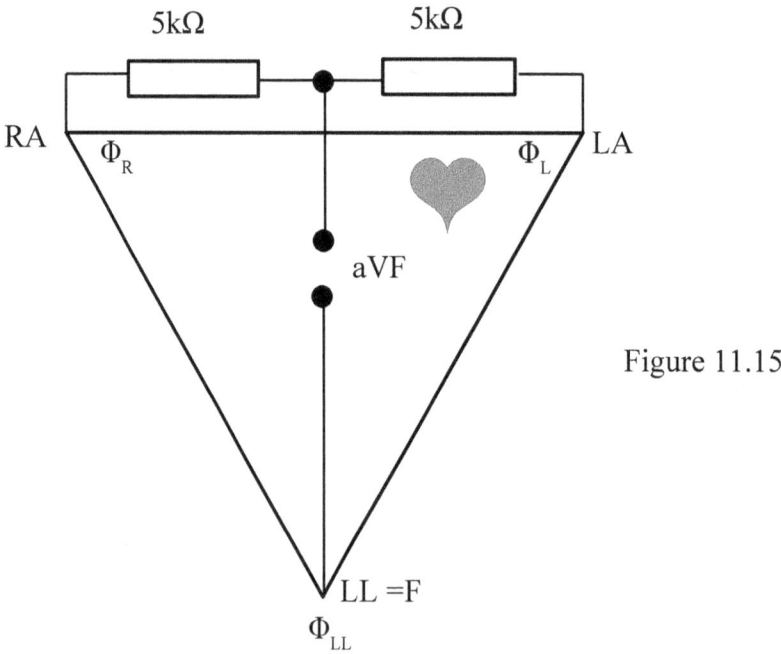

Figure 11.15

$$aVF = \Phi_{LL} - 1/2\,(\Phi_R + \Phi_L) = 3/2\,(\Phi_{LL} - V_w) \qquad (11.52)$$

Also,

$$\text{aVR} = -\left(\frac{V_i + V_{ii}}{2}\right) \qquad (11.53)$$

$$\text{aVL} = \left(V_i - \frac{V_{ii}}{2}\right) \qquad (11.54)$$

$$\text{aVF} = \left(V_{iii} + \frac{V_i}{2}\right) \qquad (11.55)$$

The electrodes for the pre-cordial leads (V_1, V_2, V_3, V_4, V_5 and V_6) are placed directly on the chest. As these leads are in close vicinity to the heart, they do not require augmentation. The pre-cordial leads view the heart's electrical activity in the horizontal plane. The heart's electrical axis in the horizontal plane is referred to as the Z axis.

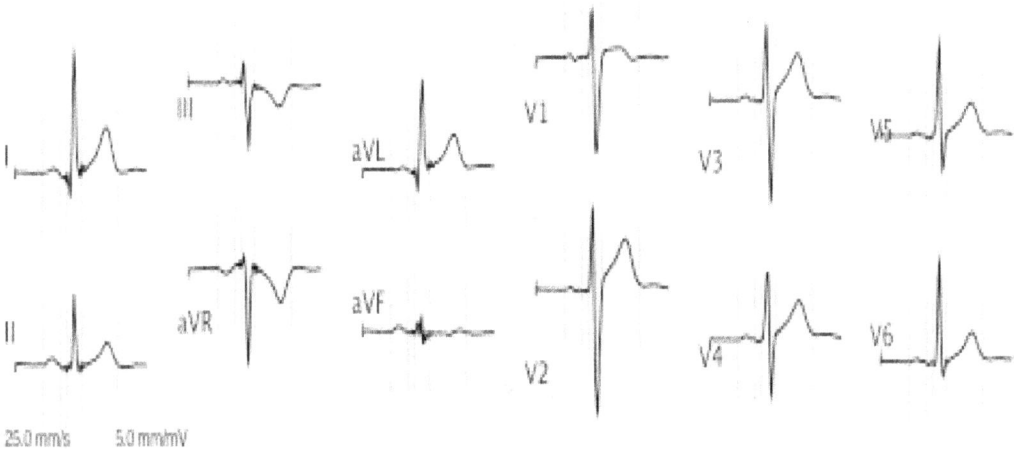

Figure 11.16 Trace of the signals from 12 leads ECG

Interpretation of ECG trace.

Waves
• P wave: represent atrial systole
• QRS complex: ventricular systole
• T wave: ventricular relaxation or diastole
• Atrial systole gets buried in the ventricular systole and therefore does not produce a waveform
• Q waves: When heart muscles are damaged the electrical current does not pass through them and instead of upright R waves, downwards Q waves are produced.

Let's look into it in more details. The first structure to be depolarised during normal sinus rhythm is the right atrium, closely followed by the left atrium. So the first electrical signal on a normal ECG originates from the atria and is known as the **P wave**. Although there is usually only one P wave in most leads of an ECG, the P wave is in fact the sum of the electrical signals from the two atria, which are usually superimposed.

There is then a short, physiological delay as the atrioventricular (AV) node slows the electrical depolarisation before it proceeds to the ventricles. This delay is responsible for the PR interval, a short period where no electrical activity is seen on the ECG, represented by a straight horizontal line also know as isoelectric line. Depolarisation of the ventricles results in usually the largest part of the ECG signal (because of the greater muscle mass in the ventricles) and this is known as the **QRS complex**.

- The Q wave is the first initial downward or 'negative' deflection
- The R wave is then the next upward deflection (provided it crosses the isoelectric line and becomes 'positive')
- The S wave is then the next deflection downwards, provided it crosses the isoelectric line to become briefly negative before returning to the isoelectric baseline.

In the case of the ventricles, there is also an electrical signal reflecting repolarisation of the myocardium. This is shown as the **ST segment** and the **T wave**. The ST segment is normally isoelectric, and the T wave in most leads is an upright deflection of variable amplitude and duration.

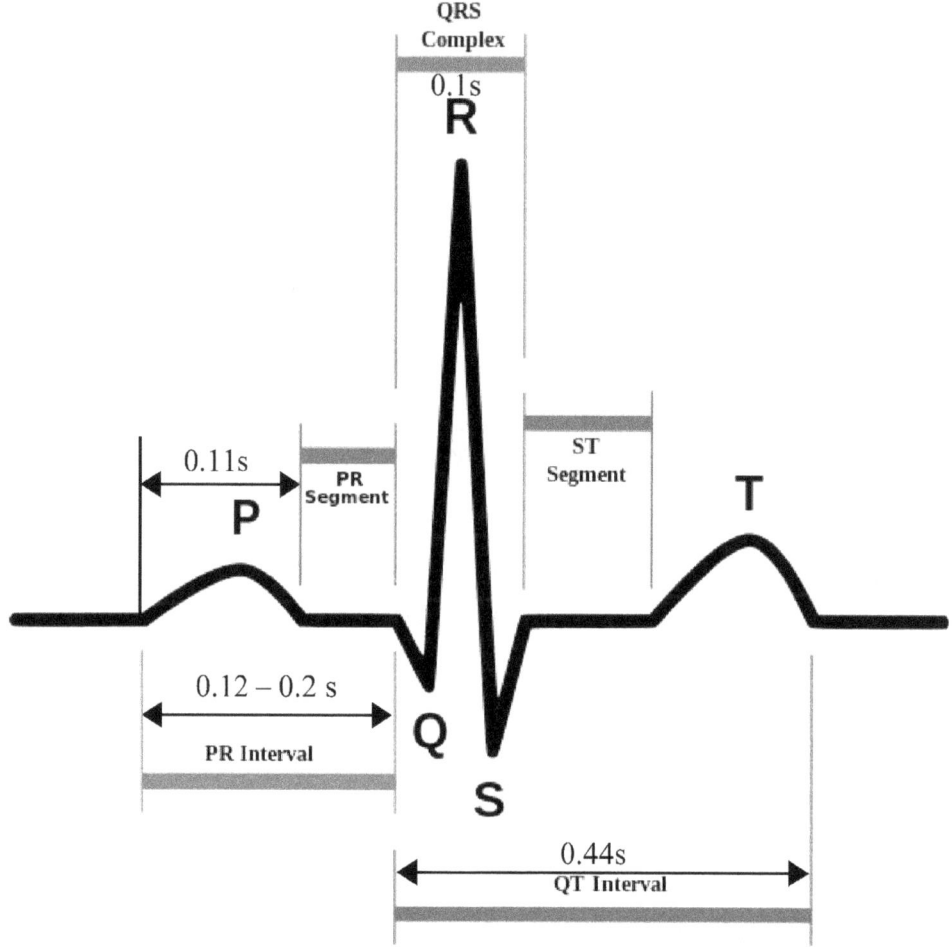

Figure 11.17 Detail of the QRS complex, showing ventricular activation time and amplitude.

Normal ECG and its intervals

The recording of an ECG on special graph paper allows the time taken, in milliseconds for the various phases of electrical depolarisation is measured. When ECG is recorded the paper speed is 25 millimetres/second so in 1 second ECG tracing covers 5 large squares or 1 large square is equal to 0.2 seconds and one small square is equal to 0.04 seconds.

P Wave: The P wave represents the spread of electrical activity over the atrium. The normal depolarization begins at the sinoatrial (SA) node near the top of the atrium. Because of the top-to-bottom, right-to-left path of the P wave, it's normally largest in lead II. The normal P wave is upright in all leads except R. When the valves between the atria and ventricles open, 70% of the blood in the atria falls through with the aid of gravity, but mainly due to suction caused by the ventricles as they expand. Atrial contraction is required only for the final 30% of the blood and therefore a relatively small muscle mass is required and only a relatively small amount of voltage is needed to contract the atria.

The P wave normally lasts less than 0.11 seconds (less the 3 small squares). An abnormally long P wave occurs whenever it takes extra time for the electrical wave to reach the entire atrium. This occurs in left atrial enlargement. The height of the P wave is normally less than 0.25 mV (2.5 small squares). An abnormally tall P wave is seen when larger amounts of electricity are moving over the atrium. This usually indicates hypertrophy of the right atrium.

P-R interval: Following the P wave is the PR segment. (NOTE: the PR segment and the PR interval are NOT the same thing.) The PR segment is not routinely measured, but may be commented on if it is depressed or elevated. During the PR segment, the electrical wave moves slowly through the atrioventricular (AV) node. This activity is not seen on the ECG. The PR interval is the time required for completion of aerial depolarization; conduction through the AV note, bundle of His, and bundle branches; and arrival at the ventricular myocardial cells. Measured from the beginning of the P wave to the first deflection of the QRS complex). Normal range 120 – 200 ms (3 – 5 small squares on ECG paper).

The PR interval may be prolonged when conduction of the electrical wave through the AV node is slow. This may be seen with degenerative disease of the node, or with digoxin, hyperkalemia, hypercalcemia, or hypothermia. The PR interval may be unusually short when conduction is rapid. A mildly short PR interval may be seen with hypokalemia or hypocalcemia. An artificially-short PR interval occurs when the QRS complex begins early, as happens with an extra conducting bundle — Wolff-Parkinson-White Syndrome (WPW).

QRS interval: After the first wave there follows a short period where the line is flat. This is the point at which the stimulus is delayed in the bundle of His to allow the atria enough time to pump all the blood into the ventricles. As the ventricles fill, the growing pressure causes the valves between the atria and ventricles to close. At this point the electrical stimulus passes from the bundle of His into the bundle branches and Purkinje fibres. The amount of electrical energy generated is recorded as a complex of 3 waves known collectively as the QRS complex. Measuring the waves vertically shows voltage. More voltage is required to cause ventricular contraction and therefore the wave is much bigger. QRS interval represents the time required for ventricular cells to depolarize. Measured from first deflection of QRS complex to end of QRS complex at isoelectric line The normal duration is 0.06 to 0.10 seconds (2.5 squares).

The diagram in figure 11.17 shows a small negative wave immediately before the large QRS complex. This is known as a Q wave and represents depolarisation in the ventricular septum. Whilst the electrical stimulus passes through the bundle of His, and before it separates down the two bundle branches, it starts to depolarise the septum from left to right. This is only a small amount of conduction (hence the Q wave is less than 2 small squares), and it travels in the opposite direction to the main conduction (right to left) so the Q wave points in the opposite direction to the large QRS complex. For example in lead I, a Q less than 1/4 of the R height, and less than one box wide, is considered normal. This is the early activation of the septum. This activation goes left — away from

lead I — and is therefore negative on the ECG. "Septal Qs" are normal in I, F, V5 and V6. The Qs are also generally innocent in lead III and lead V1 if no other abnormality is seen.
Q waves are "significant" if they are greater than 1 box in width (longer than 0.04 msec).Significant Q waves indicate either myocardial infarction or obstructive septal hypertrophy.

The first upward deflection of the QRS is called the R wave. Most of the ventricle is activated during the R wave. The R wave represents the electrical stimulus as it passes through the main portion of the ventricular walls. The wall of the ventricles are very thick due to the amount of work they have to do and, consequently, more voltage is required. This is why the R wave is by far the biggest wave generated during normal conduction. More muscle means more cells. More cells means more electricity. More electricity leads to a bigger wave. The R wave may be prolonged if the ventricle is enlarged, and may be abnormally high (indicating strong voltage) if the ventricular muscle tissue is hypertrophied. If a second upward deflection is seen, it's called an R-prime wave. R-prime waves are never normal, but indicate a problem in the ventricular conduction system.

The S wave is any downward deflection following the R wave. The S wave represents depolarisation in the Purkinje fibres. The S wave travels in the opposite direction to the large R wave, because as can be seen on the earlier picture, the Purkinje fibres spread throughout the ventricles from top to bottom and then back up through the walls of the ventricles. Like the R wave, an abnormally large S wave may indicate hypertrophy of the ventricle.

Lengthening of the QRS indicates some blockage of the electrical activity in the conducting system. This may be due to ischemia, necrosis of the conducting tissue, electrolyte abnormality, or hypothermia.

ST Segment: There is a brief period between the end of the QRS complex and the beginning of the T wave where there is no conduction and the line is flat. During this time, the ventricle is contracting, but no electricity is flowing. The ST segment is therefore usually even with the baseline. The length of the ST segment shortens with increasing heart rate. Abnormality of electrolytes may also affect the ST segment length, however measurement of the length of the ST segment alone is usually not of any clinical use. However, upward or downward shifts in the ST segment are extremely important. The deviation of the ST segment from baseline can indicate infarction or ischemia, pericarditis, electrolyte abnormality, or ventricular strain. ST segment elevation or depression is generally measured at a point two boxes beyond the QRS complex.

T Wave: The T wave represents the wave of repolarisation, as the ventricle prepares to fire again. The T wave is normally upright in leads I, II, and V3-V6. It is normally inverted in lead R. Ts are variable in the other leads (III, L, F, and V1-V2). n young children, T waves may be inverted in the right pre-cordial leads (V1 to V3). Occasionally, these T inversions persist in young adults.

T wave abnormalities may be seen with, or without ST segment abnormality. Tall T waves may be seen in hyperkalemia or very early myocardial infarction. Flat T waves occur in many conditions. Inverted T waves may be seen in both ischemia and infarction, late in pericarditis, ventricular hypertrophy, bundle branch block, and cerebral disease.

QT Interval

The QT interval is measured from the start of the QRS complex to the end of the T wave. This interval represents the duration of activation and recovery of the ventricular myocardium. A QT interval in excess of 0.44 sec is a marker of myocardial electrical instability - A lengthened QT interval is a biomarker for possible development of ventricular arrhythmia, syncope and sudden death. It is also affected by diet, gender, alcohol, time of the day, menstrual cycle and heart rate, so may require some adjusted to improve the detection of patients at increased risk of ventricular arrhythmia.

The standard clinical correction is to use *Bazett's formula*:

$$QT_B = \frac{QT}{\sqrt{RR}} \qquad (11.56)$$

where QT_B is the QT interval corrected for heart rate, and RR is the interval from the onset of one QRS complex to the onset of the next QRS complex, *measured in seconds.*

Louis Sigurd Fridericia derived an alternative correction formula using the cube-root of RR.

$$QT_F = \frac{QT}{\sqrt[3]{RR}} \qquad (11.57)$$

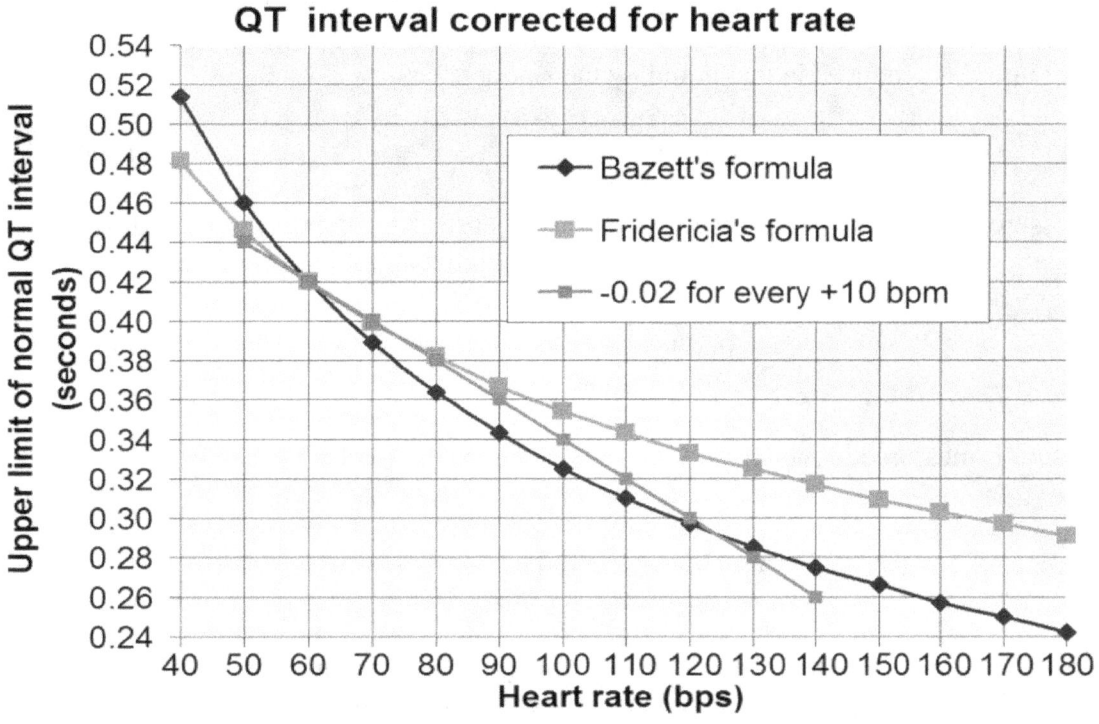

Figure 11.18 Upper limit of normal QT interval, corrected for heart rate according to *Bazett's formula and* Fridericia's formula

Hemodynamics

Hemodynamics can be defined as the physical factors that govern blood flow. Blood flow is the continuous running of blood in the cardiovascular system. Some important physiological facts on blood.

Blood account for the 7% of body mass. We take a person with 75 kg mass. So the mass of blood in his body is 5.25 kg. The density of blood is 1060 kgm^{-3}. The total volume can be calculated.

$$V = \frac{m}{\rho} = \frac{5.25\ kg}{1060\ kg\ m^{-3}} = 4.95\ mm^3 \approx 4.95\ liters$$

Flow rate is the volume of fluid that move pass a give point per unit time.

$$\text{Flow rate, } Q = \frac{V}{t} \tag{11.58}$$

Blood flow rate will be continuous under the following assumptions:

a- the blood is incompressible

b- there is no source (transfusion) or sink (leakage) of blood flow

c- vessels in which blood flow has rigid walls.

Blood vessels which transport blood away from the heart are called arteries and those vessels which transport blood towards the heart are called veins.

Due to its ability to flow, fluid exerts a force uniformly in all directions. The force exerted by a fluid at rest on any area is perpendicular to that area. When force F is applied to an area A of a liquid pressure P is produced in that liquid by the amount:

$$P = \frac{F_p}{A} \tag{11.59}$$

The pressure in a liquid will increase with depth, because as the depth is increased so is the weight of the water on top is increased. In a fluid with density ρ the pressure difference between two points separated by height h is:

$$P_2 - P_1 = \rho g h \tag{11.60}$$

The pressure is measured in mmHg and in SI units (derived) Pascal

$$1\ mmHg = 133.332\ Pa$$

Figure 11.19 shows a simplified model of the cardiovascular system. The heart is divided into two sides and each side has a pump which functions as an independent pump. The systemic arteries are a pressure reservoir that maintains blood flow during ventricular relaxation. The arterioles alter their diameter are functioning like variable resistance. Exchange between blood and cells only take place at the capillaries. Veins serve as an expandable reservoir. The systemic system carries this oxygenated blood to the body and brings deoxygenated blood back to the heart. Blood leaving the left ventricle enters the systemic arteries, shown in figure 11.19 by the elastic region. Pressure of 125mmHg (16.63kPa) created by the contraction of the left ventricle is stored in the elastic walls of the arteries and is slowly released through elastic recoil. This maintains a continuous driving pressure of blood flow during ventricular relaxation.

The mean aortic pressure is about 95mmHg (12.64kPa) in a normal individual. The mean blood pressure does not fall very much as the blood flows down the aorta and through large distributing arteries called a large systemic cycle. It is not until the small arteries and arterioles that there is a large fall in mean blood pressure. Approximately 50-70% of the pressure drop along the vasculature occurs within the small arteries and arterioles. Downstream small vessels called arterioles direct the distribution of blood to individual tissues by selectively constricting or dilating, so they are known as the site of variable resistance.

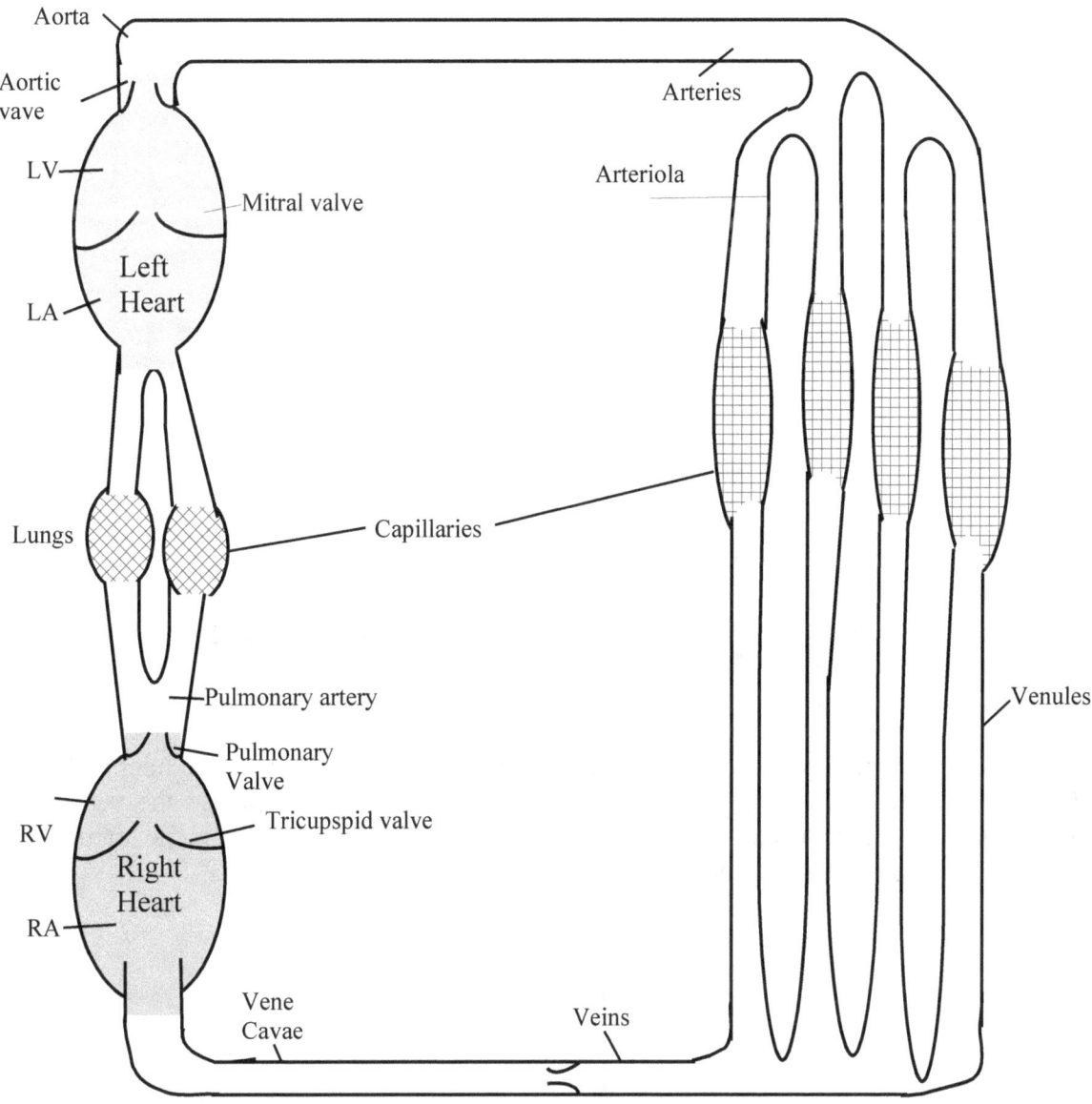

Figure 11.19 Blood circulatory system

A pulmonary system carries blood to the lungs where it becomes oxygenated and carries it back to the heart. At capillaries the mean pressure is 25-30mmHg (3.33 - 3.99kPa). The pressure falls further as blood travels into the veins and back to the heart. At right atrium blood is pumped by weak contraction at the pressure of 5-6mmHg (0.67- 0.80kPa) to the right ventricle. Right ventricle chamber compresses the blood with a pressure of 25mmHg (3.33 kPa) which pump blood through pulmonary arteries and come to left atrium where blood pumped by weak contraction with a pressure of 7-8mmHg (0.93-1.06kPa) to left ventricle.

Regarding the distribution of blood volume within the circulation, 80% is in the systemic circulation and this then divided in systemic cycle as 15% in arteries, 10% in capillaries and 75% in veins. 20% blood is in the pulmonary circulation and that also is divided in the pulmonary cycle as 46.5% in pulmonary arteries, 7% in pulmonary capillaries and 46.5% in the pulmonary veins.

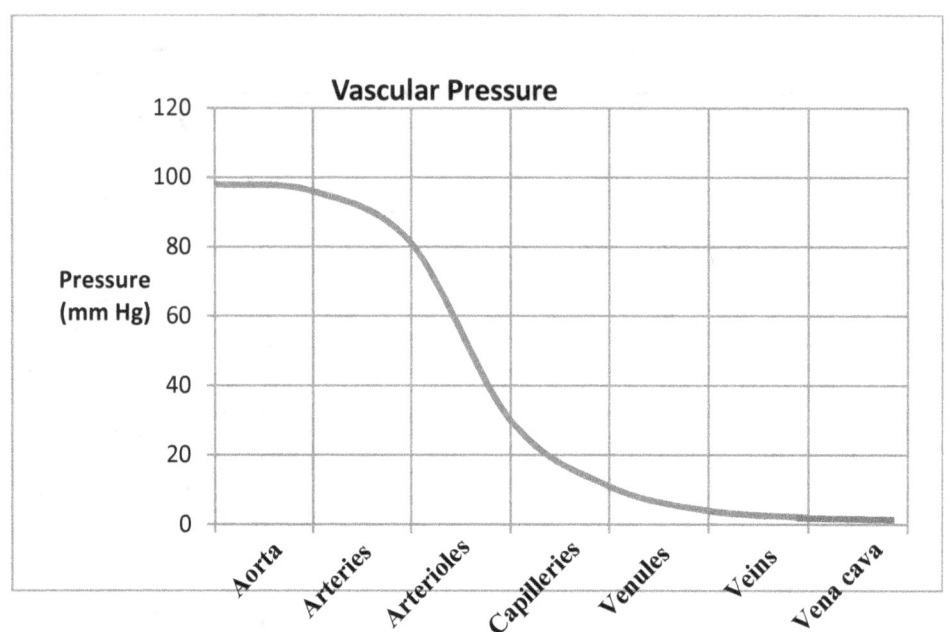

Figure 11.20 Blood pressure in various part of circulatry system

The heart can be modelled by a pump system with four valves to control the inlet and outlet of blood flow. The two atrioventricular valves which are between the atria and the ventricles, are mitral valve the and the tricuspid valve. The two semilunar valves, which are in the arteries leaving the heart, are the aortic valve and the pulmonary valve.

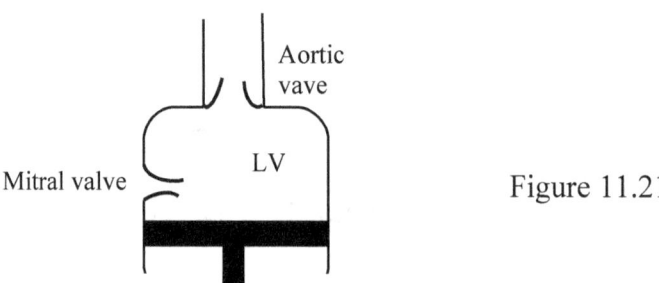

Figure 11.21

To push the blood out of LV, mitral valve will close and heart muscle will contract and blood will be squeezed out from aortic valve. The amount of work heart needs to empty the LV of a certain volume of blood can be calculated from the force muscle have to exert:

$$\Delta W = \int F\, dt = \int \frac{F}{A} A\, dr = \int \frac{F}{A} dV = \int_{V_1}^{V_2} P\, dV \qquad (11.61)$$

$V_2 - V_1 = \Delta V$ is the volume of blood pumped in each compression.

Pressure change from diastolic pressure P_0 to systolic pressure P_s and then falls back to diastolic pressure when the heart muscles relax. This can be shown graphically. See figure 11.22

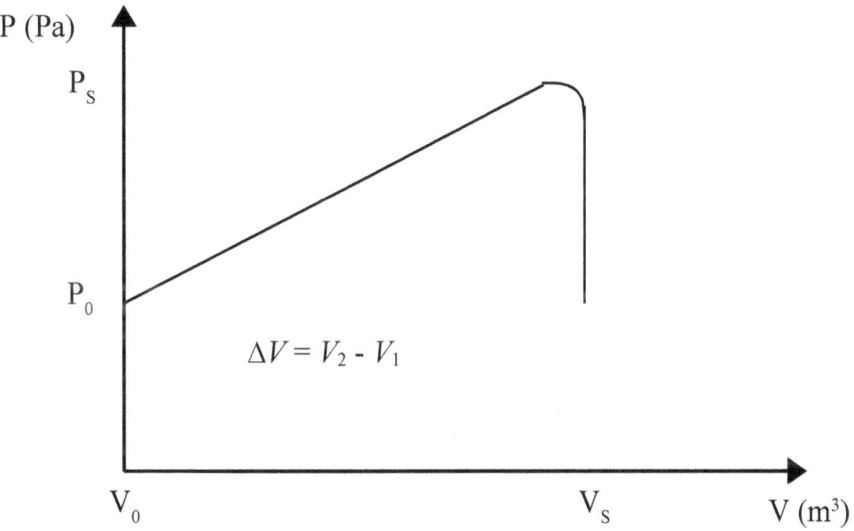

Figure 11.22 Pressure versus volume graph for blood flow through left ventricles.

The total work done is the area under this graph.

$$\Delta W = P_0 \cdot \Delta V + \frac{1}{2}(P_s - P_0) \cdot \Delta V \tag{11.62}$$

Example: Looking back on page 260, we have $P_0 = 12.64$ kPa, $P_s = 16.63$ kPa and $\Delta V = 7.5 \times 10^{-5}$ m^3. Calculate the work done by the heart.

$$\Delta W = (12.64 \times 10^3) \cdot (7.5 \times 10^{-5}) + \frac{1}{2}((16.63 - 12.64) \times 10^3) \cdot (7.5 \times 10^{-5})$$
$$= 0.948 + 0.149$$
$$= 1.09 \text{ J}$$

Flow velocity

The blood continuously flows through various parts of the body and as we know that this size of each part is different, so

$$Q_{total} = Q_{arteries} = Q_{arterioles} = Q_{capillaries} = Q_{venules} = Q_{veins}$$

The flow rate through cross-sectional area A and velocity v with length L can be written as:

$$Q = \frac{V}{t}, \text{ also velocity through length is } v = \frac{L}{t} \text{ and volume is } V = L.A$$

$$Q = \frac{L.A}{t} = vA \tag{11.63}$$

Since velocity is inversely proportional to cross-sectional area that leads to:

$$v_{artries} > v_{arterioles} > v_{capillaries} < v_{venules} < v_{veins}$$

Flow resistance

Ideal flow is frictionless. In real a fluid molecule rub against each other and therefore have a friction which slows them down. This friction between molecules is called viscous friction and is proportional to the fluid velocity and the coefficient of viscosity of that fluid. In a pipe with diameter d, the velocity of the fluid will be faster in the middle and will decrease towards the pipe wall. Such fluid flow is called *laminar* flow.

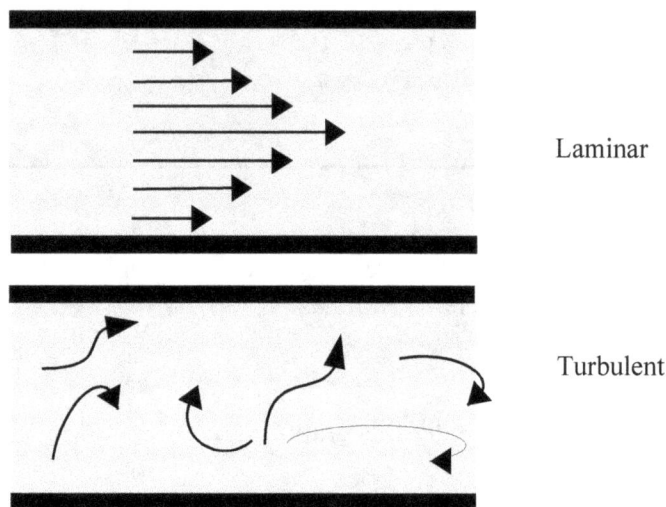

Figure 11.23 Fluid flow through a pipe, length is proportional to the velocity of fluid.

The fluid flow through the pipe due to pressure difference. Frictional force per unit area is proportional to the velocity gradient:

$$F_{viscosity} = \eta \frac{dv}{dr} \qquad (11.64)$$

where η is coefficient of viscosity (Pa.s)

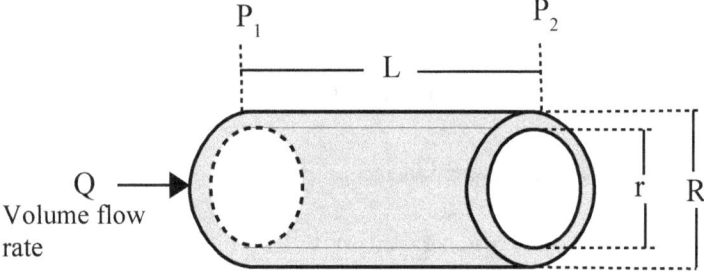

Figure 11.24 Volume fluid flow through a pipe with thickness (R-r)

The fluid outside the central region is moving slow due to viscous drag across the cylindrical surface at radius r, for a length L, pipe surface area is 2πrL. The friction force is in the opposite direction to the direction of fluid motion.

$$F_{viscosity} = -2\pi rL\eta \frac{dv}{dr} \tag{11.65}$$

In an equilibrium condition of constant velocity, where the net force goes to zero.

$$F_{viscosity} + F_{pressure} = 0 \tag{11.66a}$$

$$-2\pi rL\eta \frac{dv}{dr} + \Delta P(\pi r^2) = 0 \tag{11.66b}$$

$$\Delta P(\pi r^2) = 2\pi rL\eta \frac{dv}{dr}$$

$$\frac{dv}{dr} = \frac{\Delta P(\pi r^2)}{\eta(2\pi rL)} = \left(\frac{\Delta P}{2\eta L}\right) \cdot r \tag{11.66c}$$

integrating,
$$\int_v^0 dV = \left(\frac{\Delta P}{2\eta L}\right) \cdot \int_r^R r\, dr \tag{11.66d}$$

$$V_{(r)} = \left(\frac{\Delta P}{4\eta L}\right)[R^2 - r^2] \tag{11.66e}$$

The continuity of flow gives volume flow with varying velocities.

$$Q = \int v \cdot dA \tag{11.66f}$$

substituting 11.66e and $dA = 2\pi r\, dr$ into equation 11.66f

$$Q = \int_0^r \left(\frac{\Delta P}{4\eta L}\right)[R^2 - r^2] \cdot (2\pi r\, dr) \tag{11.66g}$$

$$Q = \left(\frac{\pi \cdot \Delta P}{2\eta L}\right) \int_0^r (R^2 r - r^3)\, dr \tag{11.66h}$$

$$Q = \left(\frac{\pi \cdot \Delta P}{2\eta L}\right)\left[\frac{r^4}{2} - \frac{r^4}{4}\right]$$

$$Q = \frac{\pi \Delta P r^4}{8\eta L} \tag{11.67}$$

This is called Poiseulle's equation.

Flow resistance R of flow rate Q is determined by the pressure difference ΔP ($P_2 - P_1$) between two points in the tube.

$$Q = \frac{\Delta P}{R} \tag{11.68}$$

equating 11.67 1nd 11.68:

$$\frac{\Delta P}{R} = \frac{\pi \Delta P r^4}{8\eta L}$$

hence,

$$R = \frac{8\eta L}{\pi r^4} \qquad (11.69)$$

A small change in radius can result in large change in resistance and flow.

Since $\qquad Q = vA$

and velocity is maximum (v_m) at the centre of the pipe, so $v = \dfrac{v_m}{2}$

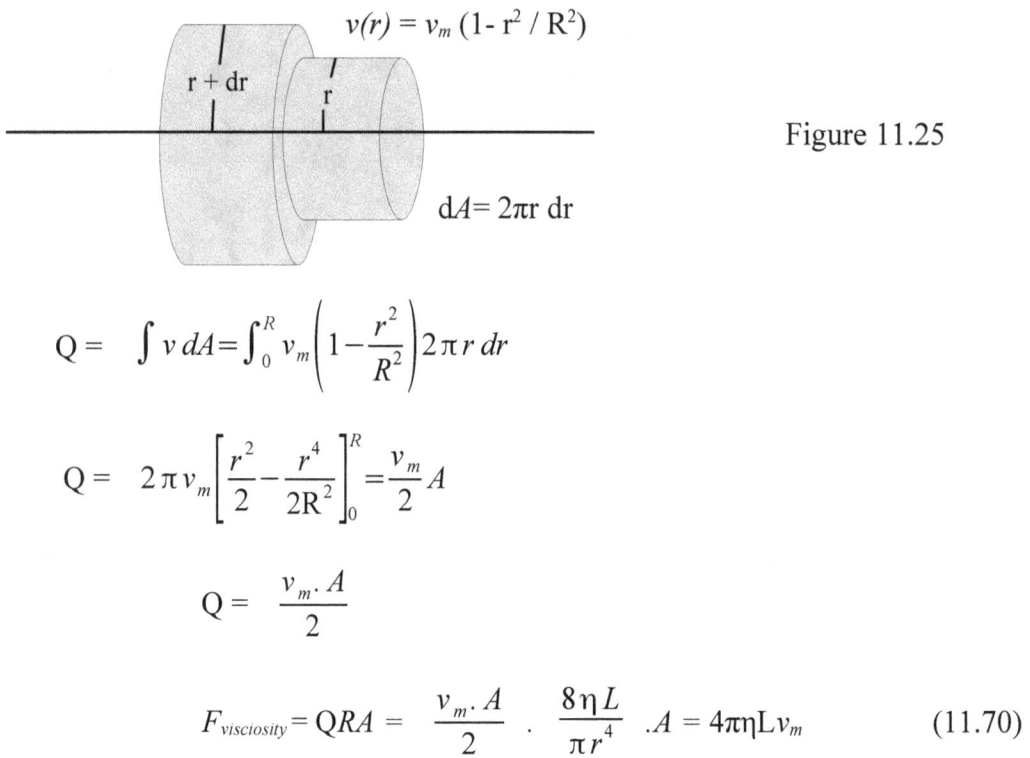

Figure 11.25

$$Q = \int v\, dA = \int_0^R v_m \left(1 - \frac{r^2}{R^2}\right) 2\pi r\, dr$$

$$Q = 2\pi v_m \left[\frac{r^2}{2} - \frac{r^4}{2R^2}\right]_0^R = \frac{v_m}{2} A$$

$$Q = \frac{v_m \cdot A}{2}$$

$$F_{visciosity} = QRA = \frac{v_m \cdot A}{2} \cdot \frac{8\eta L}{\pi r^4} \cdot A = 4\pi \eta L v_m \qquad (11.70)$$

There are two types of resistance arrangements, series and parallel. This comes out from defining resistance to flow, such that $\Delta P = QR$ for series arrangement.

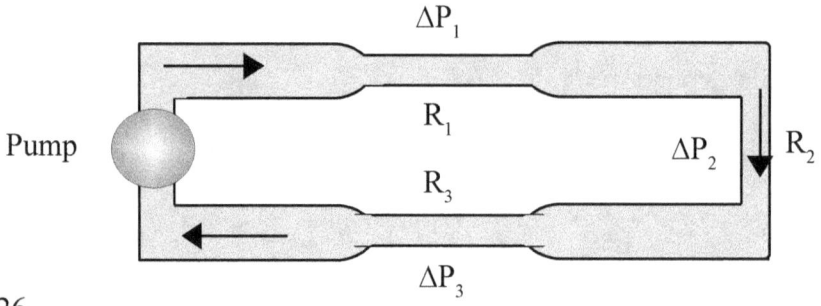

Figure 11.26

$$\Delta P = \Delta P_1 + \Delta P_2 + \Delta P_3 \qquad (11.71a)$$

$$\Delta P = QR_1 + QR_2 + QR_3 \qquad (11.71b)$$

$$\Delta P = QR_T$$

hence,

$$R_T = R_1 + R_2 + R_3 \qquad (11.71c)$$

For parallel arrangement Q will split into branches,

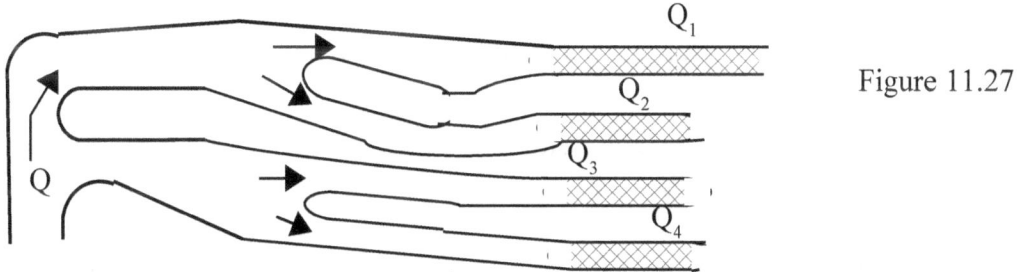

Figure 11.27

$$Q = Q_1 + Q_2 + Q_3 \qquad (11.72a)$$

$$Q = \frac{\Delta P}{R_1} + \frac{\Delta P}{R_2} + \frac{\Delta P}{R_3} \qquad (11.72b)$$

$$Q = \frac{\Delta P}{R_T}$$

$$\frac{1}{R_T} = \frac{1}{R_1} + \frac{1}{R_2} + \frac{1}{R_3} \qquad (11.72c)$$

Example: Calculate the resistance of the aorta. The length and radius of average aorta is 30cm and 1.3 cm respectively. Compare this with the resistance of capillary with 1 cm length and radius of 10μm. Take the viscosity coefficient of blood as 3.5×10^{-3} Pa s

$$R_a = \frac{8\eta L}{\pi r^4} = \frac{8(3.5 \times 10^{-3}\, Pa\, s)(0.3\, m)}{\pi (0.013)^4\, m^4} = 93.62\ \text{kPa m}^{-3}\text{s}$$

$$R_c = \frac{8\eta L}{\pi r^4} = \frac{8(3.5 \times 10^{-3}\, Pa\, s)(0.01\, m)}{\pi (10 \times 10^{-6})^4\, m^4} = 8.91 \times 10^{15}\ \text{Pa m}^{-3}\text{s}$$

Turbulent flow: In turbulent flow vortices, eddies and wakes make the flow unpredictable. Turbulent flow happens in general at high flow rates and with larger pipes. The flow condition of the fluid is characterized by Reynold Number which is defined as:

$$R = \frac{2r\rho \bar{v}}{\eta} \qquad (11.73)$$

where $\bar{v}$ is the average velocity and ρ is the density of the fluid

The flow will be laminar if Reynold number is less than 2000 and will be turbulent if this number is greater than 4000 and in between it will be in a transitional phase. A reduction in blood vessel due to fatty deposits can increase the velocity leads to transition from laminar to turbulent flow.

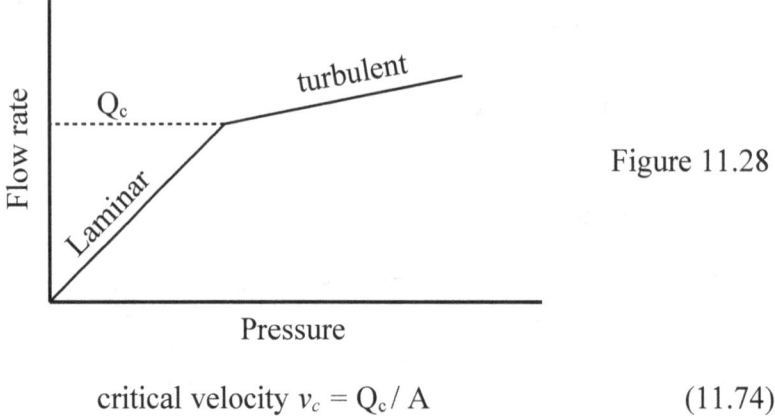

Figure 11.28

$$\text{critical velocity } v_c = Q_c / A \qquad (11.74)$$

Example: Consider the coronary artery and vein portion of the human circulatory system. The blood flow rate into the arteries is 6 cm³s⁻¹. The radius of each coronary artery is 0.1cm and 4.5 mm long. The left coronary artery is split into two circumflex arteries of each with a radius of 0.07cm and a length of 3.7mm. Calculate the velocity of the blood in each portion and total resistance of this circulatory system.

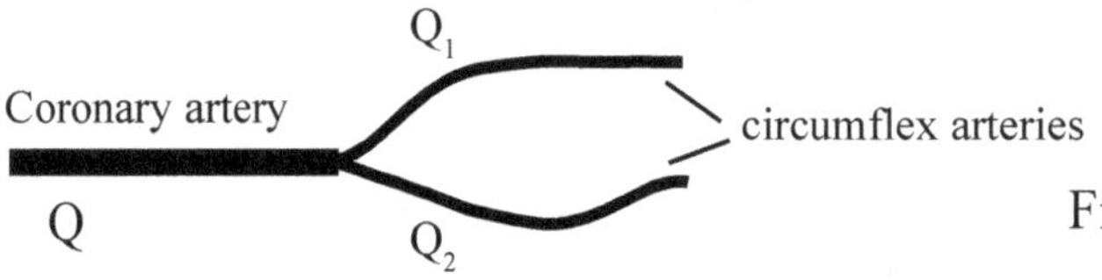

Figure 11.29

$Q = vA$ $\qquad$ Area of coronary artery $= \pi r^2 = \pi(0.001)^2 = 3.14 \times 10^{-6} \text{ m}^3$

$$v = \frac{Q}{A} = \frac{6 \times 10^{-6} \, m^3 \, s^{-1}}{3.14 \times 10^{-6} \, m^2} = 1.91 \, ms^{-1}$$

Now flow rate has split into two equal portions. Hence the velocity through each of circumflex arteries is same.

Area of coronary circumflex artery $= \pi r^2 = \pi(7 \times 10^{-4})^2 = 1.54 \times 10^{-6} \text{ m}^2$

$$v = \frac{Q}{A} = \frac{3 \times 10^{-6} \, m^3 \, s^{-1}}{1.54 \times 10^{-6} \, m^2} = 1.95 \, ms^{-1}$$

Resistance of coronary artery:

$$R = \frac{8\eta L}{\pi r^4} = \frac{8(3.5 \times 10^{-3} \, Pa\, s)(4.5 \times 10^{-3} \, m)}{\pi(1 \times 10^{-3})^4 \, m^4} = 40.11 \text{ kPa m}^{-3}\text{s}$$

Resistance of coronary circumflex artery:

$$R = \frac{8\eta L}{\pi r^4} = \frac{8(3.5 \times 10^{-3} \, Pa\, s)(3.7 \times 10^{-3} \, m)}{\pi(7 \times 10^{-4})^4 \, m^4} = 137.34 \text{ kPa m}^{-3}\text{s}$$

The total resistance of two parallel circumflex arteries:

$$R_T = \frac{(137.34 \times 10^3)(137.34 \times 10^3)}{(137.34 \times 10^3)+(137.34 \times 10^3)} = 68.67 \text{ Pa m}^{-3}\text{s}$$

so, the total resistance is $68.67 + 40.11 \times 10^3 = 40.18$ kPa m^{-3}s

Pressure and blood vessel wall

Pressure within walled vessel exerts a force on the wall. This force can be quantified by law of *LaPlac*.

For thin walled vessel:

$$T = a.P.r$$

T = wall tension (N)
P = trasmural pressure (Nm^{-2})
a = normally 2

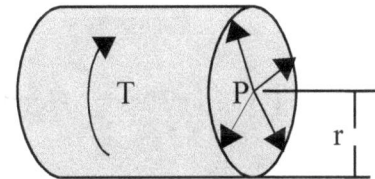

Figure 11.30a

For thick walled vessel:

$$\sigma = P\,r\,/W$$

σ = stress on wall (Nm^{-2})

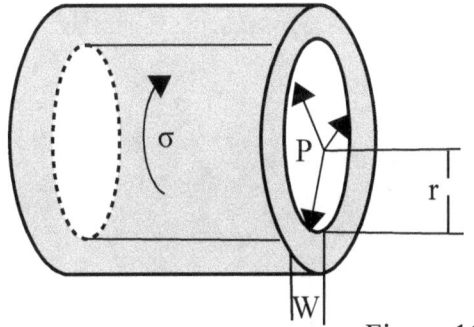

Figure 11.30b

Measuring blood velocity with ultrasound

The *Doppler effect*, named after the Austrian physicist Christian Doppler (29 November 1803 – 17 March 1853), who proposed it in 1842, is the change in frequency of a wave for an observer moving relative to its source. This Doppler effect is utilized in ultrasound applications to find blood velocity by analysing the relative frequency shift of the received signal brought by the movement of red blood cells.

The diagram in figure 11.31 shows a Doppler transducer placed on the skin and aimed at an angle, θ, towards a blood vessel, which contains blood flowing with a velocity of u m/s, at any instant. The transducer emits ultrasound waves of frequency, f_o, and echoes generated by moving reflectors in the blood, e.g. red blood cells, have a frequency, f_r. The difference between these two frequencies, Δf, is related to the velocity of the flowing reflectors through the following equation:

$$v = \frac{\Delta f\, c}{2 f_0 \cos\theta} \tag{11.75}$$

where c is the propagation speed of ultrasound in soft tissues, typically 1540 ms^{-1}

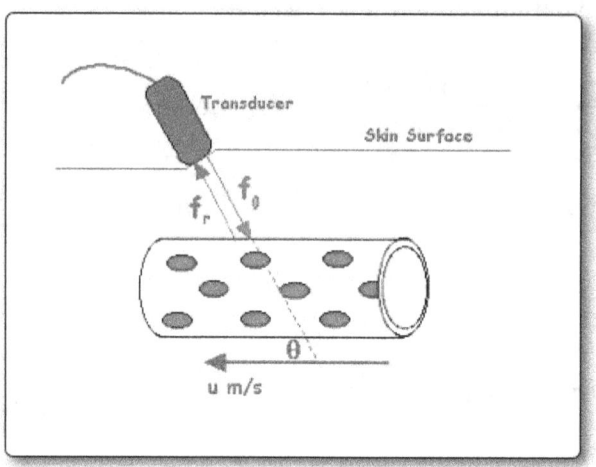

Figure 11.31 Measuring velocity of blood flow using ultrasound transducer

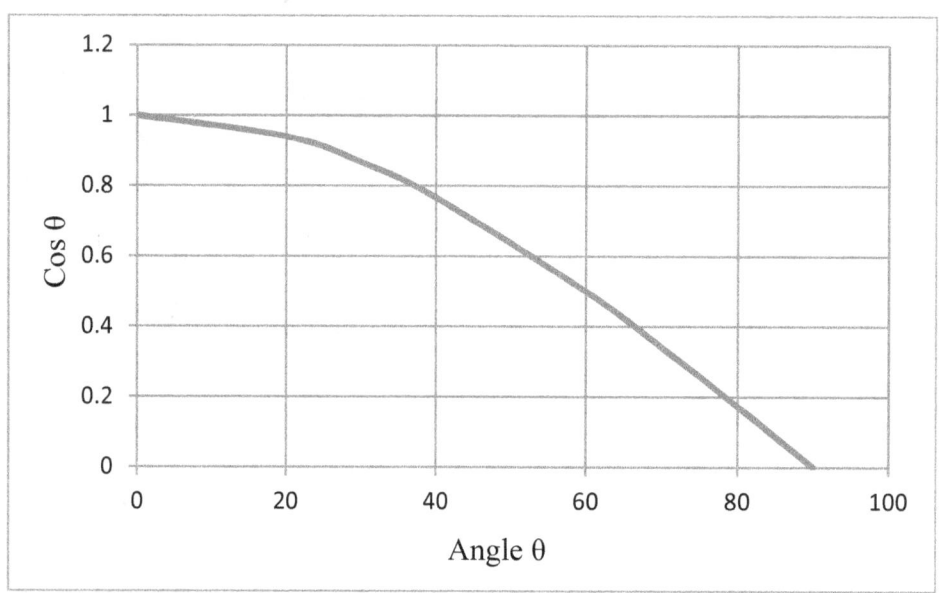

Figure 11.32 Variation of cos θ obtained against angle θ

By looking at figure 11.32 we can see that there will be no signal echo at an angle of $90°$ and best result can be achieved at an angle of $0°$ and this cannot be practical in reality. To get good result angle between $30°$ and $60°$ is desirable. At $10°$ the velocity is almost 98% of the actual value.

Example: Ultrasound transducer is placed on the arm at an angle of $40°$, The reflected signal has a frequency of 4 kHz when the transmitter frequency was 20 MHz. Calculate the blood flow velocity. Also calculate the velocity when the transducer is held at $60°$. Take the speed of the sound through arm as 1550 ms^{-1}.

$$\Delta f = (20 - 4)\ \text{MHz} = 16\ \text{MHz}$$

$$v = \frac{\Delta f\, c}{2 f_0 \cos\theta} = \frac{(4 \times 10^3)(1550)}{2(20 \times 10^6)(\cos 40°)} = 0.20\ \text{ms}^{-1}$$

and at $60°$ this speed is measured as 0.31 ms^{-1}

Exercise – 11

Q1 – What ion is responsible for the depolarization phase of the Action Potential?

a) potassium ion
b) sodium ion
d) calcium ion

Q2 – What is the status of neuron when the axon cannot be stimulated to fire?

a) threshold
b) rising phase
c) rest position
d) refractory position

Q3 – Which ions cause hyper-polarisation when they are trying to reach their equilibrium potential?

a) potassium ion
b) sodium ion
c) chlorine ion
d) calcium ion

Q4 – How do neurons alter their membrane potential?

a) they selectively alter the permeability of their membranes to ions by opening and closing gated ion channels in the membrane.
b) ion pumps on the cell membrane are activated, which pump ions outside the cell.
c) they selectively alter the permeability of their membranes to ions by changing the protein composition of their membranes

Q5 – A typical neuron has a resting membrane potential of about ……….. ?

a) -90 mV
b) -70 mV
c) 40 mV
d) 0 mV

Q6 – At what membrane voltage do neuronal voltage-gated Na^+ channels become activated?

a) -90 mV
b) -15 mV
c) -50 mV
d) 40 mV

Q7 – Which statement best describes the electrical state of a neuron at rest?

a) The outside of a neuron is more negatively charged than the inside.
b) The inside of a neuron is more negatively charged than the inside.
c) The inside and the outside of a neuron are neutral.
d) K^+ ions leak into a neuron

Q8 – Which of the following statements concerning delayed voltage-gated K^+ channels is FALSE?

a) opening of these channels does not occur until the action potential is near its peak.
b) these channels are closed when the membrane potential returns to a value close to the K+ equilibrium potential
c) these channels become fully activated when the membrane potential has shifted to -50mV.
d) without the opening of these channels, membrane repolarisation would occur more slowly.

Q9 – At the peak of the action potential, V_m is approximately +50 mV. Assuming normal intracellular and extracellular K^+ concentrations at 300K (1) calculate the driving force (in mV) that acts on K^+ ions, and (2) use the information obtained in part 1 to determine the direction in which K^+ ions will flow (i.e., into the cell or out of cell).

Q10 – Assume the following concentrations of Na^+ and K^+ ions: $[Na^+]_o$ = 145 mM, $[Na^+]_i$ = 15 mM, $[K^+]_o$ = 4 mM, and $[K^+]_i$ = 150 mM, $[Cl^-]_o$ = 110 mM, $[Cl^-]_i$ = 10 mM. Assume further that at the peak of the action potential, $p_K : p_{Na} : p_{Cl}$ is 1 : 0.05 : 0.45. Calculate V_m at the peak of the action potential at 37 °C.

Q11 – The equilibrium potentials for particular ions are determined by their concentrations inside and outside of the cell and are calculated using the Nernst equation. Calculate the equilibrium potential across the plasma membrane assuming that it is due solely to (*a*) K^+ and (*b*) Na^+. The concentrations for K^+ are 160 mM inside and 7 mM outside. The concentrations for Na^+ are 20 mM inside and 200 mM outside. Also show the direction of flow of ions.

Q12 – In the experiment the concentrations of K^+ ion on either side of a membrane that it perfectly selective for that ion and the potential difference (voltage) was measured across that membrane under equilibrium conditions as show in the table. Plot the graph of PD against log (concentration). Show that the gradient is 58.15 mV.

K^+ (mM)	V_k
1	-124.5
3	-96.8
10	-66.5
30	-38.8
100	-8.5
300	19.2

Q13 – Find the cell potential of a galvanic cell based on the following reduction half-reactions at 35 °C for cadmium reaction. $V^{\ominus}_{Cd}$ is -0.403 V and $V^{\ominus}_{Pb}$ is -0.126 V. Where $[Cd^{2+}]$ = 0.020 M and $[Pb^{2+}]$ = 0.200 M

$$Cd^{2+} + 2\,e^- \rightarrow Cd$$
$$Pb^{2+} + 2\,e^- \rightarrow Pb$$

Q14 – the typical value of capacitance for biological membrane is 1.15 µF. Calculate the value of the capacitor for a cell with diameter of 45 µm.

Q15 – Charge on capacitor is given by Q = CV. If the membrane potential is -75mV, calculate the number of K⁺ ions cross the channel. Take the capacitor value for biological membrane from Qc. If the concentration of K⁺ is 500mM, calculate the percentage change in concentration.
Hint: charge on one ion is 1.6×10^{-19} C

Q16 – Write down the parallel conductance equation for the resting potential as a function of equilibrium potentials and conductances. Use following values to calculate V_r.
Leak current $I_l = 0$
$g_k = 0.07$, $g_{Na} = 0.026$, $g_{Cl} = 0.1$ and $V_k = -72.1$ mV, $V_{na} = 52.4$ mV, $V_{Cl} = -57.2$ mV

Q17 – If all the values remain the same as in Q16, Calculate to two significant digits the value to which gK would have to be reduced in order to make V_r be -45 mV. Show your work.

Q18 – The *cardiac action potential* is a specialized _____ in the heart, with unique properties necessary for function of the electrical conduction system of the heart.

a- action potential   b- ion channel   c- photoreceptor cell   d- neuron

Q19 – This phenomenon is called calcium-induces calcium release and increases the myoplasmic free Ca^{2+} concentration causing _____.

a- muscular system   b- muscle contraction   c- cardiac muscle   d- smooth muscle

Q20 – The fast _____ can be modelled as being controlled by a number of *gates*.

a- calcium channel   b- sodium channel   c- potassium channel   d- ion channel

Q21 – What is *phase* 4 in cardiac action potential.

a- resting potential   b- action potential   c- depolarisation   d- repolarisation

Q22 – In an _____, electrical systole of the ventricles begins at the beginning of the QRS complex.

a- Cardiopulmonary resuscitation   b- Blood pressure   c- Electrocardiography   d- Circulatory system

Q23 – The frequency of the cardiac cycle is the _____.

a- Heart rate   b- blood pressure   c- pulse   d- heart beat

Q24 – Phase 0 of the ventricular cell action potential results from

a. the closing of voltage-gated K+ channels
b. the activation (opening) of voltage-gated Na+ channels
c. the activation (opening) of L-type Ca2+ channels
d. the inactivation of pacemaker channels

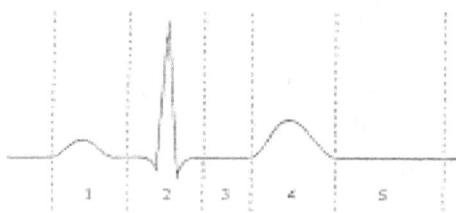

Q25 – In the figure above, the period during which the rising phase of the action potential of ventricular muscle cells will occur is represented by_____. Explain your answer.

a- 1   b- 2   c- 3   d- 4   e- 5

Q26 – The ECG of a patient with severe hyperkalemia ([K^+] = 10 mM) is likely to display (relative to normal).

a. a shortened, but larger than normal, P wave
b. a prolonged QRS complex
c. no discernible T wave
d. a decreased PR interval
e. all of the above are correct

Q27 – Diagram below shows an ECG trace for a healthy person.

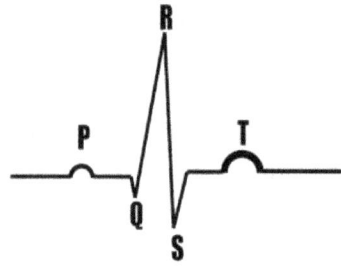

a- Add a suitable scale and unit to the potential axis.
b- Add a suitable scale to the time axis.
c- State the electrical events which give rise to the points P, R and T.

Q28 – Show that

a) - $aVL = \left(V_i - \dfrac{V_{ii}}{2} \right)$

b) - $aVF = \left(V_{ii} + \dfrac{V_i}{2} \right)$

Q29 – Use the work done by left ventricles in the heart to pump out 75ml blood from the example on page 263 and calculate how many calories are required by the heart to function. If an average person heart is only 16% efficient how many calories are required to meet that demand. BMR of average person is 1680 kcal/day. Calculate how many percent is taken by heart beat per day? Take 1 cal = 4.184 J

Q30 – To circulate the blood in the human body, heart need to pump blood against gravity. If the aortic pressure is 12.64 kPa. Calculate the blood pressure at the feet and head. Take the distance between heart and feet as 1.4 m and between heart and head as 0.4m

Q31 – Air force pilot can sometime blackout due to extra 'g' they feel due to the aircraft's certain movement. Human can blackout due to lack of oxygenated blood supply to the brain. Show by calculation how this happens?

Q32 – The blood is being transfused from a height of 1m into the patient's vein which is at 2.4 kPa pressure. The radius of the vein is 0.25 mm and is 4 cm long. How long will take to transfuse one litre blood.

If the vein radius was 0.2 mm how much longer will take to transfuse one litre blood.

Q33 – Fill in the missing words.

In the human heart, deoxygenated blood returning to the heart from the body enters the ____1____ via the venae cavae. Next, the blood passes into the ____2____, which pumps the blood through the ____3____ leading to the lungs. Oxygenated blood from the lungs flows into the ____4____ and then passes to the left ventricle, which pumps the blood out to the body via the aorta.

Q34 – What is Bradycardia ?

Q35 – Which side of the heart has deoxygenated blood?

Q36 – Name the valves on the right side of the heart?

Q37 – List the path of the electrical conduction of heart?

Q38 – List three factors governing resistance to blood flow.

Q39 – What does the heart do during:
i- systolic phase
ii- diastolic phase

Q40 – A normal PR interval is _____.

a) 0.4 second - 0.10 second, b) 0.6 second – 0.12 second, c) 0.12 second – 0.20 second

Q41 – Which wave in the ECG is caused by atrial depolarization.

Q42 – The normal duration of QRS is _____.

a) less than 0.12 seconds, b) 0.4 seconds – 0.10 seconds, c) 1.0 second – 1.5 second

Q43 – Q waves are always _____.

a) negative,   b) positive,   c) bi-phase

Q44 – If QT and spreads over 12 squares, calculate the duration of this wave on the ECG.

Appendix A

Solutions – 1

1- $E_K = \frac{1}{2} mv^2 = \frac{1}{2} \times 800 (10)^2 = 40$ kJ

2- $m = \dfrac{2E}{v^2} = \dfrac{2 \times 32800}{8^2} = 1025$ kJ

3- $v = \sqrt{\dfrac{(2E)}{m}} = \sqrt{\dfrac{(2 \times 85000)}{3000}} = 7.53\, ms^{-1}$

4- $E_P = mgh = 1200 \times 45 \times 10 = 540$ kJ

5- $m = \dfrac{E}{mg} = \dfrac{2400}{4 \times 10} = 60$ kg

6- $mg = \dfrac{E}{h} = \dfrac{60,000}{5} = 12$ kN

7- At the bottom of the hill he has kinetic energy $E_K = \frac{1}{2} mv^2 = \frac{1}{2} 70 (6)^2 = 1260$ J
To reach to the top he will use this kinetic energy to convert into potential energy $E_K = E_P$

$$mgh = 1260 \text{ J} \qquad h = \dfrac{1260}{70 \times 10} = 1.8 m$$

8- $E = \frac{1}{2} kx^2 = \frac{1}{2} 500(1.25)^2 = 390.63$ J
$= \frac{1}{2} 500(1.50)^2 = 562.50$ J

9- Thermal energy $U_t = 3/2\, kT$ and
Kinetic energy $E_K = \frac{1}{2} m\langle v^2 \rangle = \frac{1}{2} 4.65 \times 10^{-26} (850)^2 = 1.68 \times 10^{-20}$ J

$$E_K = U_t \qquad 3/2\, kT = \frac{1}{2} m\langle v^2 \rangle$$

$$T = \dfrac{(mv^2)}{3k} = \dfrac{4.65 \times 10^{-26} kg\, (850^2 m^2 s^{-2})}{3(1.38 \times 10^{-23} JK^{-1})} = 811\, K$$

10- amount per Calorie $= \dfrac{13.5 g}{67} = 0.20$ g Cal^{-1} (remember: 1kcal = 1Cal)

protein = 0.20 g Cal^{-1} × 3.45 Cal = 0.70 g
carbohydrate = 0.20 g Cal^{-1} × 45.65 Cal = 9.21 g
fat = 0.20 g Cal^{-1} × 15.38 Cal = 3.11 g
fibres = 0.20 g Cal^{-1} × 2.38 Cal = 0.48 g

11- SPL = 20 log$_{10}$ (P$_{rms}$ / P$_t$)
120 = 20 log$_{10}$ (P$_{rms}$ / P$_t$) ⟹ 120/20 = log$_{10}$ (P$_{rms}$ / P$_t$) ⟹ 6 = log$_{10}$ (P$_{rms}$ / P$_t$)
Threshold pressure P$_t$ is 20 µPa 10^6 = P$_{rms}$ / P$_t$
P$_t$ = 10^6 (20 × 10^{-6}) = 120 µPa.

12- $E = \dfrac{P^2 A}{\rho v} \cos\phi$ standard condition $\rho = 1.2$ kgm^{-3} $v = 343$ ms^{-1}

$E = \dfrac{(20)^2 \times 2}{1.2 \times 343} \cos 35^o = 1.59$ J

13- $P = VI$ $I = \dfrac{P}{V} = \dfrac{2500W}{230V} = 10.87$ A

$Q = It + 10.87 (2 \times 3600) = 78.26$ kC

14- $E = hf$ $f = \dfrac{E}{h} = \dfrac{3 \times 10^{-16} J}{6.63 \times 10^{-34} Js} = 452.49 \times 10^{15}$ Hz

Solutions – 2

1- In transverse wave particle would move perpendicular to the direction of wave motion. So particle of the medium would move both northward and southward.

2- Only longitudinal wave can travel through fluid.

3- a – amplitude   b – wavelength   c – crest   d – rest position   e – trough   f – peak-to-peak

4- a – 5m   b – 20s   c – 1/20 = 0.05 Hz   d – 4m   e – 10m   f – v = 0.05 x 4 = 1/5 = 0.2 ms^{-1}

5- decrease

6- quadruple   E α A^2 hence when amplitude double energy quadruples   50 x 2^2 = 200J

7- f = 75Hz   T = $\dfrac{1}{f} = \dfrac{1}{75} = 0.013\ seconds$

8- $\dfrac{1}{60} Hz$ as frequency is number of cycles per second.

9- $v = f\lambda$ = 1.23 x 270 = 332.1 ms^{-1}

10- By doubling the frequency, wavelength is halved so speed of wave doesn't change.

11- $\dfrac{V_i}{V_r} = \dfrac{n_r}{n_i} = \dfrac{3 \times 10^6 ms^{-1}}{1.98 \times 10^6 ms^{-1}} = \dfrac{n_r}{1}$ as $n_i = 1$

n_r = 1.54   so is salt

12- $\dfrac{sin 45^o}{sin r} = 1.08$

$r = \sin^{-1}(\dfrac{sin 45^o}{1.08}) = 40.89^o$

13- $\dfrac{\sin 42°}{\sin r} = \dfrac{2.42}{1.52} = 1.592$

$r = \sin^{-1}(\dfrac{\sin 42°}{1.592}) = 24.85°$

14- $C = \sin^{-1}(\dfrac{1}{1.44}) = 43.98°$

15- $v = \dfrac{c}{n} = \dfrac{3 \times 10^8}{2.42} = 1.24 \times 10^8\, ms^{-1}$ Also $\theta_1 = 90° - 40° = 50°$

$\phi_c = \sin^{-1}(\dfrac{1}{2.42}) = 24.4°$

$\sin \theta_2 = \dfrac{n_1}{n_2} \sin \theta_1 = \dfrac{1.33}{2.42} \sin 50°$, hence $\theta_2 = 24.9°$

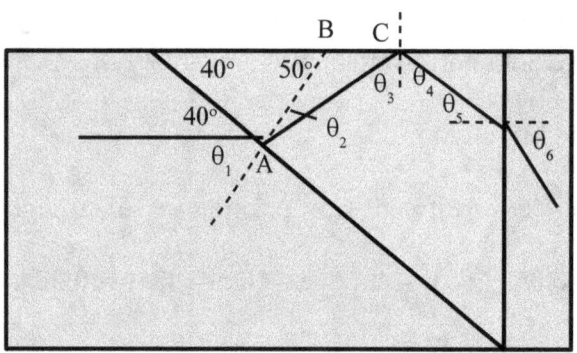

angle ABC = 180° − 50° = 130°, hence angle ABC = 180° − (130° + 24.9°) = 25.1°

so $\theta_3 = 90° - 25.1° = 64.9°$

Because the angle of incident at diamond-air boundary is 64.9° is larger than critical angle (24.4°) so the light ray is totally internally reflected.

Angle of reflection = Angle of incident
$\theta_3 = \theta_4 = 64.9°$

Therefore, $\theta_5 = 90° - \theta_4 = 25.1°$

$\sin \theta_6 = \left(\dfrac{n_d}{n_w}\right) \sin \theta_5 = \left(\dfrac{2.42}{1.33}\right) \sin 25.1° = 50.5°$

Solutions – 3

1- a) orange, b) infrared, c) radio waves, d) violet
2- a) green, b) gamma rays, c) microwave, d) UHF
3- a) orange, b) microwave, c) infrared, d) radio

4- a) electrons in atom can only occupy discrete energy levels, the ground state is the lowest energy state an electron can occupy.

b) in the excitation state electron in atom move to higher energy level by absorbing energy, where as in ionisation electron is completely removed from the atom.

c) frequency depend on energy difference of electron energy. Electron energy level have discrete values.

d) first we need to convert energy into Joule. $13.6 \times (1.6 \times 10^{-19}) = 2.18 \times 10^{-18}$ J

$$E = hf, \qquad f = \frac{2.18 \times 10^{-18} J}{6.63 \times 10^{-34} Js} = 3.28 \times 10^{15} Hz$$

5- a) $\Delta E = -0.26 \times 10^{-18}$ J $- (-0.59 \times 10^{-18}$ J$) = 0.33 \times 10^{-18}$ J

$$f = \frac{E}{h} = \frac{0.33 \times 10^{-18} J}{6.63 \times 10^{-34} Js} = 5 \times 10^{14} Hz$$

b) n = 3 and n = 2

6- $E = \dfrac{hc}{\lambda} = \dfrac{(6.63 \times 10^{-34} Js)(3 \times 10^{8} ms^{-1})}{680 \times 10^{-9} m} = 2.93 \times 10^{-19} J$

7- false

8- X-rays

9- It is reflected, refracted and diffracted. It can not be polarised.

10- $E = hf = 6.63 \times 10^{-34}$ Js $\left(\dfrac{3 \times 10^{8} ms^{-1}}{760 \times 10^{-9} m}\right) = 2.61 \times 10^{-19} J$

11- $\dfrac{2.61 \times 10^{-19} J}{1.63 \times 10^{-19} C} = 1.63$ eV

12- P = σ A T^4 = $(5.67 \times 10^{-8}$ Wm^{-2}K$^{-4}) \times (4\pi (0.02)^2$ m$^2) \times (1200$ K$)^4$
         = 59 kJs^{-1}

13- a) $\lambda = \dfrac{c}{f} = \dfrac{3 \times 10^{8} ms^{-1}}{102 \times 10^{6} Hz} = 2.94 m$

b) $E = hf = (6.63 \times 10^{-34} Js) \times (102 \times 10^{6} Hz) = 6.76 \times 10^{-26} J$

c) if n photons, each with energy E passing a particular point, then energy per second (power) is:

$$P = nhf, \qquad n = \frac{P}{hf} = \frac{120 \times 10^{3} W}{6.76 \times 10^{-26} J} = 1.78 \times 10^{30} \text{ photons per second}$$

14- $\lambda = \dfrac{c}{f} = \dfrac{3 \times 10^8 \, ms^{-1}}{2 \times 10^{14} \, Hz} = 15 \times 10^{-6} \, m = 1500 \, nm$

15- $E = hf = 6.63 \times 10^{-34} \, Js \left(\dfrac{3 \times 10^8 \, ms^{-1}}{102.6 \times 10^{-6} \, m} \right) / 1.603 \times 10^{-19} \, C = 12.09 \, eV$

16- $E = 13.6 \times (1.602 \times 10^{-19}) = 2.178 \times 10^{-18} \, J$

$$\lambda = \dfrac{ch}{E} = \dfrac{(3 \times 10^8 \, ms^{-1})(6.63 \times 10^{-34} \, Js)}{2.178 \times 10^{-18} \, J} = 91.29 \, nm$$

This wavelength is in UV region of the EM spectrum

17- $\lambda = \dfrac{c}{f} = \dfrac{3 \times 10^8 \, ms^{-1}}{5.09 \times 10^{14} \, Hz} = 5.89 \times 10^{-7} \, m = 589 \, nm$

18- $E = hf = 6.63 \times 10^{-34} \, Js \left(\dfrac{3 \times 10^8 \, ms^{-1}}{545 \times 10^{-9} \, m} \right) = 3.65 \times 10^{-19} \, J$

19- Electric and magnetic fields are oscillating perpendicular to each other and to the direction of the wave propagation.

20- Microwave and infrared are absorbed by the water in Earth's atmosphere. If we find place which is dry and above the clouds we can receive microwaves and infrared. Examples are Wyoming infrared observatory which is on the summit of Jelm mountain at an altitude of 2943m. For microwave Owens valley telescope (1222 m) and south pole telescope are good examples.
X-rays and gamma rays are totally absorbed by the ozone layer, which is very high above any mountain so only way to observe at that wavelength we need to send telescopes in the earth's orbit.

21- $f = \dfrac{c}{\lambda} = \dfrac{3 \times 10^8 \, ms^{-1}}{300 \times 10^{-9} \, m} = 1 \times 10^{15} \, Hz$

22- Gamma ray burst, Particle accelerator like CERN, Nuclear power plant and radioactive nuclei

23- $\lambda = \dfrac{c}{f} = \dfrac{3 \times 10^8 \, ms^{-1}}{88 \times 10^6 \, Hz} = 3.4 \, m$

$\lambda = \dfrac{c}{f} = \dfrac{3 \times 10^8 \, ms^{-1}}{108 \times 10^6 \, Hz} = 2.78 \, m$

24- a and b

25- $E = 3.4 \times (1.602 \times 10^{-19}) = 5.45 \times 10^{-19} \, J$

$\lambda = \dfrac{ch}{E} = \dfrac{(3 \times 10^8 \, ms^{-1})(6.63 \times 10^{-34} \, Js)}{5.45 \times 10^{-19} \, J} = 364.6 \, nm$

26- This does not depend on frequency.

$$t = \frac{d}{c} = \frac{8 \times 10^{10} \, m}{3 \times 10^8} = 267 s$$

27- $\lambda = \frac{c}{f} = \frac{3 \times 10^8 \, ms^{-1}}{95.8 \times 10^6 \, Hz} = 3.13 \, m$

28- They all travel as wave, Travel at the speed of light, Travel in vacuum, carry energy and they are transverse waves

29- see answer 19

30- see answer 20

31- $\eta_T = \eta_E + \eta_B = (½ . 8.85 \times 10^{-12} . [2 \times 10^8]^2) + (½ . [647 \times 10^{-3}]^2 / 4\pi \times 10^{-7}) = 343.56 \, kJm^{-3}$

Solutions – 4

1- $\frac{1}{f} = \frac{1}{u} + \frac{1}{v} = \frac{1}{10} - \frac{1}{15} = \frac{1}{v}$ so $v = \frac{3-2}{30} = \frac{1}{30}$ Hence $v = 30 \, cm$

+v means that the image is real and inverted.

$m = \frac{v}{u} = \frac{30}{10} = 3 \, time$ also, $m = \frac{h_i}{h_o}$ hence, $h_i = m \, h_o = 3 \times 4 = 12$ cm tall

2- a) $P = \frac{-1}{f}$ therefore, $f = \frac{-1}{P} = \frac{-1}{2.5} = -0.4 \, m = -40 \, cm$

b) $\frac{1}{v} = \frac{-1}{40} - \frac{1}{60} = \frac{(-3-2)}{120} = \frac{-5}{120}$ therefore, $v = \frac{-120}{5} = -24 \, cm$

-v means that the image is virtual and erect.

c) $m = \frac{v}{u} = \frac{24}{60} = 0.4$

d) $h_i = m \, h_o = 0.4 \times 3 = 1.2$ cm

e) [Diagram: Diverging lens showing Object, f, and image]

f) object is erect, virtual and smaller than the object.

3- a) $\frac{1}{v_1} = \frac{1}{f_1} - \frac{1}{u_1} = \frac{1}{0.5} - \frac{1}{1.5} = \frac{(3-1)}{1.5} = \frac{2}{1.5}$, $v_1 = 0.75 \, cm$ that is behind the first lens at 3.25 cm

b) $m_1 = \dfrac{v_1}{u_1} = \dfrac{0.75}{1.5} = 0.5$

c) $u_2 = 1 - 0.75 = 0.25$ cm

d) $\dfrac{1}{v_2} = \dfrac{1}{f_2} - \dfrac{1}{u_2} = \dfrac{1}{1} - \dfrac{1}{0.25} = \dfrac{(1-4)}{1} = \dfrac{-3}{1}$, $v_2 = -0.33$ cm

that is in front of the second lens at 3.17 cm

e) $m_2 = \dfrac{v_2}{u_2} = \dfrac{-0.33}{0.25} = -1.33$

f) $m_t = m_1 \times m_2 = 0.5 \times 1.33 = 0.66$

g) real, diminished and inverted

4- $\dfrac{1}{v} = \dfrac{1}{8} - \dfrac{1}{25} = \dfrac{(25-8)}{200} = \dfrac{17}{200}$ therefore, $v = \dfrac{200}{17} = 11.76$ cm

$\dfrac{v}{u} = \dfrac{h_i}{h_o}$ $\quad h_i = \dfrac{v}{u} \times h_o = \dfrac{11.76}{25} \times 10 = 4.71$ cm

5- $\dfrac{1}{v} = \dfrac{-1}{10} - \dfrac{1}{35} = \dfrac{(-7-2)}{70} = \dfrac{-9}{70}$ therefore, $v = \dfrac{-70}{9} = -7.8$ cm

$h_i = \dfrac{v}{u} \times h_o = \dfrac{7.8}{35} \times 10 = 2.2$ cm

6- <u>thick</u>, <u>thin</u>, <u>thin</u>, <u>thick</u>

7- $\dfrac{1}{v} = \dfrac{1}{20} - \dfrac{1}{60} = \dfrac{(3-1)}{60} = \dfrac{2}{60}$ therefore, $v = \dfrac{60}{2} = 30$ cm

$h_i = \dfrac{v}{u} \times h_o = \dfrac{-30}{60} \times 20 = -10$ cm

$m = \dfrac{v}{u} = \dfrac{10}{20} = 0.5$

8- Second lens must form a virtual image at 20cm to the left, therefore $v_2 = -20$ cm

$\dfrac{1}{u_2} = \dfrac{1}{f_2} - \dfrac{1}{v_2} = \dfrac{1}{25} + \dfrac{1}{20} = \dfrac{(4+5)}{100} = \dfrac{9}{100}$, $u_2 = \dfrac{100}{9} = 11.11$ cm

The object for second lens comes from the image of the first lens. Therefore, the image distance for the first lens is
$v_1 = 60$ cm $- u_2 = 60 - 11.11 = 48.89$ cm

The distance of the original object must be located to the left of the first lens.

$$\frac{1}{u_1} = \frac{1}{f_1} - \frac{1}{v_1} = \frac{1}{12} - \frac{1}{48.89}, u_1 = 15.90 \, cm$$

$$m_t = m_1 m_2 = \left(\frac{v_1}{u_1}\right)\left(\frac{v_2}{u_2}\right) = \left(\frac{48.89}{15.90}\right)\left(\frac{-20}{11.11}\right) = -5.54$$

9- $\frac{1}{v_1} = \frac{1}{f_1} - \frac{1}{u_1} = \frac{1}{45} - \frac{1}{60}, v_1 = 180 \, cm \quad m_1 = \frac{v_1}{u_1} = \frac{180}{60} = 3$

$\frac{1}{f_2} = \frac{1}{u_2} + \frac{1}{v_2} = \frac{1}{30} + \frac{1}{195}, f_2 = 26 \, cm \quad m_2 = \frac{v_2}{u_2} = \frac{195}{30} = 6.5$

$m_t = m_1 m_2 = 3 \times 6.5 = 19.5$

$m_t = \frac{h_i}{h_o}$, hence $h_o = \frac{h_i}{m_t} = \frac{70}{19.5} = 3.59 \, cm$

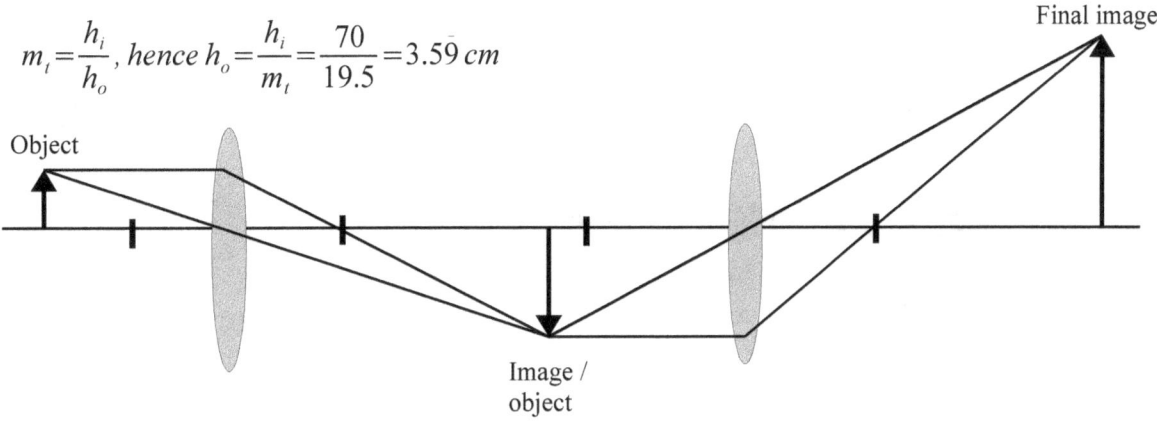

10- a) $f = -f$ and $u = -3f$ \quad Substituting in $\frac{1}{f} = \frac{1}{v} - \frac{1}{u}$

$\frac{-1}{f} = \frac{1}{v} - \frac{1}{-3f}, \frac{1}{v} = \frac{1}{-3f} - \frac{1}{f} = \frac{-1-3}{3f} = \frac{-4}{3f}, v = \frac{-3}{4}f$ \quad The image is virtual and upright

b) $\frac{-1}{f} = \frac{1}{v} - \frac{1}{-f}, \frac{1}{v} = \frac{1}{-f} - \frac{1}{f} = \frac{-1-1}{f} = \frac{-2}{f}, v = \frac{-1}{2}f$ \quad The image is virtual and upright

c) $\frac{-1}{f} = \frac{1}{v} - \frac{1}{-1/2f}, \frac{1}{v} = \frac{1}{-f} - \frac{2}{f} = \frac{-1-2}{f} = \frac{-3}{f}, v = \frac{-1}{3}f$ \quad The image is virtual and upright

11- $\frac{1}{f} = \frac{1}{v} + \frac{1}{u}, \frac{1}{v} = \frac{1}{f} - \frac{1}{v} = \frac{1}{36} - \frac{1}{54} = \frac{2-1}{108} = \frac{1}{108}, v = 108 \, cm \quad m = \frac{-v}{u} = \frac{-108}{54} = -2$

image is real, inverted and enlarged

12- Erect, virtual, smaller, between lens and focal length, same side as object

Solutions – 5

1- B   2- C   3-C   4-A   5-C   6-B

7- $^{198}_{86}Rn \rightarrow ^{4}_{2}He + ^{194}_{84}Po$

8- $^{26}_{14}Si \rightarrow ^{26}_{13}Al + ^{0}_{+1}e$

9- $^{12}_{6}C + ^{238}_{92}U \rightarrow ^{246}_{98}Cf + 4\,^{1}_{0}n$

10- i) ½ x ½ x1000 = 250 cps

ii) $\lambda = \dfrac{\ln 2}{t_{1/2}} = \dfrac{0.693}{(30 \times 60)} = 3.85 \times 10^{-4}\,s^{-1}$

$N = N_0 e^{-\lambda t}$ $\dfrac{25}{1000} = e^{-(3.85 \times 10^{-4})t}$

$\ln 0.25 = -(3.85 \times 10^{-4})t$
-3.69 = -3.85 x 10⁻⁴t

$t = \dfrac{-3.69}{-3.85 \times 10^{-4}} = 9581.5\,sec = 160\,min$

11- 4 half-lives = 4 x 5730 = 22920 years

12- $\lambda = \dfrac{\ln 2}{t_{1/2}} = \dfrac{0.693}{(6 \times 3600)} = 3.21 \times 10^{-5}\,s^{-1}$

$\dfrac{dN}{dt} = -\lambda N$ $N = \dfrac{\frac{dN}{dt}}{\lambda} = \dfrac{3.5 \times 10^8}{3.21 \times 10^{-5}} = 1.1 \times 10^{13}$

mass of N nuclei = $\dfrac{N \times mass\,number}{Avogadro's\,Number} = \dfrac{(1.1 \times 10^{13})(99)}{6.023 \times 10^{23}} = 1.79 \times 10^{-9}\,g = 1.79\,ng$

13- $\dfrac{1}{2^n}$ where n is the number of half-lives $\dfrac{60}{5} = 12$ hence n =12

$\dfrac{1}{2^{12}} = \dfrac{1}{4094}$ remains

14- $N = N_0 e^{-\lambda t}$ $32 = 64 e^{-(0.00043)t}$ $0.5 = e^{-(0.00043)t}$

ln 0.5 = -0.00043t

-0.693 = -0.00043t   t = 1611 years

15-

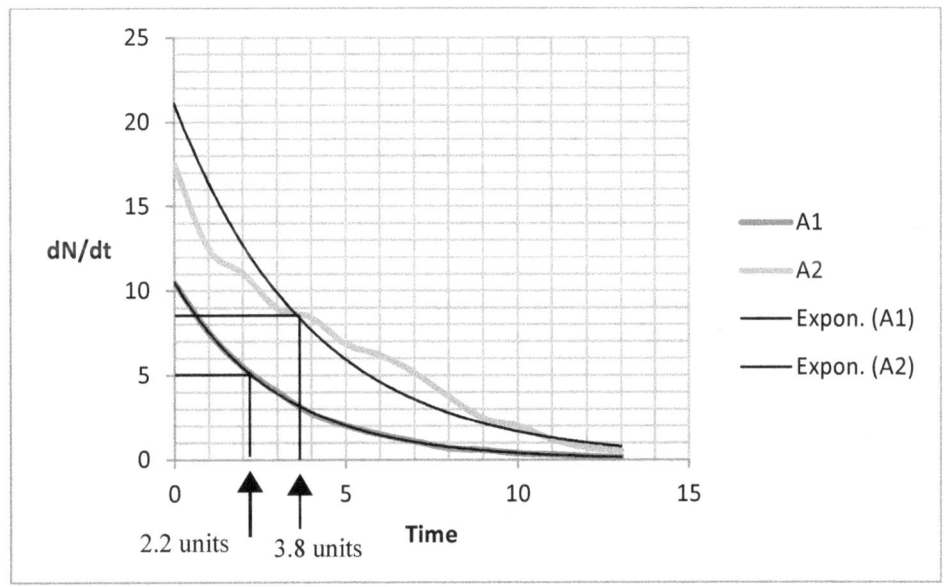

The half life of A1 is 2.2 unit time and A2 is 3.8 unit time.

16- a) Gamma rays are more appropriate as they are very penetrating.

b) i) $\lambda = \dfrac{\ln 2}{t_{1/2}} = \dfrac{0.693}{5.3} = 0.13 \, years^{-1}$

ii) $\dfrac{dN}{dt} = -\lambda N \quad \dfrac{dN}{dt} = -0.13 \times (1 \times 10^{-3}) = 130.78 \, \mu g \, years^{-1}$

iii) $A = A_0 e^{-\lambda t} \quad 1.04 \times 10^{10} = 4.16 \times 10^{10} e^{-0.13 t}$

$\dfrac{1.04 \times 10^{10}}{4.16 \times 10^{10}} = e^{-0.13t}$ taking ln of both side

-1.386 = -0.13t   hence,   t = 10.6 years

17-   A = 238 – 8(4) = 206
        Z = 92 + 6 – 8(2) = 82

18- $\lambda = \dfrac{\ln 2}{t_{1/2}} = \dfrac{0.693}{5} = 0.14 \, days^{-1}$

$A = A_0 e^{-\lambda t} \quad \dfrac{A}{A_0} = e^{-(0.14 \times (15/24))} = 0.92 = 92\%$

loss of activity is 8%

19- i)
$$A = A_0 e^{-\lambda t} \quad 300 = 1500 e^{-(24 \times 3600 \lambda)} \quad \ln 0.2 = -\lambda 86400$$

$$\lambda = 1.86 \times 10^{-5} \text{ s}^{-1}$$

ii) $t_{1/2} = 0.693/\lambda = 0.693/1.86 \times 10^{-5} = 10.34 \text{hrs}$

iii) $A = \lambda N \quad 1500 = 1.86 \times 10^{-5} N \quad$ hence $N = \dfrac{1500}{1.86 \times 10^{-5}} = 80.65 \times 10^6$

20- $0.5g = (50g)(0.5)^n$

$\dfrac{0.5}{50} = 0.5^n \qquad 0.01 = (0.5)^n$

$\log_{10} 0.01 = n \log_{10} 0.5$

$n = \log_{10} 0.01 / \log_{10} 0.5 = 6.64$ days

Solutions – 6

1- d) as all the other requires one number to be constant and other two number to be variable

2- a) Gamma and X-rays of the same energy have identical interaction with matter. The only difference is in production. Gamma ray are produced from the nucleus of the atom and x-ray from the outer orbital electrons.

3- b) daughter can't decay until its formed. Since the rate of formation of the daughter is slow, in the equilibrium mixture the daughter will appear to decay with the parent's half-life. When separate from the mixture (after elusion) the daughter decay with its characteristic half-life.

4- d)

5- d)

6- c) $\dfrac{10 \, mrem}{1.66 \, mrem \, hr^{-1}} = 6.02 \, hrs$ so 6 hours are required to reach to this level but it is reduced by two half-lives or by factor of 4 so time will increase by factor of 4 i.e 24 hrs

7-c) $\lambda = \dfrac{\ln 2}{t_{1/2}} = \dfrac{0.693}{73} = 9.495 \times 10^{-3} \, hrs^{-1}$

$A = A_0 e^{-\lambda t} \quad A = 10 e^{-(9.495 \times 10^{-3} \times 98)} = 3.94$

8- b) $\dfrac{15000}{3} = 5000 \text{cpm}$

9- d) 1mCi = 10^{-3} Ci x 3.7 x 10^{10} dpc/Ci = 3.7 x 10^7 dps

10- b) When one half value layer is removed from between the source and the detector, the intensity reading doubles.

$$I_0 = 10 \text{ mR/h} \qquad 2^2 = 4 \qquad = 40 \text{mR/h}$$

11- a) 40 hrs/wk x 56 wk/year = 2240 hrs/year
Dose rate received is 5600 mR / 2240 = 2.5 mR/hr in one year

By inverse square law 40mR/hr x $(10 \text{ cm})^2$ = 2.5 mR/hr $(d)^2$

$$d^2 = 4000 \text{ mRhr}^{-1}\text{cm}^2 / 2.5 \text{ mRhr}^{-1} = 1600 \text{ cm}^2$$

$$d = 40 \text{ cm}$$

12- a), b), d)

13- c) The actual time required to reach max. activity level in the Mo/Tc generator is 23 hr.

14- a) in β^{-1} decay Z number of the decay always increase by one unit to counter balance the -1 charge on the emitter electron.

15- Equivalent dose from slow neutrons

$$H_{neutrons} = DW_R$$
$$= 475 \times 10^{-6} \times 3$$
$$= 1.43 \times 10^{-3} \text{ Gy}$$

Equivalent dose from gamma rays

$$H_{gamm\ rays} = DW_R$$
$$= 3 \times 10^{-3} \times 1$$
$$= 3 \times 10^{-3} \text{ Gy}$$

$H_{total} = H_{neutrons} + H_{gamm\ rays} = 4.43 \times 10^{-3}$ Gy

16-

Equivalent dose rate / μSvh^{-1}	25	12.5	6.25	3.13
Half-value thickness	0	1	2	3

Three half value thickness of lead reduce the equivalent dose rate to the required value.

3 x half-value thickness = 120mm
half-value = 120/3 = 40 mm

17- Each half layer is sufficient thickness to reduce radiation by a factor of two. To reduce the activity by $\frac{1}{2^4} = \frac{1}{16}$. This lead plate is equal to 4 half value layer

Activity kilo per second	48	24	12	6	3
Half value layer	0	1	2	3	4

18- 60 halves twice to reach 15 so the thickness
$\quad$ = 12 x 2
$\quad$ = 24 mm

19- $A = \dfrac{N}{t} = \dfrac{30000}{5 \times 60} = 100 Bq$

20- The energy absorbed per kilogram of mass of absorbing tissue is the absorbed dose. Unit is Gray.

21- $D = \dfrac{E}{m} = \dfrac{7}{0.25} = 28 Gy$

22- $H = DW_R = (80 \times 10^{-3}) \times 10 = 800$ mSv

23- $W_R = \dfrac{H_m}{E} = \dfrac{700 \times 10^{-3} \times 15 \times 10^{-3}}{5 \times 10^{-3}} = 2.1$

24- $t_e = \dfrac{t_p \times t_b}{t_p + t_b} = \dfrac{138 \times 40}{138 + 40} = \dfrac{5520}{178} = 31 \ days$

25- $\dfrac{1}{t_p} = \dfrac{1}{t_e} - \dfrac{1}{t_b} = \dfrac{1}{3.075} + \dfrac{1}{4} = \dfrac{1}{0.0752} = 13.3 \ hrs$

26- Use of formula $I = k \dfrac{I_0}{x^2}$ or $\dfrac{I_1}{I_2} = \left(\dfrac{r_2}{r_1}\right)^2$

$\dfrac{\left(25 - \left(\dfrac{120}{60}\right)\right)}{\left(I - \left(\dfrac{120}{60}\right)\right)} = \left(\dfrac{30}{20}\right)^2 \qquad I = 12.2$ cps

27- _ it only emits gamma rays – least ionizing and hence cause little damage

_ short half-life – so will not remain active in the body after use

_ it has long enough half-life to remain active during diagnosis

_ substance has low toxicity that can be tolerated by the body

_ can be prepared on site

28- e) efficiency and resolution are inversely proportional to energy. High energy collimeters have more lead and fewer holes than low energy collimeter.

29- $\frac{300mm}{512} = 0.59 \, mm/pixel$

30- A PMT work by causing an increasingly avalanche of electron. The PMT are quite large and may have a diameter of few centimetres. So if each flash of light is detected by one photomultiplier the resolution of the image would be fairly poor. A light guide, which is made of transparent plastic, spreads the light from each flash around many photomultiplier. By comparing the intensity of each signal you can work out where on the crystal the flash of light occurred.

Solutions – 7

6- b)

7- a)

8- ligaments

9- (E x 3.5) = (22 x 15) + (12 x 35)

$$E = \frac{750}{3.5} = 214.29 \, N$$

Resolving vertically
R + 22 + 12 = 214.29
R = 214.29 – 34
= 180.29 N

10- (E x 3.5) = (22 x 15) + (24 x 35)

$$E = \frac{11700}{3.5} = 334.29 \, N$$

11- See the example on page 125

12- See the example on page 134

13- a) a   b) b   c) c

14- a) F   b) F   c) F   d) T   e) T   f) T   g) F   h) F   I) T

15- 1) a   2) c   3) d   4) c

16- (i) – compressive stress = T cos 40° = 574.53 N

shear stress = T sin 40° = 482.10 N

(ii) – Remember that the three forces body weight W, E and S are in equilibrium

Resolving forces perpendicular to the column:

E sin 12° – 750 sin 70° = 0

therefore,

$$E = 3390 \text{ N}$$

Resolving forces parallel to the column

$$S - E \cos 12° - 750 \cos 70°$$

therefore,

$$S = 3572 \text{ N}$$

This force is almost 5 times the body weight and compresses the disc by almost 20%.

(iii) – Now, there are four forces and S acts at a smaller angle to the axis of column. This has two components, S_c and S_s.

Taking moment at S

$$E \times \tfrac{2}{3} L \sin 12° = 750 \times \tfrac{2}{3} L \sin 70° + 300 L \sin 70°$$

$$0.139 E = 751.76 \text{ N}$$

$$E = 5408 \text{ N}$$

Resolving forces parallel to the column

$$S_c = E \cos 12° + 750 \cos 70° + 300 \cos 70°$$

$$S_c = 5828 \text{ N}$$

Resolving force perpendicular to the column

$$S_s + 750 \sin 70° + 300 \sin 70° - 5408 \sin 12°$$

$$S_s = 1124.38 - 986.68 = 137.70 \text{ N}$$

The total reaction is

$$S^2 = (S_c)^2 + (S_s)^2 = (5828)^2 + (137.7)^2$$
$$S = 5830 \text{ N}$$

$$\theta = \tan^{-1}\left(\frac{137.7}{5828}\right) = 1.35°$$

Solutions – 8

10- $Q_C = h A (T_S - T_F)$
$= 2000 \times 1 \times (50 - 20)$
$= 60 \text{ kJs}^{-1}$

11- $Q/A = h (T_S - T_F)$
$= 10 (30 - (-30))$
$= 60 \text{ Cal/m}^2 \cdot \text{hr}$

12- a) homeostasis
b) thermoregulation
c) skin
d) vasodilation
e) evaporation

13- a) i   b) i   c) ii   d) iii   e) ii

14- ii

15- i

16- ii

17- iii

18- i

19 – The hypothalamus signals the sweat glands to begin producing perspiration when the body temperature rises above normal.

20- Sweat

21- (a) Respiration for *muscular* activity; respiration inefficient/releases waste heat/all energy 'ends up as 'heat'

(b) Larger surface area: volume ratio, or less fat under skin; more rapid / more heat loss from body surface.

(c) Humidity reduces diffusion gradient / less difference in water potential; less evaporation of sweat;
less cooling due to use of heat energy for evaporation of sweat

(d) Temperature receptors stimulated in; hypothalamus; heat loss centre stimulated; nerve impulses to sweat glands; increase rate of / start sweat production; nerve impulses to skin arterioles; vasodilation.

Solutions – 9

1- a) Pupil   b) Retina   c) Iris   d) Vitreous Humour   e) Optic nerve   f) Eye lens

g) Fovea   h) Rods   I) Aqueous Humour   J) Suspensory ligaments   k) Sclera

L) Astigmatism

2- Both the rods and the cones are light receptors. The rods function in dark (Black & White vision) and cones function in colour vision.

3- Accommodation is the ability of the eye see sharply all objects which are located at different distances in front of the eye. An adult can change an object in focus at large distance to an object at 7 cm in about 350 milliseconds. For this to occur eye lens need to change it focal length or its power and it can change up to 15 Dioptres in this short period.

Ciliary muscles change the power of the lens. To focus on distant objects the ciliary muscles relax making the eye lens thin and this will increase the focal length of the lens so eye can focus distant object sharply on the retina. To focus on nearby objects the ciliary muscles contract making the eye lens thick. This decreases the focal length of the lens so eye can focus near objects.

4- Iris regulates the amount of light entering the eye by contracting its opening called pupil.

5- A positive or convex lens. It increases the power of the eye lens system and move the focal point from behind the retina to the surface of the retina to make a sharp image.

6- a) position furthest from the eye that an object may be viewed clearly.
   b) diverging or concave lens.

   c) $P = \dfrac{1}{f} = \dfrac{1}{\infty} - \dfrac{1}{12} = -12D$

   d) identify v as near point

   $\dfrac{1}{u} - \dfrac{1}{0.15} = \dfrac{-1}{1.2}$ or $\dfrac{1}{u} = \dfrac{1}{0.15} - \dfrac{1}{1.2} = 5.83\,\text{m}^{-1}$

   $u = 0.17\,\text{m}$

7- a) focal length of the lens is adjusted by ciliary muscles to form sharp on the retinal

   b) first iris closes and size of pupil will decreases. Then retina or rods/cones become less sensitive.

   c) Long sightedness or hypermetropia

   d) i- eye ball too short,   2 Lens too weak or thin

   Converging or convex lens is used to correct the sight.

   e) $\dfrac{1}{f} = \dfrac{1}{0.25} + \dfrac{1}{-0.8}$

   $P = 2.75D$

   f) $\dfrac{1}{u} + \dfrac{1}{\infty} = 2.75$

   $u = 0.36\,\text{m}$

8- a) objects can be distinguished as separate.

b) diameter or size of pupil and wavelength of light.
separation of detectors on retina
number of detectors per unit area

e) $\theta = \tan^{-1}\left(\dfrac{2.41 \times 10^{14}}{2.37 \times 10^{17}}\right) = 5.8 \times 10^{-2} \, degree$

f) yes, because angular separation is greater than resolving power (minimum angle).

9- a) (i) Dioptres D or m^{-1}

   (ii) concave or diverging lens

   (iii) power of lens

   b) (i) shortsighted

   (ii)

   meeting on retina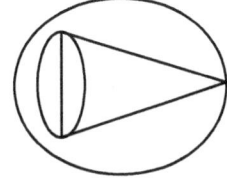

   meeting before retina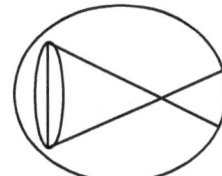

   c) $\dfrac{1}{\infty} + \dfrac{1}{v} = -2.75$ thus $v = 0.36$ m

10 a) (i) closer point or distance to eye at which an image is in focus.

   (ii) furthest point or distance from eye at which image is in focus.

   b) (i) shortsighted

   (ii) eyeball too short or eye lens too strong or cornea is too curved.

   (iii) 1. u = infinity
        v = -400 cm

       2. power = -0.25 D

(iv) $f = -400$ cm

$v = -25$ cm

near point distance is 26.7 cm

11- converging lens with positive power is required.

$d_0 = 25$ cm $\quad\quad d_1 = -73$ cm

$$f = \frac{d_0 \times d_1}{d_0 + d_1} = \frac{25 \times (-73)}{25 - 73} = +38 \, cm$$

$$p = \frac{1}{f} = \frac{1}{0.38} = 2.6 \, D$$

12- $\quad d_0 = \infty \quad\quad d_1 = -500$ cm $= -5$ m

$$P = \frac{1}{f} = \frac{1}{d_0} + \frac{1}{d_1} = \frac{1}{\infty} + \frac{1}{-5} = -0.2 \, D$$

$d_0 = 25$ cm $\quad\quad d_1 = -73$ cm

$$P = \frac{1}{f} = \frac{1}{d_0} + \frac{1}{d_1} = \frac{1}{0.25} + \frac{1}{-0.75} = +2.75 \, D$$

Solutions – 10

1- eustachian tube

2- 20 Hz to 20 kHz

3- true

4- The equal loudness contour shows the difference between sensitivity to sound pressure levels at particular frequencies.

5- Intensity – the energy of the sound wave arriving at the ear per second m².

Intensity – level is a comparison of intensity with a reference value. The reference value is a threshold of hearing.

Intensity is a measured in Wm^{-2}, intensity level is measured in dB.

6- $35 dB = 10 \log_{10} \dfrac{I_1}{10^{-12}}$

$(10^{3.5}) \times (10^{-12}) = I_1 \quad\quad\quad\quad I_1 = 3.16 \times 10^{-9} \, Wm^{-2}$

$$75\text{dB} = 10 \log_{10} \frac{I_2}{10^{-12}}$$

$(10^{7.5}) \times (10^{-12}) = I_2$ $\qquad\qquad I_2 = 3.16 \times 10^{-5}\ \text{Wm}^{-2}$

$\dfrac{I_2}{I_1}$ factor of 10000

7- Response of human ear is non-linear or logarithmic and range of audible sound intensities are very large.

8- $\text{dB} = 10 \log_{10} \dfrac{I}{I_0}$ rearranging

$\log_{10} \dfrac{I}{I_0} = \dfrac{1}{10}\ \text{dB}$ this is a straight line equation $y = mx$

the gradient is 1/10 or 0.1

9- $90 = 10 \log_{10} \dfrac{I}{10^{-12}}$

$I = 10^9 \times 10^{-12} = 1 \times 10^{-3}\ \text{Wm}^{-2}$

now, $I = \dfrac{P}{A}$ where P is power or energy per second

energy per second $= IA = 2\pi (2)^2 \times (1 \times 10^{-3}) = 25\ \text{mJs}^{-1}$

10- to localize sound the ear to ear distance must be equal to the wavelength of the sound. The average separation of ears is 15 cm.

$$f = \frac{v}{\lambda} = \frac{343\ ms^{-1}}{0.15} = 2.5\ kHz$$

11- c

12- b, The decibel scale is mathematically related to a measurable quantity. Where as phon scale is subjective and depends on persons perception of how load a sound is.

13- b, The point with coordinates 60 dB and 100 Hz is in between the 30 phon line and the 40 phon line. So the phon ratio is between 30 and 40 phon.

14- c, A 10 phon increase in the sound level is most often perceived as a doubling of loudness. A 40 phon sound is assigned a sone rate of 1. A 50 phon sound is 10 phons greater; so its sone rating is two times greater than 1. Thus a sone rating of 2. 60 phon will be 4 sone. 70 phon sound be twice as a 60 phon sound having a rating of 8 sone.

15- b, 10^{-2} Wm^{-2} is 100 in dB. Figure 10.3 this dB can be combined with 200Hz to determine the phon ratio. The coordinate point with value of 200 Hz and 100Hz lies above 90 phon line. An estimate is that it is an approximately 100 phon. From figure 10.2 100 Phone sound has sone rating of 64.

16- a)

17- b)

18- a)

19- $\Delta L = L_2 - L_1$ where $L_1 = 10 \log_{10} \left(\dfrac{I_1}{I_0}\right)$ $L_2 = 10 \log_{10} \left(\dfrac{I_2}{I_0}\right)$

$= 10 \log_{10} \left(\dfrac{I_2}{I_0}\right) - 10 \log_{10} \left(\dfrac{I_1}{I_0}\right)$

but $\log_{10} A - \log_{10} B = \log_{10} \dfrac{A}{B}$

hence

$$\Delta L = 10\left\{\log_{10}\left(\dfrac{I_2}{I_0}\right) - \log_{10}\left(\dfrac{I_1}{I_0}\right)\right\}$$

$$= 10\{(\log_{10} I_2 - \log_{10} I_0)\} - 10\{(\log_{10} I_1 - \log_{10} I_0)\}$$

$$= 10\{(\log_{10} I_2 - \log_{10} I_1)\} = 10 \log_{10} \left(\dfrac{I_2}{I_1}\right)$$

$I_2 = 2 I_1$ $\Delta L = L_2 - L_1 = 10 \log_{10}(4) = 10 \times 0.602 \approx 6$ dB

20- $L_1 = 10 \log_{10} \left(\dfrac{I_1}{I_0}\right) = 70$ $L_2 = 10 \log_{10} \left(\dfrac{I_2}{I_0}\right) = 80$

$\log_{10} \left(\dfrac{I_1}{I_0}\right) = 7$ $\log_{10} \left(\dfrac{I_2}{I_0}\right) = 8$

$10^7 = \left(\dfrac{I_1}{I_0}\right)$ $10^8 = \left(\dfrac{I_2}{I_0}\right)$

$I_1 = I_0 (10^7) = (10^{-12})(10^7)$ $I_2 = I_0 (10^8) = (10^{-12})(10^8)$
$= 10^{-6}$ Wm^{-2} $= 10^{-4}$ Wm^{-2}

$I_1 = 0.01 I_2$ or $I_2 = 100 I_1$

21- $\quad L_{sum} = I_{total} = I_1 + I_2$
$\qquad\qquad\qquad = I_1 + 100I_1$
$\qquad\qquad\qquad = 101I_1$

$\qquad L_{sum} = 10\log_{10}\left(\dfrac{101I_1}{I_0}\right)$

$\qquad\qquad \approx 80\text{ dB}$

22- Intensity I_1 is due to blender, $\quad 80 = 10\log_{10}\left(\dfrac{I_1}{I_0}\right) \implies I_1 = 10^8 I_0$

Intensity I_2 is due to dough maker, $78 = 10\log_{10}\left(\dfrac{I_2}{I_0}\right) \implies I_2 = 6.3 \times 10^7 I_0$

$\qquad L_{sum} = I_{total} = I_1 + I_2$

$\qquad\qquad\qquad = 10^8 I_0 + 6.3 \times 10^7 I_0$
$\qquad\qquad\qquad = 1.63 \times 10^8 I_0$

$\qquad L_{sum} = 10\log_{10}\left(\dfrac{1.63 \times 10^8 I_0}{I_0}\right) = 82\text{ dB}$

23- P = IA a) $5 \times 10^{-5} \times 10^{-12} = 5 \times 10^{-17}\text{ Js}^{-1}$

$\qquad\qquad$ b) $5 \times 10^{-5} \times 1 = 5 \times 10^{-5}\text{ Js}^{-1}$

24- $\;L = ¼\lambda$
$\quad\lambda = 4L = 4(2.6 \times 10^{-2}) = 0.104\text{ m}$

$\qquad f = \dfrac{v}{\lambda} = \dfrac{340\, ms^{-1}}{0.104\, m} = 3.27\, kHz$

25- using $\;v = \sqrt{\dfrac{E}{\rho}}$
$\qquad 1144\text{ ms}^{-1}$
$\qquad 458\text{ ms}^{-1}$
$\qquad 3780\text{ ms}^{-1}$
$\qquad 1529\text{ ms}^{-1}$

26- $\dfrac{I_1}{I_2} = \dfrac{P}{4\pi r_1^2} \div \dfrac{P}{4\pi r_2^2} = \dfrac{r_2^2}{r_1^2} = \dfrac{(800)^2}{(400)^2} = 16$

$\qquad I_1 = 16 I_2 = 16 \times 1.1 = 17.6\, Wm^{-2}$

27- $\quad 1 = 10 \log_{10} \left(\dfrac{I}{I_0}\right)$

$\left(\dfrac{I}{I_0}\right) = 10^{0.1} = 1.26$ or increase by 26%

28- a) $L_1 = 10 \log_{10} \left(\dfrac{I_1}{I_0}\right) \qquad\qquad 60 = 10 \log_{10} \left(\dfrac{I_1}{I_0}\right)$

$\left(\dfrac{I_1}{I_0}\right) = 10^6 \qquad\qquad I_1 = 10^6 I_0 = (10^6)(10^{-12}) = 10^{-6}$ Wm^{-2}

b) $I_2 = 2I_1 = 2 \times 10^{-6}$ Wm^{-2}

$L_2 = 10 \log_{10} \left(\dfrac{I_2}{I_0}\right) = 10 \log_{10} \left(\dfrac{2I_1}{I_0}\right) \implies L_2 = 10 \log_{10} 2 + 10 \log_{10} \left(\dfrac{10^{-6}}{10^{-12}}\right)$

$L_2 = 3 + 60 = 63$ dB

29- Individually they all produce sound level L correspond to intensity I. When all 4 sings their total intensity is 4I with sound level of 80 dB.

80 dB $= 10 \log_{10} \left(\dfrac{4I}{I_0}\right)$

80 dB $= 10 \log_{10} 4 + 10 \log_{10} \left(\dfrac{I}{I_0}\right)$

80 dB $= 10 \log_{10} 4 + L$

80 dB $= 6$ dB $+ L$

hence

$L = 80$ dB $- 6$ dB $= 74$ dB

30- 3

Solution – 11

Q1 – b) When the membrane potential reaches threshold, Voltage-gated sodium ions open. This allows sodium to rush into the cell, resulting in the depolarization of the cell.

Q2 – d) During the Refractory Period, all of the voltage-gated Na channels are either open or inactive. At this point, they cannot be re-stimulated to fire.

Q3 – a) During repolarisation, potassium ions leave the cell to try to reach it's equilibrium potential, resulting in the hyper-polarisation phase.

Q4 – a

Q5 – b

Q6 – c

Q7 – a

Q8 – c

Q9 – Using Nernst equation 11.11 and from table 1 $K^+_i = 140$ and $K^+_o = 5$

$$V = \frac{RT}{zF} \ln \frac{c_1}{c_2} = 0.0259 \ln \frac{140}{5} = -86.14 \text{ mV}$$

$$V_{DF} = V_m - V_{eq} = 50 \text{ mV} - (-86.14 \text{ mV}) = 136.14 \text{ mV}$$

Looking at table 2 show that ion's movement is out of the cell across the plasma membrane

Q10 – using equation 11.7

$$V_m = \frac{8.314 \times 310}{96485} \ln \left(\frac{1 \times 4 + 0.05 \times 145 + 0.45 \times 10}{1 \times 150 + 0.05 \times 15 + 0.45 \times 110} \right)$$
$$V_m = 0.0267 \ln 0.0786$$
$$V_m = -67.92 \text{ mV}$$

Q11 – a). The equilibrium potential for K^+ is $V = 58$ mV x $\log_{10}$ 7 mM/160 mM or 58 x (-1.359) = –78.82 mV. Ion flows inward.

b). The equilibrium potential for Na^+ is $V = 58$ mV x $\log_{10}$ 200 mM/20 mM or 58 x 1 or 58 mV. Ion flows outward.

Q12 –

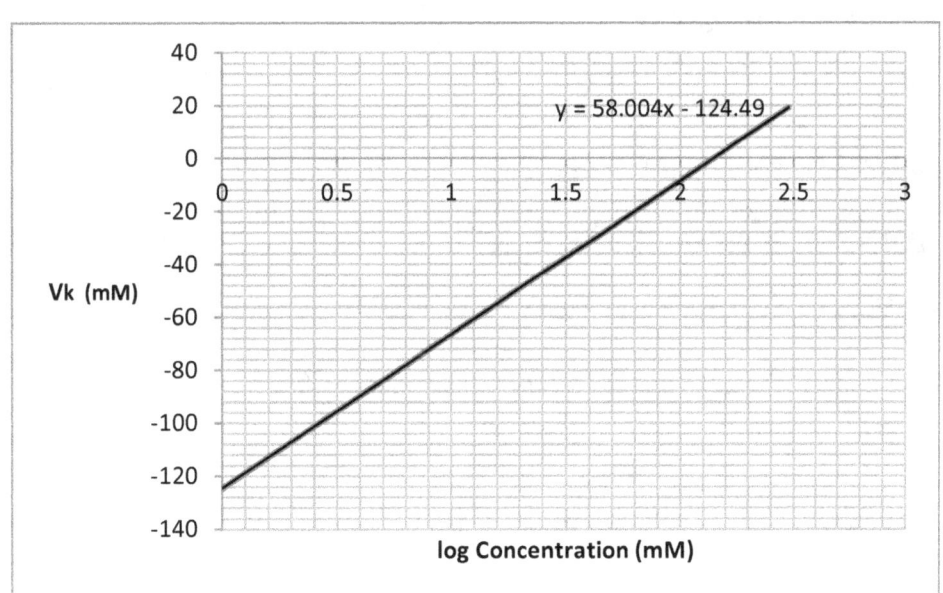

Q13 – The first step is to determine the cell reaction and total cell potential. For this reaction to be galvanic, the cadmium reaction must be the oxidation reaction. $Cd \rightarrow Cd^{2+} + 2e^-$
$V_{Cd}^\ominus = +0.403$ V and $Pb^{2+} + 2e^- \rightarrow Pb$ $V_{Pb}^\ominus = -0.126$ V

The total cell reaction is:
$$Pb^{2+}(aq) + Cd(s) \rightarrow Cd^{2+}(aq) + Pb(s)$$

$$V_{total}^\ominus = V_{Cd}^\ominus + V_{Pb}^\ominus = 0.403 + (-0.126) = 0.277 V$$

also $\dfrac{RT}{zF} = \dfrac{8.314 \times 303}{2 \times 9.648 \times 10^4} = 13.27\,mV$

$V_{cell} = V_{total}^\ominus - 13.27 \log_{10} \dfrac{C_{iCd}}{C_{oPb}}$ $V_{cell} = 0.277 - 13.27 \log_{10} \dfrac{0.02}{0.2} = 0.3 V$

Q14 – Capacitor = Capacitance x surface area of sphere (considering cell is spherical)

$= (1.15 \times 10^{-6}\,F/cm^2) \times (4\pi\,(2.25\times10^{-3})^2)$
$= 73.16\,pF$

Q15 – $Q = CV$

$Q = (73.16 \times 10^{-12})(75 \times 10^{-3})$
$= 5.49 \times 10^{-12} C$

No of ions moving across = $\dfrac{5.49 \times 10^{-12}}{1.6 \times 10^{-19}} = 34{,}293{,}750$ ions

The volume of the cell is $4/3\,\pi r^3 = 4/3\,\pi\,(22.5\times10^{-6})^3 = 4.77 \times 10^{-14}\,m^3 = 4.47 \times 10^{-11}$ litres with 400mM concentration it contains $(4.47 \times 10^{-11})(500 \times 10^{-3}) = 2.39 \times 10^{-11}$ moles of K^+ ions. 1 mole consists of 6.03×10^{23} ions so the total K^+ ions are 1.44×10^{13}.
if 34,293,750 leave that side of membrane this reduce the concentration by 0.00024%

Q16 – $I_{leak} = 0 = g_K(V - V_K) + g_{Na}(V - V_{Na}) + g_{Cl}(V - V_{cl})$

$0 = g_K V - g_K V_K + g_{Na} V - g_{Na} V_{Na} + g_{Cl} V - g_{Cl} V_{cl}$

$0 = V(g_K + g_{Na} + g_{Cl}) - g_K V_K - g_{Na} V_{Na} - g_{Cl} V_{cl}$

$V(g_K + g_{Na} + g_{Cl}) = g_K V_K + g_{Na} V_{Na} + g_{Cl} V_{cl}$

$V = g_K V_K + g_{Na} V_{Na} + g_{Cl} V_{Cl} / g_K + g_{Na} + g_{Cl}$

$V = 0.07 \times -72.1 + 0.026 \times 52.4 + 0.1 \times -57.2 / 0.07 + 0.026 + 0.1$

$V = -47.8\,mV$

Q17 - $I_{leak} = 0 = g_K(V - V_K) + g_{Na}(V - V_{Na}) + g_{Cl}(V - V_{cl})$

$$g_K (V - V_K) = -g_{Na}(V - V_{Na}) - g_{Cl}(V - V_{Cl})$$

$$g_K = -g_{Na}(V - V_{Na}) - g_{Cl}(V - V_{Cl}) / (V - V_K)$$

$$g_K = -0.026 (-45 - 52.4) - 0.1 (-45 - -57.2) / (-45 - -72.1)$$

$$g_K = 0.0484 \; \mu S$$

Q18 – a, Q19 – b, Q20 – b, Q21 – a, Q22 – c, Q23 – a, Q24 – b

Q25 – b, Period 2 contains the QRS complex. The QRS complex is the period when the action potential spreads through the muscle cells of the ventricle, beginning as the action potential propagates from the AV node into the bundle of His and ending with the depolarization of the cells at the epicardial surface of the ventricles.

Q26 – b, A patient with severe hyperkalemia ($[K^+]$ = 10 mM) will have decreased action potential propagation velocity in the heart and elsewhere. This is primarily due to 'resting inactivation' of voltage-gated Na+ channels due to membrane depolarization. This will result in a prolonged P wave (and hence a prolonged PR interval, which is defined as the time from the start of the P wave to the start – or peak – of the QRS complex) and a prolonged QRS complex. The T wave is enlarged ("peaked") by hyperkalemia.

Q27 – a- 0 marked where line meets axis with maximum value of 1 and unit as mV
b- uniform scale starts at 0 and has value 0.7 (0.9 to 0.5) at end of T wave.
c- P - depolarisation of atria
R - depolarisation of ventricles (and repolarisation of atria)
T - repolarisation of ventricles

Q28 – a) $aVL = (\Phi_L - \Phi_R) - \dfrac{(\Phi_{LL} - \Phi_R)}{2}$

$2aVL = 2\Phi_L - 2\Phi_R - \Phi_{LL} + \Phi_R = 2\Phi_L - \Phi_{LL} - \Phi_R$

$aVL = \Phi_L - \dfrac{1}{2}(\Phi_R + \Phi_{LL})$

b) $aVF = \Phi_{LL} - \Phi_L + \dfrac{\Phi_L - \Phi_R}{2}$

$2aVF = 2\Phi_{LL} - 2\Phi_L + \Phi_L - \Phi_R = 2\Phi_{LL} - \Phi_L - \Phi_R$

$aVF = \Phi_{LL} - \dfrac{1}{2}(\Phi_L + \Phi_R)$

Q29 – Average heart beat rate is 70 bpm, although it varies with age and activity. Hence heart beat once every 0.86 seconds.

Power = energy / time = 1.09J / 0.86 s = 1.27 W

1.27 W = 0.261 cal / sec = 22.51 kcal / day

as heart is only 16% efficient, hence total calories required by heart are $\dfrac{22.51}{0.16}$ = 140.68 kcal /day

This is equal to 8.45% of BMR.

Q30 – using equation 11.62

$P_F = \rho.g.h = 1060 \times 9.81 \times 1.40 = 14.56$ kPa

Pressure at feet $= P_F + P_A = (14.56 + 12.64)$ kPa $= 27.20$ kPa

$P_H = \rho.g.h = 1060 \times 9.81 \times 0.41 = 4.26$ kPa

Pressure at head $= P_A - P_H = 12.64 - 4.26 = 8.38$ kPa

Q31 –

Lets assume they experience extra 2.5g force on them, so their body will be feel 3g as one is always present.

The distance between heart and brain is on average is 40 cm. To send the blood to the head the pressure must be less than the aortic pressure.

$P_H = 1060 \times (3.5 \times 9.81) \times 0.4 = 14.55$ kPa

This is higher than aortic pressure so brain will not receive oxygenated blood and pilot will blackout and if not recovered soon can lead to death.

Q32 – using the flow rate equation $Q = \dfrac{\pi \Delta P r^4}{8 \eta L}$

$Q = \dfrac{\pi[(1060 \times 9.81 \times 1) - 2400](0.00025)^4}{8(4 \times 10^{-3})(0.04)} = 7.7 \times 10^{-8}$ m³/s

$Q = 0.077$ cm³/s

At this flow rate blood will take 13 seconds ($13 \times 0.077 = 1$) for each cm³ to flow into the vein. It will take 3.6 hours for a litre of blood to be transfused.

1000 cm³ = 1 litre, hence $13 \times 1000 = 13000$ seconds. 1 hour = 3600 seconds so $13000 / 3600 = 3.61$ hour

If r is now reduced to 0.2 mm then the flow rate will decrease by the factor $(0.2/0.25)^4 = 0.41$. The time to take for new transfusion time is $3.61 / 0.41 = 8.81$ hours. So decreasing radius by factor of 0.8 will increase the time by 2.5 time

Q33 – 1- right atrium; 2 - right ventricle; 3 - pulmonary artery; 4 - left atrium

Q34 – When heart beat is slower than 60 bpm.

Q35 – right   Q36 – tricuspid and pulmonic

Q37 – SA   AV   Bundle of HIS   Bundle Branches   Perkinje fibres

Q38 – 1- length, 2- viscosity, 3- radius   Q39 – i- contract   ii – relaxes

Q40 – c)   Q41 – P wave

Q42 – a)   Q43 – b)

Q44) each square is equal to 40 millisecond.

Time in 12 squares = 12×40 ms = 480 ms or 0.48 second

Index

Abductors 120
Absolute temperature 31
Absorption spectrum 32,34
Absorped dose 97,99
Absorptiometry 143
Accelerator 83
Action potential 161
Activity 74
   rate 77
Adductor 120
Adiabatic 203
   index 203
   pressure 203
   volume 203
Adiposity 149
Adrenaline 145
Aerobic 138,146
Agonist 121
Alpha Particles 63,64,65,66,73
   decay 63,66
Alumina column 88,90,91
Americium-241 65
Amino acid 141
Ampáre 8
Amplitude 17,20,107
Amylase 138
Anaerobic 138
Annihilation 70
Anode 73,106,107
Antagonistic 119
Anterior chamber 163,169,171
Antioxidant 146
Aorta 244,245
Aqueous humour 169,171
Arteries 243,244,245,260,261,262
Arterioles 244,260,261
Arthritis 139,168
Astigmatism 165,196
Atomic mass 65,66,70
Atomic number 63,66,70,71
Atomic structure 19
Atrioventricular node – AV 248,256,257
Atrium 243,245,257
Auditory canal 211,212,213,214,215
Avalanche 73
Avogadro's number 77
Axon 161,162,163

Balmer series 34,35
Basal Metabolic Rate 141,142,143,144,145,146
Basal metabolism 140,144
Basilar membrane 218,219,220

Bazett's formula 259
Becquerel, Henri 61,65
Becquerel 97
Bending 130,132
Beta decay 65,66
Beta minus 62,66,74
Beta plus 62,66,68
Biceps 119,120,123
Biological
   effects 64,69,72
   half life 96
Biopotential 231
Black body radiation 31,32
Blind spot 180
Blood 243,244,245,259,260,261,262
Blood stream 138,139,140
Body mass index 146
Body temperature 142,145
Bohr, Niels 32
Boltzmann constant 5,237
Boltzmann distribution 236
Brachial Muscle 117,120
Brachytherapy 70,72,102
Brain 138,142,159
Buffer 27
Bulk modulus 19,202

Caffeine 146
Calories 6,144,145
Cancer 72,141
   cell 65
   lung 64
Capillaries 243,244,245,261,262
Carbohydrate 6,15,137,140,146
Carbon dating 77
Cardiovascular 138,145,148,243
Carl, Anderson 68
Cataract 169,170
Cathode 73,106
C-band 40
Centre of curvature 49
Cerebral cortex 211
Chain reaction 9,85
Charge 8,31,35,37,63,65,67
Charm quark 67
Chemical energy 5,119
Chemotherapy 101
Cholesterol 140
Choroid 163,173
Chromatography 88,89

Chromatic aberration 189
Chromosome 72,168
Chronic 72,140,145,168
Circulation 243

pulmonary 243,262
systemic 243,244,262
Cladding 27
Cochlear 217,218,219
  fluid 218
  duct 219
  tube 219
  implant 224,225
Collagen fibres 122,168
Collagen fibrils 165
Collimator 103,104,105
Compressive stress 129
Compton scattering 71
Concave 49,54,56
Conductance 234,235,242,246,247
Conduction 150,151
Cones 162,174,175,176
Conjunctivitis, allergic 166
Conjunctiva 166,167
Constipation 145
Conservation of energy 1,11,67
Contrast 188
Convection 151,152
  force 152
  natural 152
Converge 49
Convex 49,50,56
Cooling 152
Core, fibre optics 27
Cornea 163,164,165,166,180,182
  disorder 164,165
  transplant 164
Cosmic rays 77
Coulomb 8
Covalent bond 5
Cow, moly 88,91
Crest 16
Critical angle 26
Crystal 61,139
Cunningham formula 144
Curie, 98
  Irene 68
  Jean 68
  Marie 61
  Pierre 61
Current 8,37,61,65, 234,235
Cyclotron 84,85

Daughter, radionuclide 93
Decay 61,63,66,67,68
  chain 71
  constant 75,86,92
Decibel 7,208
Degenerate 66
Dendrites 160,161,232,234,243
Dendron 160
Density 18,19,71,201

Depolarisation 162,232,233,240,245,255,257
Depression 145
Depth
  field 188
  focus 187,188
Detection circuitry 108
Deuterium 9
Diabetes 138,139,146,170
Digestion 138
Diagnose 101
Diagnostic 72,83,100,101
Diarrhoea 145
Diet 140,145
Dioptres 163,182,183
Disaccharides 137,138
Dirac, Paul 68
Diverge 49
DNA 42
Doppler effect 269
Dorsiflexor 119
Dose 72,102
Down quark 67,68
D shaped 83,84
Dynamic
  contraction 120
  range 205,206
Dynodes 107

Ear
  canal 213,214,215,218
  external (outer) 211, 212,213
  inner 212,218
  middle 212,215,216,218
Eccentric contraction 121
Effective half life 97
Effort 113,114,116,117,118
Einstein, Albert 9,10
  energy equation 63
Einthover's Law 253
Elastic
  potential energy 2,3
  property 18,19
Electrocardiogram, ECG 247,248,249,250,255,256,258
Elbow, joints 117,118,119,122,123
Electron 8,66,73
  multiplier 106
  volt 35
Electrical energy 8
Electric field 37,38,74
Electrodes 73,84
Electromagnetic 10,32,62,63
  spectrum 31,38
Electrometer 61
Eluate 89,90,91
Elution 87,91,94,95
Emmetropic 182
Emissivity 153
Emission

continuum 31,32,33
discrete 31
Enamel 139
Endolymph 218,219
Energy
 conversion 12
 level 32,33,34,35
Enzyme 138,139
Epinephrine 145
Equilibrium 16,31,92,93,94,95
Equivalent dose 97,98
Excitable cell 231,233,234
Exercise 145
Exponential equation 75
Expose 61,72
 exposure 61,99
Extension 119
Extensors 119
Eustachian tube 215,216,217,218
Evaporation 151,154
 perspiration 150,154
Eye 41, 159,162-183
 anatomy 163
 accommodation 182,183
 pink 166
 resolution 183,185,186

Fajan
 Kazimierz 66
 Soddy law 66
Faraday constant 236,237
Faraday, Michael 37
Farsighted 182,195
Fat 6,137,139,140,144,146
Fatty acid 139,140
Fermi, Enrico 67
Fermi golden rule 67
Fever 150
Feynman diagram 67,68,69
Fibre
 optics 27
 scope 28
First class lever 113,134
Fission 9,85,86
Fitzhugh-Naguma model 233
Fleming's left hand rule 74,84
Flexion 119
Flexor 120
Flow 264
 laminar 264,268
 turbulent 264,267,268
Fluorescence 61,69
Flux
 neutron 86
 rate 153
Focal length 49,56,58,182,183,193

Follicular 150
Frequency 17,18,35,36
Fructose 137,138
Fulcrum 113,114,115,116,117,118,119
Fusion 9

Gamma
 camera 103,105
 decay 70
 rays 45,63,70,104
Geiger- Muller 73
 detector 72
Gel generator 88,90
Generator 83,86
Glands 149
Glucose 5,6,137,138,139
Gluteal 125
Glycerol 139
Goldman-Hodgkin-Katz equation 235
Gout 139
Graded potential 161
Gravitational potential energy 1
Gray 98,99
Ground state 32

Hahn, Otto 67
Hair cells 211,219,220
 inner cells 219
 outer cells 219
Half-life 75,86,87,91,92,93,94
 physical 95
 effective 96
Hamstring 121
Hard x-rays 43
Harris-Benedict equation 142
Hearing loss 221
 acquire 223
 conductive 217,221
 congenital 222
 sensorineural 222
 threshold 205
Hearing aid 223
 behind the ear 223
 in the ear 223
Heart 243,244,245,247,248,250,255
Heat 5
Heat capacity ratio 203
Heavy water 85
Helium 63
Hemodynamics 259
Herschel, William 41
Hertz, Heinrich 37,39
Hip 126

Hodgkin-Huxley model 233,234,240,242,243
Homeostasis 149
Hook's law 2
Hormone 144,145

Human hearing 201,206,208,211
Humerus 117
Hume's method for LBM 143
Humidity 19,154
Hydrogen spectrum 34
Hygroscopic 105
Hyperopia 182,195,197
Hyper-polarization 233
Hyperthermia 148
Hypothermia 148
Hypothalamus 142,148,149,150,151

Ibn Sahl 21
Ideal gas 5
Immune system 150
Impedance
  acoustic 214
  characteristic 214
Incident ray 21,25
Incus 215
Index finger 187
Inertial property 18
Inflammatory 139,165
Infra-red 41
Ingestion 139
Insulin 138,139,145
Intravenous 100,101,102
in vitro 100
in vivo 88,100
Ion channel 161,234,241,243,247
  voltage-gated 161,233,234
  pump 161,234
Ionisation 61
  chamber 61
Ionising radiation 98,99
Iris 163,168
Iritis 168
Isobaric nuclide 66
Isokinetic 121
Isomeric transition 71
Isometric muscle contraction 121
Isotonic contraction 121
Isotonic nuclide 66
Isotope 63,68,83,84

Jacket, fibre optics 27
James's method for LBM 143
Katch-McArdle formula 143
K capture 69
Kerb cycle 146
Kidney 139,140
Kinetic energy 3,4,11,63
Kirchhoff's law 235,252
Klystron 40

K shell 69
Krypton-83 69

Labelling 100
Lactose 137,138
Latent heat of vaporisation 154
Law of
  radioactive decay 74
  reflection 25
  refraction 21
L-band 40
Lead 62
  augmented voltage 251,253,254
  bipolar 250
  chest 251
  standard limb 251
  unipolar 250
  12 lead ECG 248,250
Lean body mass 143
Lens 49,50,54,55,56,57,163,169,170,183,191,193
  concave 49,54,56,193
  convex 49,50,56,195
  equation 55
Lethargy 145
Lever 113,114,115,116,117
  arm length 118
Lift(ing) 131
Ligaments 121,122,123
Lipid 100
Liver 101,138,138,139,140
Load 114,115,116,117
Longitudinal wave 16
Lordosis 129
Loudness 207
Leak potassium channel 231,232,233
Louis Sigurd Fridericia formula 259
Low back pain 132
Lumbosacral disc 128,131,132
Lyman series 34,35

Magnetic field 37,38,74,84
Magnetron 40
Malleus 215
Mass deficit 63
Max, James Clarke 37
Max, Planck 31
Mechanical
  advantage 113,118
  disadvantage 113,114,116
  leverage 118
Medical imaging 72
Meitner, Lise 67
Metabolic 83
  equivalent task (MET) 148
  rate 140,141,142,144,145,146,147
Metabolise 139

Metabolism 6,140,142,145,148
metal 8,19
Metastases 101
Metastable 71
Microwave 40
Moderation 85
Molybdenum-99 83,86,88,89,90,91
Monosaccharides 137,138,139
Multi-mode fibre 28
Muscles 113,117,118,119,120,121
Muscular system 118
Myopic 182,190

Nearsighted 165,171,182
Nernst equation 235,236,237
Neurons 160,161,231,232
Neurotransmitter 161,162,171,211,220
Neutron 9,61,62,63,67
  activation 83,86
  deficiency 83
Neutrino 67,68,69
Newton's
  cooling law 152
  second law 4
  Laplace equation 201
Night vision 41
Nodes of Ranvier 161
Non-invasive 100
Non thermal radiation 32
Normal 21,23,25,26
Nucleus 61,62,63,66
Nuclear
  energy 9
  equation 63,64
  isomer 71
  medicine 72,83,86
  reactor 83,85,86

Obesity 138
Occipital lobe 159
Odourless 139
Organic 139
Organ of Corti 218,219
Orsted, Hans 37
Oscillating 37
Ossicles 215,216
Overshoot phase 233
Oxidized 138
Ozone layer 42,45

Pair production 71
Parent nuclide 92,93
Parietal lobe 159
Paschen series 34,35
Pauli, Wolfgang 67
Peak to peak 17

Period 17,18
Permeability of free space 38
Permittivity of free space 38
Perspiration 154
Perilymph 218,219
PET 70,85
Phosphorus 68,69
Photo
  cathode 106,107
  cell 107
  electric effect 71
  graphic plate 42,43,44,61
  multiplier tube 105,106,107
  receptor cell 174,175,180
  synthesis 77,137
Photon 31,33,35,37,106
Phons 207,208
Physical half life 96
Pinna 211,212
Pivot 113
Plantarflexor 119
Point of insertion 118
Poiseulle's equation 265
Polarize 232,233
Polarity 84
Polonium 61
Positron 9,68,70,83,85
Potential
  action 231,232,233,239
  cardiac action 245,247
  difference 8
  equilibrium 238
  energy 1,2,8
  membrane 231,232,233,235,236,238,240,245
  Nernst 234,235,238
  resting 231
Power 8,9,209
  of accommodation 183
  of lens 58,182,183,194
Presbycusis 226
Presbyopia 182
Pressure 7,202,203,204,205,206
Principal
  axis 49
  focus 49
  plane 49

Pronation 122
Protein 6,138,141,147
Proton 9,61,62,63,67
  deficient 83
Pulse height analysis 107
Pupil 163,168,169

Purkinje fibres 257,258
Pyrogen 86

Quality control 87
Quantum 37,70

307

theory 67
Quantized 32
Quark 62,67

Radar 40
Radiation 150,153
Radiant energy 10
Radio
   active nucli(de) 63,71,74,83
   activity 61,66
   isotope 68,70,100,101,102,102
   nuclide 61,101,102,103
   pharmaceutical 83,96,100,101,102,110
   telescope 39
   therapy 65
   waves 39
Radium 61,63
Ray diagram 50
Reactor, nuclear 83,85
Refraction 20,21,24,50
   law 21
Refractive
   index 21,22,24,26,28,30,189
   power 163, 182
Rem 98,99
Renal system 65
Resonance 212
Respiration 138
Resistance 113,116
Resistor 107
Resolution 104
Resting phase 232
Rest metabolism rate 143
Retina 169,174
Retinal
   detachment 172,173
   scan 177
Reynold number 268
Rising phase 232
Rods 162,174,175,176
Root mean square speed 5
Roentgen, Wilhelm 43
Roentgen 100
Rutherford, Ernest 65

Saline solution 88,91
Sankey energy diagram 13
Saturated 140
Scala
   media 218
   tympani 217,218,22
   vestibulli 217,218,219
Scintigraphy 103
Scintillation crystal 105
Sclera 166,167,174

Second class lever 114,117
Secular equilibrium 93
Shear stress 128,129,131,201
Semicircular 84
Shivering 148
Sievert 98,99
Sign conventions 56
Single mode fibre 28
Sinoatrial node SA 246,247,248,257
Skeletal muscles 118,119,149
Skin cancer 42
Smoke detector 65
Snellius, Willebroad 21
Snell's law 21
Snellen test 183
Soddy, Frederick 65
Sodium iodide 105
Soft x-ray 43
Solvent extraction 88
Sones 207
Sound 16,18
   energy 6,214
   intensity 20,25
   level 206
   pressure level 7,205,206
   threshold level 20
   wave 6,7,8,19,201,202,215

Speed
   of light 23,38
   of sound 6,19,201,203,204,214
   of wave 18
Spinal cord 127
Standing waves 212
Stapes 215,219
Static contraction 120
Stefan-Boltzmann law 31
Stereocilia 219,220
Sterile 86,88
Stochastic 61
Stoma 163,165,192
Strange quark 67
Stress 19,145
Strong nuclear force 62
Sublimation 88,89
Sucrose 137,138
Sugar 137,138,139
Supination 122
Sweat 148,149
Sweet 137,138
Synapse 160, 161, 162, 175
Synergist 120
Tears 163,166,167
Technetium-99 72,83,87,90,111,112
Tectorial membrane 219,220
Tellurium 63
Temperature 4,18,30,148,149,150,151,152,153,154
   absolute 203
   regulating 148,149
Temporal lobe 159

Tendons 118,121,122
Therapeutic 101,102
Thermalisation 85
Thermal
  conductance 151
  conductivity 151
  emission 31
  energy 4,5
Thermogenesis 140,146
Thermoregulation 149,154
Third class lever 116,123
Thomson J. J 65
Thorium 63
Tinnitus 226
Torque 123,133
Total internal reflection 26,27
Tracer radioactive 72,82
Transient equilibrium 94,95
Transmutation 66
Transverse waves 16
Trough 17
Tricep 119
Triglycerides 139
Tumour 70,72,82,102
Tympanic membrane 213,216

Ulna 117,122
Ultra violet 42,43
Ultrasound 269,270
Universal gas constant 204,236
Unsaturated 140
Up quark 67,68
Uranium 9,61,63,65
Uric acid 139

Vacuum 84,153
Valve 244
  aortic 244
  bicuspid 245, 261
  mitral 244
Vasoconstriction 148
Vertebral column 127,128,129
Vein 243,260,262
Venae cavae 244
Ventricles 243,244,245,257,260,262
Ventral stream 159
Venules 243
Vestibular 217,219
Vibration 201,211,215,219,220
Virtual image 53
Viscosity, coefficient 264
Vision
  acuity 180,183,184
  blurred 165,167,169
  mesopic 176
  photopic 175
  scotopic 176
Visual cortex 159
Vitamin 139
Vitreous humour 171
Volatility 88

Wave 255
  P 255,257
  PR interval 257
  Q 255,256,258
  QRS 255,256,257,258
  QT interval 258
  R 256,258
  S 256, 258
  ST 256, 258

Wavelength 16,17,18,176,212
Weak force 67
Weight loss 144,145
Window
  round 217
  oval 219
Work done 1

X-rays 43,44,45
Xylose 138

Yield 86

Zonules 169,182

Acknowledgements

Acknowledgement is made to the courtesy of all who kindly given permission for reproduction of the diagrams. The author has attempted to trace the copyright holders of all material reproduced in this publication and apologize to copyright holders if permission to publish in this form has not been obtained.

Figure 1.8 Photographer Reinraum 2009-03-20 PD ; Figure 1.9 Bowman by Pearson Scott Foresman 2008 PD; Figure 3.1 spectral continous by it: Utente 2007; Figure 3.2 Iron spectrum by Yttriomai 2009; Figure 3.9 Pe Roberto Landeu de Moura -1904 PD; Figure 3.10 National astronomy and ionospahere centre, Cornell USA, PD; Figure 3.11 http://wmap.gsfc.nasa.gov/media/101080 PD; Figure 3.12 frankelectronic. en. made-in-china. com; Figure 3.13 flickr.com 2009; Figure 3.14 Mouagip / DNA_UV_mutation gif:NASA/ David Herring, PD; Figure 3.15 Wilhelm Rontgen NASA. PD; Figure 3.16 medical. acaltechnology.com, smithsdetection. com ; Figure 5.2 Napy1Kenobi, derivative work: sjlegg, 2009-05-05 Creative common attribution share alike 3 unported; Figure 5.9 Joel Holdsworth 2007-03-09 PD; Figure 5.11 Inductiveload 2007-05-10 PD; Figure 5.12 US custom and border protection bureau 03-2006 PD; Figure 5.14 Dirk Hunniger, derivative work by Nevdka 2009-06-26 Creative common attribution share alike 3 unported; Figure 6.3 Kieran Maher, Nuclear medicine lab. Melbourne, 1990 PD; Figure 6.9 Fizzy 2005-11-08, Creative common attribution share alike 3 unported; Figure 6.12 Biddharth Banolhu, university of Illinois, college of Engineering; Figure 6.14 Arturo1299 Creative common attribtion-share alike 2.5; Figure 7.3, 7.7 & 7.10 The book of exposition Edited by Homer Heath Nugent, Rensselaer Polytechnic Institute 1922; Figure 7.11b and 7.13 www.medtrng.com Figure 7.11a http://www.badmintoncentral.com/forums/attachment.php?attachmentid=38674&stc=1&d=1199591070; Figure 7.12 Gray anatomy plate 418 PD; Figure 7.15 patient.co.uk; Figure 7.19 For men of understanding [for mechanism in our body] by Harun Yahya ISBN-1-84200-003-9; Figure 7.20 dynamic chiropractic 2012(the bio-mechanics of); Figure 7.21 Carpal Tunnel Ex, Creative common universal public domain dedication; Figure 8.2 Gichtfussim Roentgenbild by Hellerhoff, Creative common attribution share alike 3 unported; Figure 8.3 www. Pharmpedia.com/image/4/44/20-6jpg Creative common attribution 2.5 generic licence; Figure 8.4 Medhero88 and M.kumorniczak Creative common attribution share alike 3 unported: Figure8.4 Encyclopedia Britanica 12th edition PD; Figure 9.1 epilepsy foundation of America; Figure 9.2 Robert H Lourtz, Eric R, Kandel, Perception & motion, depth & form; Figure 9.3 Quasar Jarosz GNU free documentation licence version 1.2, Creative common attribution share alike 3 unported; Figure 9.4a synapse_illustration2_tweaked: Nrets, derivative work: Looie496, Creative common attribution share alike 3 unported 9.4b quia.com, nervous system; Figure 9.5 Sunshineeconnelly, Creative common attribution share alike 3 unported; Figure 9.7 Lady Weaxzezz, GNU free documentation licence ver.1.2; Figure 9.8 Jonathan Trobe, M.D., University of Michigan Kellogg Eye centre, Creative common attribution share alike 3 unported; Figure 9.9 iris Gray plate 878 PD; Figure 9.10 iritis by Paul SK 2011-02-09, GNU free documentation licence ver.1.2; Figure 9.11 Rakesh Ahuja MD, GNU free documentation licence ver.1.2, Creative common attribution share alike 3 unported; Figure 9.12 Nation Eye institute, National institute of health USA ref: EDA08 www.nei.nih.gov/ photo/eydis/index.asp PD; Figure 9.13 Gray plate 881 PD; Figure 9.14 James Cook, catalase.com; Figure 9.15 Gnuplot vector version by Qef, PD. Figure 9.16 chm. Davidson.edu; Figure 9.17 Nation Eye institute, National institute of health USA; Figure 9.18 Morleyj PD; Figure 9.19 Acuity human eye by Hans-Werner Hunziker, Creative common attribution share alike 3 unported; Figure 9.20 symptoms treatment.org; Figure 9.21 eyecairo.net; Figure 9.29 US navy photo by mass communication specialist 2nd class Chad A. Bascom   ID 100217-N-7032B-023 PD; Figure 9.30 Bill C, GNU free documentation licence ver.1.2; Figure 9.31 Edouard Spooner at Flickr, Creative common attribution share alike 3 unported; Figure 9.32-33 Transtutors .com; Figure 9.34 Tallfred; Figure 9.35 H.Hahn GNU free documentation licence ver.1.2, Creative common attribution share alike 3 unported; Figure 10.3 replaygain.hydrogenaudio.org; Figure 10.4 cohandsandvoices.org; Figure 10.5 by Covalent PD; Figure 10.7 Source http//training.seer.gov/ module_anatomy/ popup-illustrations/ images /illu_auditory_ossicles.jpg, PD; Figure 10.9 earsite.com; Figure 10.11 cnx.org; 10.12 Fred the oyster, Creative common attribution share alike 3 unported; Figure 10.13 physiologyonline.physiology.org; Figure 10.14 Michael D. Mann The nervous system in action; Figure 10.15 Earth 64 net PD; Figure 10.16 Edwtie, GNU free documentation licence ver.1.2, Creative common attribution share alike 3 unported; ; Figure 10.17 National institute of health – USA department of health and human services, PD; Figure 11.3 Nrets, Creative common attribution share alike 3 unported; Figure 11.5 unknown, GNU free documentation licence ver.1.2, Creative common attribution share alike 3 unported; Figure 11.6 Patrick J. Lynch, Mikael Haggstran, GNU free documentation licence ver.1.2, Creative common attribution share alike 3 unported; Figure 11.7 Mnokel modified by Silvia3, Creative common attribution share alike 3 unported; Figure 11.8 Richard Weston, myotonicdystrophy.com: Figure 11.9 Markus Kuhn modified by Stannered, PD: Figure 11.11 Madhero88 PD; Figure 11.12 Mikael Haggstrom, Creative common attribution share alike 3 unported PD;   Figure 11.16 Ske, GNU free documentation licence ver.1.2, Creative common attribution share alike 3 unported; Figure 11.17 Agateller (Anthony Atkielski) PD; Figure 11.18 Mikael Haggstrom PD; Figure 11.31Kieran Maher PD.

PD: Public domain

www.ingramcontent.com/pod-product-compliance
Lightning Source LLC
Chambersburg PA
CBHW081045170526
45158CB00006B/1859